D0484091

HANDBOOK OF DRUG INTERACTIONS

HANDBOOK OF DRUG INTERACTIONS

GERALD SWIDLER

ENDO LABORATORIES

Los Angeles, California

Wiley-Interscience

A DIVISION OF JOHN WILEY & SONS, INC.

NEW YORK • LONDON • SYDNEY • TORONTO

Notice Concerning Trademark or Patent Rights. The listing or discussion in this book of any drug in respect to which patent or trademark rights may exist shall not be deemed, and is not intended as, a grant of, or authority to exercise, or an infringement of, any right or privilege protected by such patent or trademark.

Library of Congress Catalog Card Number: 73–155775

ISBN 0–471–83975–2

Printed in the United States of America.

10 9 8 7 6 5 4 3 2 1

PREFACE

This compilation was begun because I believed that there was a need by the physician and pharmacist for additional help in the safe treatment of patients and for the protection of the physician against possible legal action, especially when more than one drug for the patient must be prescribed.

I have based this book on three premises: the pharmacological complexity and potency of many pharmaceuticals on the market, the time or lack of it on the part of the medical and pharmaceutical professions necessary to research each pharmaceutical preparation or combination of products, and the difficulty of correlating the material.

The *Handbook of Drug Interactions* was compiled from information supplied to the medical profession by the pharmaceutical manufacturers in the form of package inserts, literature, and advertisements, the *Physicians Desk Reference*, medical and pharmaceutical journals, and pharmacology texts. Whenever trade-marked names are used, the language of the manufacturer has been adhered to. When animal studies were used, they are so designated.

Since this book does not intend to masquerade as a pharmacology text, no effort has been made to include the mechanisms for all drug interactions. Many of them are so pharmacologically or physiologically complex that I felt I would be doing the reader a disservice by trying to condense this information when a pharmacology text would give complete understanding. The mechanisms not listed for other interactions are either unknown or self-evident. In several instances the mechanism for a drug interaction listed in one section of the book differs from or contradicts a similar drug interaction mechanism in another. This contradiction or difference occurs because the source material and the information supplied by the drug manufacturer do not agree or because a new study appeared which in my opinion was so good that I felt obligated to include it; e.g., Anticoagulants + Chloral Hydrate.

This book has a dictionary format in that all entries are arranged alphabetically. This was done to make it easier to locate a particular drug or drug interaction. The interaction is explained in an encyclopedic format, which to conserve reading time presents the information as concisely as possible.

The statement "See listed under the following agent(s)" following a main entry indicates crossfiling. This was done to make the reader aware that a particular drug can interact with any number of drugs, differing not only in name but in pharmacologic action, and to encourage him to look up that drug and all other drugs with which it may interact.

The notation "See" or "See also" refers the reader to a similar mechanism for a drug interaction, a similar grouping of drug interactions under a pharmacologic entry, or a breakdown of a fixed-combination drug into its individual components, in which case the reader can then look up each of the ingredients. I have in most instances interpolated the generic name to a trade-marked name for that drug if the drug is marketed by only one manufacturer, even if it differs from the manufacturer of the combination drug.

For ease in locating a particular drug the index is in two sections. The General Index, pages 351 through 370, is a list of all agents found in the book. The generic names of single-ingredient drugs are given next to them in parentheses. The Generic Name Index, pages 371 through 384, is a listing of drugs in the book by generic or chemical name with the trade-marked name(s) listed beneath.

All entries in the book are crossfiled for ease and rapidity in locating a particular drug.

I make no claim to have covered every pharmaceutical preparation available to the physician or found on the pharmacy shelf. Unfortunately, lack of information, either in the form of definitive studies or from the manufacturer, forced me to omit material from this edition. The study of drug interactions is relatively new and there is much we do not know.

GERALD SWIDLER

West Covina, California
February, 1971

CONTENTS

HANDBOOK OF DRUG INTERACTIONS

A

ACETATE FABRICS + Cresatin: see Cresatin + Acetate Fabrics

ACETYLCHOLINE: see listed under the following agents:

Isordil
Pentritol
Sorbitrate

ACETYLSALICYLIC ACID: see listed under the following agents:

Levoprome
Liquamar

ACHROMYCIN

+ Antacids: see Achromycin + Food
+ Drugs, High Calcium Content: see Achromycin + Food
+ Food: Absorption of Achromycin is impaired by concomitant administration of high calcium content drugs such as some antacids and foods and some dairy products such as milk. Oral forms of tetracycline should be given one hour before or two hours after meals. Mechanism: Tetracyclines are chelating agents that form insoluble chelates with multivalent cations (Ca^{++}, Mg^{++}, Al^{+++}) and gastrointestinal absorption is inhibited. (1, 55, 81)
+ Milk and Dairy Products: see Achromycin + Food
+ Trisoralen: see Trisoralen

ACHROMYCIN Parenteral

+ Calcium-Containing Solutions: The use of solutions containing calcium should be avoided, since they tend to form precipitates (especially in neutral or alkaline solution) and therefore should not be used unless necessary. Ringer's Injection U.S.P. and Lactated Ringer's Injection U.S.P. can be used with caution, however, for the calcium ion content in these diluents does not normally precipitate tetracycline in an acid media. (1)
+ Hemastix: see Hemastix

ACIDIC AGENTS

+ Alkaline Media: The following drugs which are weak acids will not be absorbed well in an alkaline media if, for example, the patient is taking antacids at the same time. Mechanism: Unionized drugs generally are more lipid soluble and poorly water soluble. The ionized drugs are poorly lipid soluble and are water soluble. The membranes in the body (cell membrane, blood vessels, stomach

ACIDIC AGENTS (continued)

+ Alkaline Media (continued): and intestinal lining) are lipoidal and therefore uncharged forms of drugs pass freely through them; charged forms transfer across cell membranes relatively slowly. Any drug that influences the pH of the fluid in a given compartment will influence absorption of a drug from there. The proportion of ionization is greater when acidic agents are in an alkaline media and therefore less lipoid soluble. (50, 55, 124)

Butazolidin	Furadantin	Salicylic Acid
Coumarins	NegGram	Sulfonamides

+ Alkanizing Agents (Urinary): Weak acids are excreted at a higher clearance in highly alkaline urine. Mechanism: The proportion of ionization is greater when acidic agents are in an alkaline media and therefore more soluble in water and less lipoid soluble. This assumes that the compound has not been primarily metabolized. (50, 55, 124) Drugs that are known to show this phenomenon of pH-dependent excretion include the weak acids:

Furadantin	Phenobarbital	Sulfonamides
NegGram	Salicylic Acid	

ACIDIC DRUGS: see listed under the following agents:
Antacids
Paveril Phosphate

ACIDIC MEDIA + Alkaline Agents: see Alkaline Agents + Acidic Media

ACIDIC SOLUTIONS: see listed under the following agents:
Brevital Sodium
Keflin
Premarin I. V.

ACIDIFYING AGENTS (Urinary) + Alkaline Agents: see Alkaline Agents + Acidifying Agents (Urinary)

ACIDITY: see listed under the following agents:
Mandelamine
Thrombin

ACIDS: see listed under the following agents:
Merthiolate
Thrombin, Topical

ACNE MIXED UBA + Antiseptic: If any of the antiseptic used to cleanse the rubber stopper is carried into the vial by the needle, denaturation of the proteins may occur and thus the special advantage of the antigen will be lost. (1)

ACRIDINE + Trisoralen: see Trisoralen

ACTH: see Adrenal Corticosteroids, Adrenocorticosteroids, Corticosteroids, and Steroids

ACTH: see listed under the following agents:

Anhydron
Anticoagulants
Biavax
Brucellergen
Dicumarol
Diuril
Esidrix
Hedulin
HydroDIURIL
Measles Virus Vaccine, Inactivated, Aluminum Phosphate Adsorbed
Metahydrin
Naqua
Naturetin
Panwarfin
Poliomyelitis Vaccine (Lilly)
Rabies Vaccine (Duck Embryo), Dried Killed Virus
Smallpox Vaccine, Dried
Tetanus Toxoid, Alum Precipitated
Tetra-Solgen
Tri-Solgen
Typhoid Vaccine
Typhus Vaccine

ACTHAR + Diabetes: Corticotrophin may aggravate diabetes mellitus so that higher insulin dosage may become necessary or manifestations of latent diabetes mellitus may be precipitated. (1)

HP ACTHAR GEL: see Acthar

ACTOSPAR

+ Oxytocics: In patients refractory to Actospar, at least three hours should elapse between oxytocics; e.g., synthetic oxytocin (Syntocinon) is administered to prevent potential synergistic action between oxytocin and Actospar. (1)

+ Syntocinon: see Actospar + Oxytocics

ACUSUL + Sulfonylureas, Hypoglycemic: Sulfonamide therapy may potentiate the hypoglycemic action of sulfonylureas. Mechanism: Displacement of the sulfonylurea from its protein binding site. (1)

ADRENAL CORTICOSTEROIDS: see listed under the following agents:

Hygroton
Measles Virus Vaccine, Inactivated, Aluminum Phosphate Adsorbed
Poliomyelitis Vaccine (Lilly)
Rabies Vaccine (Duck Embryo), Dried Killed Virus
Tetanus Toxoid, Alum Precipitated
Tetra-Solgen
Tri-Solgen
Typhoid Vaccine
Typhus Vaccine

ADRENAL CORTICOSTEROIDS + Diabetes: Close observation of diabetic patients and sometimes an increase in insulin requirements may be necessary due to the hypoglycemic effects of adrenal corticosteroids. The gluconeogenesis, when not compensated by an adequate insulin output, leads to hyperglycemia and glycosuria. (2)

ADRENERGIC AGENTS

+ Diabetes: Because of the glycogenolytic effect, adrenergic agents should be very carefully administered to diabetic patients. (2)
+ Isoniazid: see Isoniazid + Adrenergic Agents
+ Monoamine Oxidase Inhibitors: Adrenergic agents should not be given to patients receiving monoamine oxidase inhibitors, since this combination may precipitate a hypertensive crisis. For mechanism see MAO Inhibitors + Sympathomimetic Drugs. (2)

ADRENERGIC NEURON BLOCKING AGENTS: see listed under the following agents:

Pertofrane
Tofranil

ADRENOCORTICAL STEROIDS: see listed under the following agents:

Anhydron
Aquatag
Diuretics, Thiazide
Enduron
Esidrix
Exna
HydroDIURIL
Hydromox
Metahydrin
Naqua
Renese
Saluron
Smallpox Vaccine

ADRENOCORTICOSTEROIDS: see listed under the following agents:

Dicumarol
Digitalis
DPT Vaccine

ADRENOCORTICOSTEROIDS

+ Anticoagulants: Concomitant use may cause a reversal of hypoprothrombinemic state.
+ Diabetes: Diabetic patients frequently need an increase in insulin dosage if they are taking adrenocorticosteroids.
+ Digitalis: Potassium depletion due to the adrenocorticosteroid may precipitate digitalis intoxication and arrhythmias without actual overdosage of the digitalis.
+ DPT Vaccine: Concomitant administration should be avoided because the steroids may interfere with antibody response.
+ Measles Virus Vaccine, Inactivated: Adrenocorticosteroids may suppress the antibody response to the vaccine; therefore, if possible,

ADRENOCORTICOSTEROIDS (continued)

+ Measles Virus Vaccine, Inactivated (continued): it would seem advisable to avoid administration of the vaccine concomitantly with these hormones.
+ Measles Virus Vaccine, Live, Attenuated: In patients under treatment with therapeutic doses of steroids, vaccination against smallpox and attacks of measles, chicken pox, and other acute contagious diseases has resulted in serious or fatal illness. At present it is not definitely established that the administration of a live attenuated measles virus vaccine has the same potential hazard. Therefore, until further definitive data are available, the physician should weigh carefully the use of this vaccine in patients under treatment with steroids. Most pharmaceutical companies consider concomitant usage contraindicated.
+ Mumps Vaccine: see Adrenocorticosteroids + Rabies Vaccine
+ Rabies Vaccine: Adrenocorticosteroids may reduce host resistance to certain infectious agents by suppression of antibody response or by other still poorly understood mechanisms. Therefore they should not be administered after exposure to infectious agents (mumps, rabies, tetanus) for which no satisfactory antimicrobial therapy is available. To do so may alter the host-parasite relationship enough to cause severe or fatal illness in spite of the prophylactic administration of a vaccine.
+ Smallpox Vaccine: see Adrenocorticosteroids + Measles Virus Vaccine, Live, Attenuated
+ Stoxil: In superficial infections Stoxil should not be used in combination with steroids. In deep infections, if such combined therapy is judged necessary, it must be employed with caution and the patient must be observed.
+ Tetra-Solgen: Concomitant administration should be avoided because the steroids may interfere with antibody response.
+ Thiazide Diuretics: Hypokalemia is more likely to develop or a preexistent potassium deficiency may be aggravated during periods of brisk diuresis.
+ Tri-Solgen: Concomitant administration should be avoided because the steroids may interfere with antibody response.

ADRENOLYTIC AGENTS + Aramine: see Aramine + Adrenolytic Agents

ADRENOSEM + Antihistamines: Antihistamines tend to inhibit the efficacy of Adrenosem and should be discontinued 48 hours before Adrenosem is started. If conditions do not permit sufficient time for withdrawal of antihistamines, the usual dose of Adrenosem should be increased to

ADRENOSEM + Antihistamines (continued): 2 cc (10 mg) for the initial injection. (1)

ADRESTAT + Antihistamines: Antihistamines may inactivate Adrestat. If a patient has received antihistamines, at least 12 hours should be allowed to elapse before starting therapy. In emergencies, however, Adrestat may be used if the dose is doubled and repeated in two or three hours if necessary. (1)

AEROLONE COMPOUND: see Isoproterenol

AIR + Surital: see Surital + Air

AKRINOL + Soap: All soap must be removed by thorough rinsing and followed by thorough drying with a towel before Akrinol cream is applied. Soap can drastically reduce the antifungal activity of Akrinol. (1, 2)

ALCOHOL: see listed under the following agents:

Ansolysen
Antabuse
Anticoagulants
Antiemetics, Phenothiazine-type
Antihistamines
Anhydron
Anti-Nausea Suprettes
Apresoline
Atarax
Aventyl
Benzodiazepines
Bristuron
Compazine
Coronary Vasodilators (Nitrates)
Coumadin
Dalmane
Dartal
Deprol
Diabinese
Dicumarol
Disophrol
Diuretics, Thiazide
Diuril
Doriden
Dymelor
Edecrin
Elavil
En-Chlor
Equanil
Esidrix
Eutonyl
Flagyl
Furoxone
Haldol
Harmonyl
HydroDIURIL
Hydromox
Hygroton
Hypoglycemic Drugs, Oral
Insulin
Inversine
Ismelin
Isordil
Lenetran
Levoprome
Librium
Lipan
Listica
MAO Inhibitor
Marplan
Matulane
Mellaril
Meprobamate
Miltown

ALCOHOL (continued):

Nardil
Navane
Niamid
Noludar
Orinase
Pacatal
Parest
Parnate
Pentritol
Periactin
Permitil
Pertofrane
Phenobarbital
Phenothiazine
Placidyl
Plaquenil
Plasmanate
PMB-200
PMB-400
Probanthine with Dartal
Proketazine
Prolixin
Quaalude
Quiactin
Repoise
Seconal Sodium
Serax
Serentil
Seromycin
Sinequan
Sintrom
Solacen
Somnofac
Somnos
Sopor
Sorbitrate
Sparine
Stelazine
Striatran
Sulfonylurea, Hypoglycemic Oral
Symmetrel
Tacaryl
Taractan
Temaril
Thorazine
Tindal
Tofranil
Tolinase
Torecan
Trancopal
Trepidone
Trilafon
Tybatran
Ultran
Valium
Valmid
Vesprin
Vistaril

ALCOHOL

+ Barbiturates: see Alcohol + CNS Depressants. Mechanism: The inhibition of drug-metabolizing enzymes by alcohol may contribute to the increased sensitivity of inebriated persons to barbiturates. (47)

+ Chlorpromazine: see Alcohol + CNS Depressants

+ CNS Depressants: The synergism of the narcotic action of alcohol with a number of central nervous system depressant drugs may produce coma or even death by respiratory depression. In this group are barbiturates, meprobamate, chlorpromazine, and similar phenothiazine derivatives. (26)

ALCOHOL (continued)

+ Diphenylhydantoin: It has been demonstrated that heavy drinking speeds up the metabolism of diphenylhydantoin. Mechanism: It is postulated that this effect is the result of stimulation of the hepatic microsomal enzyme system in the human liver responsible for the metabolizing of diphenylhydantoin. (45)

+ Insulin: Alcohol has a glucose-lowering action that augments the hypoglycemic effects of insulin and may cause severe hypoglycemia and irreversible neurological changes. Studies have shown that alcohol inhibits the formation of new glucose from amino acids and other precursors. The interference with gluconeogenesis has been attributed to an increase in the ratio of reduced to oxidized nicotinamide adenine dinucleotide ($NADH_2$/NAD) within the hepatic cell during the oxidation of alcohol. Increases in the $NADH_2$/NAD ratio inhibits the entrance of glycerol, lactic acid, and specific amino acids into the metabolic pathways through which these metabolites are converted to glucose. (49)

+ Meprobamate: see Alcohol + CNS Depressants

+ Nitroglycerin: If nitroglycerin and alcohol are given at approximately the same time, an Antabuse-alcohol type reaction may occur. (26)

+ Orinase: It has been demonstrated that heavy drinking speeds up the metabolism of tolbutamide. Mechanism: It is postulated that this effect is the result of stimulation by alcohol of a hepatic microsomal enzyme system in the human liver responsible for the metabolism of Orinase. (35, 45) See Alcohol + Sulfonylurea, Hypoglycemic.

+ Phenothiazines: see Alcohol + CNS Depressants

+ Sulfonylurea, Hypoglycemic: The oral antidiabetic drugs of the sulfonylurea series such as Orinase and Diabinese when used concomitantly with alcohol may cause an Antabuse-alcohol reaction but milder. Mechanism: The reaction is probably due to a similar mechanism which occurs with Antabuse-alcohol. Alcohol is oxidized in the body to acetaldehyde, acetic acid, and carbon dioxide. Antabuse blocks the enzyme system responsible for the conversion of acetaldehyde into acetate and large amounts of acetaldehyde accumulates in the blood. This evokes a fall in blood pressure, gastrointestinal distress, and the faintness characteristic of the Antabuse-alcohol syndrome. (26, 41) See Alcohol + Orinase.

+ Warfarin: It has been demonstrated that heavy drinking speeds up the metabolism of Warfarin. Mechanism: It is postulated that this effect is the result of stimulation by alcohol of the hepatic microsomal enzyme system in the human liver responsible for the metabolizing of Warfarin. (45)

ALDACTAZIDE: see Aldactone and Hydrochlorothiazide

ALDACTAZIDE
+ Diabetes: Thiazides may decrease glucose tolerance, thus temporarily exaggerating abnormalities of glucose metabolism in diabetic patients or causing them to appear in those latent diabetes. Aldactazide may have similar effects on glucose tolerance. (1)
+ Sodium: The most common electrolyte disturbance encountered with Aldactazide therapy is dilutional hyponatremia. Administration of sodium in this situation is usually contraindicated. (1)

ALDACTONE: see listed under the following agents:
Dyazide
Dyrenium
Lasix

ALDACTONE
+ Antihypertensive Agents: Aldactone may potentiate the action of other antihypertensive drugs. The dosage of such drugs, particularly the ganglionic blocking agents, should be reduced at least 50% when Aldactone is added to the regimen. (1)
+ Diuretics, Mercurial: see Aldactone + Diuretics, Thiazide
+ Diuretics, Thiazide: Aldactone exerts a true synergistic effect when administered concomitantly with thiazide or organic mercurial diuretics. When administered in combination with thiazide or organic mercurial diuretics, the potassium loss induced by these diuretics is offset. Supplemental administration of potassium is contraindicated during such therapy unless glucocorticoids are administered. When Aldactone is used concomitantly with thiazide or organic mercurial diuretics, it has a synergistic effect on sodium excretion. The possible development of hyponatremia, manifested by dryness of the mouth, thirst, lethargy, drowsiness, and a low serum sodium, must be considered. (1, 2)
+ Ganglionic Blocking Agents: see Aldactone + Antihypertensive Agents
+ Glucocorticoids: see Aldactone + Diuretics, Thiazide and Aldactone + Potassium
+ Potassium: Potassium supplementation, either in the form of medication or as a potassium rich diet, is not indicated unless a glucocorticoid is also given. Such supplementation to Aldactone alone may cause hyperkalemia in rare instances. (1, 50)

ALDOCLOR: see Aldomet and Diuril

ALDOMET: see listed under the following agents:

Dopar	Nardil
Esidrix	Naturetin
Eutonyl	Parnate
Larodopa	Raudixin
Levodopa	Rauwiloid
Levophed	Sympathomimetics, Direct-Acting
MAO Inhibitor	Unitensin

ALDOMET

+ Anesthesia: Concomitant usage may prevent the surgical patient from compensating for the hypotensive challenge of anesthesia. (55) Mechanism: see Ismelin + Anesthesia.
+ Antihypertensive Agents: Therapy with Aldomet may be initiated in most patients already under treatment with antihypertensive agents by terminating all previous medication except thiazide drugs which may be continued. By gradually decreasing the dosage of ganglionic blocking agents and Ismelin and gradually adding Aldomet a smooth transition in therapy can be accomplished. (1) See also Aldomet + Drugs.
+ Aramine: Aldomet may mildly potentiate the action of Aramine. (55) Mechanism: see Aldomet + Wyamine.
+ Diuretics, Thiazide: When a thiazide diuretic is given simultaneously, the antihypertensive effect is enhanced. In addition, the thiazide counteracts the retention of sodium and the increased plasma volume that often occur after long-term administration of Aldomet. (1, 2)
+ Drugs: When Aldomet is used in combination with other drugs, potentiation of the action may occur. See also Aldomet + Diuretics, Thiazide and Aldomet + Antihypertensive Agents. (1)
+ Ganglionic Blocking Agents: see Aldomet + Antihypertensive Agents
+ Ismelin: see Aldomet + Antihypertensive Agents
+ Levophed: see Aldomet + Sympathomimetics, Direct-Acting
+ Methamphetamine: Methamphetamine may inhibit the effects of Aldomet to some extent. (55)
+ Sympathomimetics, Direct-Acting: Direct-acting vasopressors such as Levophed given to counteract severe hypotension in a patient who has taken Aldomet, will induce a much greater response than would be expected under normal circumstances. Mechanism: Aldomet works to prevent the uptake of Levophed into inactivation sites and therefore potentiates the effect of the vasopressor. (55)
+ Sympathomimetics, Indirect-Acting: see Aldomet + Aramine and Aldomet + Wyamine
+ Wyamine: Aldomet may mildly potentiate the action of Wyamine. (55) Mechanism: It is postulated that Aldomet is metabolized to alpha

ALDOMET (continued)

+ Wyamine (continued): methylnorepinephrine (a false neurotransmitter) which displaces norepinephrine and is stored in its place in the nerve endings. Alpha methylnorepinephrine may be less potent than norepinephrine or is more tightly bound and less susceptible to release by nerve stimulation. The indirect-acting sympathomimetics act by releasing this false neurotransmitter which may be the cause of the potentiation. (104)

ALDORIL: see Aldomet and HydroDIURIL

ALKALIES + Thrombin, Topical: see Thrombin, Topical + Alkalies

ALKALINE AGENTS

+ Acidic Media: The following drugs which are weak bases will not be absorbed well in an acidic media. Mechanism: Unionized drugs generally are more lipid soluble and poorly water soluble. The ionized drugs are poorly lipid soluble and are water soluble. The membranes in the body (cell membrane, blood vessels, stomach and intestinal lining) are lipoidal and therefore uncharged forms of drugs pass freely through them; charged forms transfer across cell membranes relatively slowly. Any drug that influences the pH of the fluid in a given compartment will influence absorption of a drug from there. The proportion of ionization is greater when alkaline agents are in an acidic media and therefore less lipoid soluble. (50, 55, 124)

  Amphetamine
  Antipyrine
  Aralen
  Inversine
  Meperidine
  Procaine
  Quinidine
  Theophylline

+ Acidifying Agents (Urinary): Weak bases are excreted at a higher clearance in highly acidic urine. Mechanism: The proportion of ionization is greater when alkaline agents are in an acidic media and therefore are more water soluble and less lipoid soluble. This assumes that the compound has not been primarily metabolized. (50, 55, 124) Drugs that are known to show the phenomenon of pH-dependent excretion include the weak bases:

  Amphetamine
  Antipyrine
  Aralen
  Atabrine
  Elavil
  Inversine
  Levo-Dromoran
  Meperidine
  Nicotine
  Procaine
  Quinine
  Theophylline
  Tofranil

ALKALINE EARTH METALS + Keflin: see Keflin + Alkaline Earth Metals

ALKALINE MEDIA + Acidic Agents: see Acidic Agents + Alkaline Media

ALKALINE SOLUTIONS: see listed under the following agents:

Anectine	Ritalin
Dopram	Succinylcholine

ALKALIZING DRUGS + Mandelamine: see Mandelamine + Alkalizing Drugs

ALKALIZING FOODS + Mandelamine: see Mandelamine + Alkalizing Foods

ALKANIZING AGENTS (URINARY) + Acidic Agents: see Acidic Agents + Alkanizing Agents (Urinary)

ALKERAN

+ Alkylating Agents: Alkeran should not be given if other similar chemotherapeutic agents have been recently administered or if the neutrophil and/or platelet counts are depressed. (1, 2)

+ Radiation Therapy: Alkeran must be given cautiously to patients receiving radiotherapy, since this treatment may seriously increase susceptibility to the toxic effects of Alkeran. Previously well-tolerated doses of Alkeran may cause severe depression of the bone marrow when skeletal radiation therapy is given concomitantly. (1, 2)

ALKON-T: Alkon-T contains methaqualone; see Quaalude or Sopor

ALKYLATING AGENTS: see listed under the following agents:

Alkeran	Measles Virus Vaccine, Live, Attenuated
Attenuvax, Lyovac	Mumpsvax, Lyovac
Biavax	M-Vac Measles Virus Vaccine, Live, Attenuated
Cendevax	Pfizer-Vax Measles-L
Fluorouracil	Rubeovax, Lyovac
Lirugen	

ALKYLATING AGENTS

+ Antineoplastic Chemotherapy: The alkylating agents should be used with caution after other chemotherapy because of the danger of damaging the bone marrow irreversibly. (2)

+ Radiation Therapy: The alkylating agents should be used with caution after a course of radiation therapy because of the danger of damaging the bone marrow irreversibly. (2)

ALPHADROL + Diabetes: Alphadrol should be used with great caution in patients with diabetes mellitus. Disturbances of glucose metabolism occur only infrequently with doses under 6 mg a day. (2)

ALUMINUM HYDROXIDE GEL: see listed under the following agents:

Rencal	Terramycin	Vibramycin
Rondomycin	Tetracyn	
Signemycin	Urobiotic	

ALVODINE

+ Barbiturates: see Alvodine + Tranquilizers
+ Muscle Relaxants: see Alvodine + Tranquilizers
+ Tranquilizers: If tranquilizers, muscle relaxants, or barbiturates have been used or are contemplated, due consideration should be given to their possible additive effect on respiration and circulation. (1)

AMBENYL EXPECTORANT: see Ambodryl and Benadryl

AMBODRYL

+ Hypnotics: Hypnotics, sedatives, or tranquilizers, if used with Ambodryl hydrochloride, should be prescribed with caution because of possible additive effect. (1)
+ Sedatives: see Ambodryl + Hypnotics
+ Tranquilizers: see Ambodryl + Hypnotics

AMINET SUPPOSITORIES + Ephedrine: Avoid concurrent use of ephedrine and Aminet suppositories. (1)

AMINOBENZOIC ACID + Trisoralen: see Trisoralen

AMINOPHYLLINE: see listed under the following agents:

Organophosphate Pesticides	Quibron
Protopam	Theo-Organidin

AMINOSALICYLIC ACID: see Para-Aminosalicylic Acid

AMINOSOL

+ Antibiotics: Antibiotics or other drugs should never be added to Aminosol solutions because of physical incompatibility. (1)
+ Drugs: see Aminosol + Antibiotics

AMMONIUM CHLORIDE: see listed under the following agents:

Aralen	Diutensin	Thiomerin
Atabrine	Plaquenil	

AMMONIUM CHLORIDE

+ Amphetamine: see Ammonium Chloride + Weak Bases
+ Meperidine: see Ammonium Chloride + Weak Bases
+ Weak Bases: Weak bases such as amphetamine or meperidine will be excreted more rapidly in a patient whose urine has been acidified with ammonium chloride. If the drug is excreted more rapidly, its effect is of shorter duration. (55)

AMPHAPLEX + MAO Inhibitors: Concomitant administration of a monoamine oxidase inhibitor is contraindicated. Certain monoamine oxidase inhibitors may potentiate the action of Amphaplex. (1) Mechanism: see MAO Inhibitors + Sympathomimetics, Indirect-Acting.

AMPHETAMINES: see Alkaline Agents + Acidic Media and Alkaline Agents + Acidifying Agents (Urinary)

AMPHETAMINES: see listed under the following agents:

- Ammonium Chloride
- Antidepressants, Tricyclic
- Darvon
- Eutonyl
- Furoxone
- Ismelin
- Meperidine
- Nardil
- Norpramin
- Parnate
- Pertofrane
- Reserpine
- Tofranil

AMPHETAMINES

+ Ismelin: Amphetamines may decrease the hypotensive effect of Ismelin. (1) Mechanism: see Ismelin + Amphetamines.
+ MAO Inhibitors: Concomitant use is contraindicated for it may cause severe headache, hypertension, cardiac arrhythmias, chest pain, intracranial bleeding, and circulatory failure. (1) Mechanism: see MAO Inhibitor + Sympathomimetics, Indirect-Acting.

AMYTAL SODIUM STERILE

+ Barbiturates: see Amytal Sodium Sterile + CNS Depressants
+ CNS Depressants: Caution should be exercised in the administration of Amytal Sodium to patients who have received opiates, tranquilizers, central nervous system depressants, or other barbiturates concurrently. The additive effect may cause respiratory depression and interfere with the cough reflex. (1)
+ Opiates: see Amytal Sodium Sterile + CNS Depressants
+ Tranquilizers: see Amytal Sodium Sterile + CNS Depressants

ANABOLIC STEROIDS: see listed under the following agents:

- Anticoagulants
- Coumadin

ANALEPTIC AGENTS: see listed under the following agents:

- Darvon
- Quaalude

ANALEPTIC AGENTS + Narcotics: Analeptic agents when used in large doses are capable of causing convulsive seizures in narcotized patients. The inherent convulsive properties of narcotics and the identical action of analeptics narrow the therapeutic margin of the analeptic agents. (4)

ANALEXIN: see listed under the following agents:

Anticoagulants
Coumadin
Diphenylhydantoin
Orinase
Panwarfin

ANALEXIN

+ Anticoagulants: Patients receiving oral anticoagulants (Dicumarol, warfarin sodium phenindione) in addition to Analexin should be carefully followed for possible excessive hypoprothrombinemia. Mechanism: Analexin potentiates the action of anticoagulants by inhibiting the metabolizing hepatic microsomal enzymes. (1, 2, 3, 29, 55)
+ Dicumarol: see Analexin + Anticoagulants
+ Phenindione: see Analexin + Anticoagulants (3)
+ Warfarin Sodium: see Analexin + Anticoagulants

ANALGESIC DEPRESSANT + Largon: see Largon + Analgesic Depressant

ANALGESICS: see listed under the following agents:

Antiemetics, Phenothiazine-type
Compazine
Eutonyl
Haldol
Largon
Levoprome
Navane
Niamid
Numorphan
Parest
Permital
Phenergan
Phenothiazine
Prolixin
Quaalude
Regitine
Repoise
Seconal Sodium
Somnofac
Sopor
Sparine
Tacaryl
Taractan
Temaril
Thorazine
Tindal
Trilafon
Vesprin

Analgesics

+ Barbiturates: see Analgesics + Respiratory Depressants
+ Mercurial Diuretics: Although impaired renal function is not necessarily a contraindication for their use, the potent analgesics may sometimes have adverse effects on renal function, such as decreased urinary output and interference with the action of mercurial diuretics. (2)
+ Respiratory Depressants: It is imperative to use reduced doses of potent analgesics when given to patients who are receiving tranquilizers, barbiturates, or other agents that increase respiratory depression. (2)
+ Tranquilizers: see Analgesics + Respiratory Depressants

ANANASE + Anticoagulants: Ananase should be used with caution in patients with abnormalities of the blood-clotting mechanism. (1)

ANDROGENS + Phenobarbital: see Phenobarbital + Androgens

ANECTINE: see Succinylcholine

ANECTINE

+ Alkaline Solutions: Succinylcholine is rapidly hydrolized by alkaline solutions and therefore loses potency rapidly if mixed with thiopental sodium (Pentothal Sodium). Such mixtures, if used at all, must be used within a few minutes of preparation; however, separate injections of the drug are preferable. (1)
+ Antibiotics: Certain antibiotics, such as neomycin, streptomycin, Kantrex, Coly-Mycin, Polymixin B, and possibly bacitracin, act as weak depolarizing agents and cause a transmission defect at the myoneural junction. On rare occasions, therefore, they may also potentiate the action of muscle relaxants and prolong recovery time following induced muscle paralysis. (1, 33, 110)
+ Anticholinesterase Drugs: Patients treated with or exposed to organic phosphate compounds (e.g., Phospholine Iodide, insecticides) and other anticholinesterases (e.g., Neostigmine, Physostigmine, Tensilon) may have decreased plasma cholinesterase activity that may intensify and prolong the action of Anectine. (1)
+ Bacitracin: see Anectine + Antibiotics
+ Coly-Mycin M: see Anectine + Antibiotics
+ Dihydrostreptomycin: see Anectine + Antibiotics
+ Insecticides, Polyphosphate: see Anectine + Anticholinesterase Drugs. Patients suffering from polyphosphate insecticide poisoning may possibly have a decreased plasma-cholinesterase activity that may intensify and prolong the action of Anectine. (1)
+ Kantrex: see Anectine + Antibiotics
+ Neomycin: see Anectine + Antibiotics
+ Neostigmine: see Anectine + Anticholinesterase Drugs
+ Pentothal Sodium: see Anectine + Alkaline Solutions
+ Phospholine Iodide: see Anectine + Anticholinesterase Drugs
+ Physostigmine: see Anectine + Anticholinesterase Drugs and Anectine + Tensilon
+ Polymixin B: see Anectine + Antibiotics
+ Procaine: Procaine prolongs and intensifies the action of succinylcholine. (1)
+ Streptomycin: see Anectine + Antibiotics
+ Tensilon: On rare occasions, when succinylcholine is given over a long period of time, the characteristic depolarization block of the

ANECTINE (continued)

+ Tensilon (continued): myoneural junction changes to a nondepolarizing block that causes prolonged respiratory depression or apnea similar to the blockage induced by tubocurarine. Under these circumstances small repeated doses of Physostigmine and Tensilon may shorten the action of Anectine. (1) See also Anectine + Anticholinesterase Drugs.

ANESTHESIA: see Anesthesia, Caudal, General, Lumbar-Block, Regional, Spinal, Surgical, and Topical and Anesthetics, Anesthetics, General, and Anesthetics, Local

ANESTHESIA: see listed under the following agents:

Aldomet
Cremomycin
Diuretics, Thiazide
Ecolid
Emivan
Enduronyl
Esidrix
Esimil
Exna
Harmonyl
Inversine
Ismelin
Kantrex
Mio-Pressin
Naqua
Neomycin
Ostensin
Permitil
Raudixin
Rauwolfia Alkaloids
Sandril
Serpasil
Singoserp
Thrombolysin
Torecan

ANESTHESIA

+ Corticosteroids: Severe hypotension and even circulatory collapse have occurred during routine anesthesia of patients who have been recently treated with corticosteroids, rauwolfia alkaloids, and certain phenothiazine tranquilizers. (2)
+ Phenothiazines: see Anesthesia + Corticosteroids
+ Rauwolfia Alkaloids: see Anesthesia + Corticosteroids

ANESTHESIA, CAUDAL + Ergotrate Maleate Inj.: see Ergotrate Maleate Inj. + Anesthesia, Caudal

ANESTHESIA, GENERAL: see listed under the following agents:

Bristuron
Reserpine
Saluron

ANESTHESIA, LUMBAR-BLOCK + Panwarfin: see Panwarfin + Anesthesia, Lumbar-Block

ANESTHESIA, REGIONAL: see listed under the following agents:

Ergotrate Maleate Inj.
Panwarfin

ANESTHESIA, SPINAL: see listed under the following agents:

Ergotrate Maleate Inj.
Vesprin

ANESTHESIA, SURGICAL + Asmolin: see Asmolin + Anesthesia, Surgical

ANESTHESIA, TOPICAL + Carcholin: see Carcholin + Anesthesia, Topical

ANESTHETICS: see also Anesthesia

ANESTHETICS: see listed under the following agents:

Anticoagulants
Coumadin
Dopram
Eskaserp
Eutonyl
Haldol
Inderal
Leritine
Lorfan
Magnesium Sulfate Inj.
Mellaril
Moderil
Narcotic Antagonist
Naturetin
Numorphan
Parnate
Phenothiazine
Prinadol
Prolixin
Regitine
Repoise
Serentil
Singoserp
Sintrom
Stelazine
Taractan
Thorazine
Torecan
Tromexan
Unitensin
Vesprin

ANESTHETICS, GENERAL: see listed under the following agents:

Antiemetics, Phenothiazine-type
Apodol
Compazine
Dartal
Eutonyl
Harmonyl
Ismelin
Laradopa
Levoprome
Mellaril
Navane
Pacatal
Permitil
Phenothiazine
Proketazine
Prolixin
Stelazine
Taractan
Tindal
Trilafon
Vesprin

ANESTHETICS, LOCAL: see listed under the following agents:

Eutonyl
Hydeltrasol Inj.

ANHYDRON: see Diuretics, Thiazide

ANHYDRON

+ ACTH: see Anhydron + Adrenocortical Steroids
+ Adrenocortical Steroids: Special precaution should be taken against

ANHYDRON (continued)

+ Adrenocortical Steroids (continued): potassium depletion when Anhydron is used by patients receiving adrenocortical steroids, or ACTH. (1, 2)
+ Alcohol: Orthostatic hypotension may be potentiated when Anhydron is combined with alcohol, barbiturates, or narcotics. (1)
+ Antihypertensive Agents: Since Anhydron augments the action of antihypertensive agents, ganglionic blocking agents, reserpine, and the veratrum alkaloids, their dosage should be reduced, perhaps by 50% of the usually recommended dosage, at the start of treatment and carefully readjusted upward or downward according to the patient's response and need. The dosage of the latter agents should be reduced because their potentiating effect may cause a fall in orthostatic blood pressure. (1, 2)
+ Barbiturates: see Anhydron + Alcohol
+ Diabetes: Although diabetes is not a contraindication to the use of Anhydron, the drug can produce a further decrease in glucose tolerance. (1, 2)
+ Digitalis: Special precaution should be taken against potassium depletion when Anhydron is used by patents receiving digitalis. (1, 2)
+ Ganglionic Blocking Agents: see Anhydron + Antihypertensive Agents
+ Narcotics: see Anhydron + Alcohol
+ Pressor Amines: Since it has been reported that thiazide compounds decrease arterial responsiveness of pressor amines, great care should be exercised in administering Anhydron to patients who are to undergo anesthesia and surgery. It is advisable to discontinue therapy with Anhydron for one week or so before elective surgery. (1, 2)
+ Reserpine: see Anhydron + Antihypertensive Agents
+ Tubocurarine: Thiazide compounds have been reported to enhance the effect of tubocurarine, great caution should be exercised in administering Anhydron to patients who are to be treated with curare or its derivatives. (1)
+ Veratrum Alkaloids: see Anhydron + Antihypertensive Agents

ANIONIC COMPOUNDS + Benasept Vaginal Jelly: see Benasept Vaginal Jelly

ANORECTICS: see Anorexiants

ANORECTICS + Furoxone: see Furoxone + Anorectics

ANOREXIANTS: see also Appetite Suppressants and Anorectics

ANOREXIANTS: see listed under the following agents:

Eutonyl Marplan MAO Inhibitor

ANSOLYSEN: see Ganglionic Blocking Agents

ANSOLYSEN: see listed under the following agents:
MAO Inhibitors
Nardil
Parnate

ANSOLYSEN
+ Alcohol: Alcohol, salt depletion, and occasionally exercise will potentiate Ansolysen. (1)
+ Anticholinergics: Concomitant administration may cause an exaggerated response to Ansolysen, particularly if constipation occurs. Combined use may also cause an exaggerated response to the anticholinergic. (1)
+ Exercise: see Ansolysen + Alcohol
+ Food: Once oral absorption of Ansolysen has taken place, the hypotensive effect of Ansolysen is enhanced by the ingestion of large amounts of food; this is probably related to splanchnic vasodilation associated with the digestive process. Because of this potentiation, dosage schedules should be arranged so that food is taken at a time when maximum drug effect is wearing off. (1)
+ Laxatives, Bulk: Plain bulk-type laxatives are not recommended for the treatment of constipation due to Ansolysen. (1)
+ Reserpine: When reserpine is used in conjunction with Ansolysen, there is frequently a reduction in the incidence of side effects and of Ansolysen requirements. (1)
+ Salt: see Ansolysen + Alcohol

ANTABUSE
+ Alcohol: The patient taking Antabuse must be warned that as long as he is taking this drug the ingestion of alcohol in any form will make him ill and endanger his life. The patient must learn to avoid disguised forms of alcohol such as sauces, fermented vinegar, cough syrups, and even aftershave lotions and back rubs. (111)
+ Brevital Sodium: see Brevital Sodium + Antabuse
+ Diabetes: Great care should be taken if Antabuse is to be administered in the presence of diabetes mellitus. (1)
+ Digitalis: A fall in serum potassium concentration has been demonstrated during the drug-alcohol reaction. Therefore potassium levels should be monitored particularly when patients are on digitalis therapy. Corrective measures should be taken as required.(1)
+ Paraldehyde: Antabuse should not be given to patients who have been recently treated with paraldehyde, and, conversely, paraldehyde

ANTABUSE (continued)

+ Paraldehyde (continued): should not be given to patients who have been treated with Antabuse. (1)

ANTACIDS: see also Aluminum Hydroxide Gel

ANTACIDS: see listed under the following agents:

Achromycin
Aureomycin
Dulcolax
Thrombin

ANTACIDS

+ Acidic Drugs: The following drugs, which are weak acide such as salicylic acid, Butazolidin, Furadantin, NegGram, and sulfonamides, will not be absorbed well in an alkaline media if, for example, the patient is taking antacids at the same time. Mechanism: Unionized drugs generally are lipid soluble and poorly water soluble. The ionized drugs are poorly lipid soluble and are water soluble. The membranes in the body (cell, blood vessels, and stomach and intestinal lining) are lipoidal and therefore uncharged forms of the drug pass freely through them; charged (ionized) forms transfer across cell membranes relatively slowly. The proportion of ionized drug is dependent on the pH of the fluid in a given compartment; therefore acidic drugs will be ionized in an alkaline media but not in an acidic media. (50, 55)
+ Butazolidin: see Antacids + Acidic Drugs
+ Furadantin: see Antacids + Acidic Drugs
+ NegGram: see Antacids + Acidic Drugs
+ Salicylic Acid: see Antacids + Acidic Drugs
+ Sulfonamides: see Antacids + Acidic Drugs

ANTHRACENE + Trisoralen: see Trisoralen

ANTIBACTERIAL AGENTS: see listed under the following agents:

Fungizone
Kantrex Inj.

ANTIBIOTICS: see listed under the following agents:

Aminosol
Anectine
Anticoagulants
Coly-Mycin M Inj.
Coumadin
Dicumarol
Flaxedil
Fungizone I.V.
Haldrone
Hedulin
Ilopan
Inversine
Loridine
Mucomyst
Mylaxen
Neomycin

ANTIBIOTICS (continued)

Panwarfin
Quelicin Chloride Inj.
Sintrom
Succinylcholine
Tromexan
Tubocurarine
Tubocurarine Chloride Inj.

ANTIBIOTICS, BACTERIOSTATIC + Penicillin: see Penicillin + Antibiotic, Bacteriostatic

ANTIBIOTICS, CURARE-LIKE: The following antibiotics have neuromuscular blocking-type action: Neomycin, Streptomycin, Kanamycin (Kantrex), Bacitracin, Colistin (Coly-Mycin), Vancomycin (Vancocin, Vinactane, Viocin), Dihydrostreptomycin, and Polymixin-B.

ANTIBIOTICS, CURARE-LIKE + Magnesium Sulfate: see Magnesium Sulfate Injection + Antibiotics, Curare-Like

ANTIBIOTICS, NEPHROTOXIC: see listed under the following agents:

Coly-Mycin M Inj.
Loridine

ANTIBIOTICS, NEUROTOXIC: see listed under the following agent:

Coly-Mycin M. Inj.

ANTIBIOTICS, OTOTOXIC: see listed under the following agents:

Mycifradin
Streptomycin

ANTICHOLINERGICS: see listed under the following agents:

Ansolysen
Aventyl
Deprol
Navane
Norpramin
Pertofrane
Phenothiazine
Phospholine Iodide
Tofranil
Vivactil
Vontrol

ANTICHOLINESTERASE AGENTS: see listed under the following agents:

Anectine
Ilopan
Phospholine Iodide
Sucostrin

ANTICHOLINESTERASE AGENTS

\+ Barbiturates: Although barbiturates are potentiated by the anticholinesterases, they may be used cautiously in the treatment of convulsions. (2)

\+ Organic Phosphorous Insecticides: Patients who have been on long-term treatment with anticholinesterase agents should avoid exposure

ANTICHOLINESTERASE AGENTS (continued)

+ Organic Phosphorous Insecticides (continued): to an atmosphere or area in which organic phosphorous insecticides have been used. (2)

+ Succinylcholine: In patients who have been on long-term treatment with anticholinesterase agents the administration of succinylcholine during general anesthesia may result in prolonged apnea. (2)

ANTICHOLINESTERASE PESTICIDES + Phospholine Iodide: see Phospholine Iodide + Anticholinesterase Pesticides

ANTICOAGULANTS: see listed under the following agents:

Adrenocorticosteroids
Alcohol
Analexin
Ananase
AquaMEPHYTON
Aspirin
Atarax
Atromid-S
Choloxin
Chymolase
Chymoral
Cuemid
Diphenylhydantoin
Dipyrone
Doriden
Euthroid
Fragicap-K
Griseofulvin
Haldol
Konakion
Librium
Loridine
Mephyton
Mithracin
Narone
Noctec
Orenzyme
Pabalate
Papase
Phenobarbital
Ponstel
Pyrilgin
Questran
Sulfasuxidine
Sulfathalidine
Synkavite
Tandearil
Thrombolysin
Thyrolar
Varidase
Vistaril
Zactrin Compound

ANTICOAGULANTS: The following agents may potentiate the hypoprothrombinemic effects of anticoagulants. The dose of the anticoagulant may have to be decreased; frequent prothrombin determinations are recommended until new anticoagulant dosage is established. See also individual agents.

Butazolidin
Diphenylhydantoin
Indocin
Tandearil

Mechanism: These agents will displace the anticoagulant from protein binding sites in blood plasma, thus making increased quantities of free anticoagulant available. (2, 9, 10, 11, 65)

ANTICOAGULANTS (continued, agents that may potentiate anticoagulants)

Aspirin Quinidine Quinine Salicylates	Mechanism: These agents depress prothrombin formation in the liver and the effect is additive with the anticoagulant. (2, 9, 10, 33, 60, 65)
Analexin	Mechanism: This agent potentiates the action of anticoagulants by inhibiting the metabolizing enzymes in the liver. (2, 65)
Antibiotics Aureomycin Chloromycetin Penicillin Sulfonamides	Mechanism: Antibiotics that suppress the organisms constituting the normal bacterial flora of the intestine and reduce production of vitamin K may increase anticoagulant activity. (2, 10, 11, 33, 60, 65)
Atromid-S	Mechanism: The increased sensitivity to anticoagulants that patients show when receiving Atromid-S may be the result of diminished availability of vitamin K present in plasma triglycerides, which are reduced. (10, 65)
Choloxin	Mechanism: The potentiation of anticoagulants by Choloxin has been theorized as the result of increasing the affinity of the anticoagulant for the repressor substance, a protein, inactivated by vitamin K, which governs the synthesis of clotting factors II, VII, IX, and X. Presumably vitamin K and the structurally similar coumarins compete for sites on the repressor protein. When these sites are filled by the anticoagulant, the repressor protein function is enhanced and the synthesis of clotting factors is inhibited. (10, 55)
Cuemid Questran	Mechanism: Chronic use of cholestyramine (Cuemid and Questran) may cause a hypoprothrombinemia associated with nutritional deficience of vitamin K. See also Questran. (1)
ACTH	
Alcohol (2, 11)	see also Alcohol + Anticoagulants
Anabolic Steroids (11)	
Anesthetics (60)	

ANTICOAGULANTS (continued, agents that may potentiate anticoagulants)

Atarax (112)
Carbon Tetrachloride (10)
Dianabol (11, 60)
Dipyrone
Drugs affecting Blood Elements
Drugs with Hepatotoxic Action (33, 60)
Low Choline
Low Cystine
Low Vitamin C
Methylthiouracil
Narcotics, Prolonged
Nilevar (10, 11, 55)
Ponstel (113)
Some Radio Compounds
Vistaril (112)

Chloral Hydrate	Several days of repeated doses of chloral hydrate administered to patients receiving oral anticoagulants will cause a potentiation of the hypoprothrombinemia. A reduction in the dose of the anticoagulant should be anticipated if chloral hydrate is given to the patient maintained on oral anticoagulants. Mechanism: Displacement of the anticoagulant from plasma protein binding sites. The displacement is believed to be caused by trichloracetic acid, the metabolite of chloral hydrate. (155, 156)

ANTICOAGULANTS: The following agents may cause a reversal of the hypoprothrombinemic effect of the anticoagulant necessitating an increase in the dose of the anticoagulant to maintain therapeutic range of anticoagulation. Caution is advised if the inhibiting agent is discontinued so as not to cause a severe hypoprothrombinemia.

Barbiturates (2, 9, 10, 54, 63, 103) Butabarbital (43) Doriden (9, 55) Griseofulvin Heptabarbital (31, 68) Meprobamate Phenobarbital (8, 11, 55, 64, 83, 103)	Mechanism: These preparations depress the response to anticoagulants by increasing the synthesis of drug metabolizing enzymes in the liver leading to more rapid inactivation of the anticoagulant.
Contraceptives, Oral	Mechanism: Oral contraceptives may inhibit the effects of anticoagulants. It is postulated that

ANTICOAGULANTS (continued, agents that may cause a reversal)

Contraceptives, Oral (continued) oral contraceptives have a vitamin K-like activity. (10, 19)

Antihistamines

Corticosteroids (2, 27)

Diet High in Vitamin K (alfalfa, cabbage, cauliflower, carrots, corn, fish, fish oils, liver, mushrooms, oats, peas, port, potatoes, soybeans, spinach, strawberry, tomato (green), tomato (ripe), wheat, wheat bran, wheat germ

Mineral Oil

Vitamin K

ANTICOAGULANTS + Estrogens: Estrogens antagonize the hypoprothrombinemic effect of the coumarin anticoagulants by increasing the vitamin K dependent coagulation factors. (154)

ANTICOAGULANTS, COUMARIN-TYPE: see listed under the following agents:

Analexin
Butazolidin
Dianabol
Tandearil

ANTICOAGULANTS, INDANDIONE + Dianabol: see Dianabol + Anticoagulants, Indandione

ANTICONVULSANTS: see listed under the following agents:

Dartal
Dilantin
Enduronyl
Gemonil
Haldol
Harmonyl
Moderil
Nardil
Navane
Rauwolfia Alkaloids
Ritalin
Singoserp
Sparine
Trilafon
Valium

ANTIDEPRESSANT DRUGS + Matulane: see Matulane + Antidepressant Drugs

ANTIDEPRESSANTS: see listed under the following agents:

Elavil
Eutonyl
Ismelin
MAO Inhibitor
Marplan
Parnate
Placidyl
Valium

ANTIDEPRESSANTS, TRICYCLIC: see also under individual agents: Aventyl, Elavil, Norpramin, Pertofrane, Tofranil, Vivactil

ANTIDEPRESSANTS, TRICYCLIC: see listed under the following agents:

Meperidine
Nardil
Niamid
Phenothiazines

ANTIDEPRESSANTS, TRICYCLIC

+ Amphetamines: The effects of amphetamine are augmented by the tricyclic antidepressants in animals and should be considered a possibility in humans. (59) Mechanism: It is postulated that the tricyclic antidepressants may interfere with the metabolism (hydoxylation in the liver) of the amphetamine, thus increasing the level of circulating amphetamine. (114)
+ Atropine: The effects of atropine are augmented by the tricyclic antidepressants. Mechanism: The tricyclic antidepressants have anticholinergic activity which is additive to atropine. (55)
+ Benzodiazepines: see Benzodiazepines + Antidepressants, Tricyclic
+ Ismelin: The use of the tricyclic antidepressants will often negate the antihypertensive effects of agents in the Ismelin class. Mechanism: The tricyclic antidepressants will prevent the uptake of Ismelin at the adrenergic nerve endings. (55)
+ MAO Inhibitors: The administration of tricyclic antidepressants with or shortly after MAO Inhibitors have caused severe reactions and even fatalities. Symptoms have ranged from dizziness, nausea, and excitation to coma, hyperpyrexia, convulsions, tremor, delerium, seizures, rigidity, and circulatory collapse. (50, 55, 115) Mechanism: see MAO Inhibitors + Antidepressants, Tricyclic.

ANTIEMETICS

+ Carbocaine: see Carbocaine + Antiemetics
+ Streptomycin: The antiemetic agents should be used cautiously, for they mask the presence of underlying organic abnormalities or toxic effects of such drugs as Streptomycin. (2)

ANTIEMETICS, PHENOTHIAZINE-TYPE

+ Alcohol: see Antiemetics, Phenothiazine-Type + CNS Depressants
+ Analgesics: see Antiemetics, Phenothiazine-Type + CNS Depressants
+ Anesthetics, General: see Antiemetics, Phenothiazine-Type + CNS Depressants
+ Barbiturates: see Antiemetics, Phenothiazine-Type + CNS Depressants
+ CNS Depressants: The additive or potentiating effect of the phenothiazine compounds on the action of other central nervous system depressants should be considered before an antiemetic of this type is given to patients under the influence of alcohol, barbiturates, potent

ANTIEMETICS, PHENOTHIAZINE-TYPE (continued)

+ CNS Depressants (continued): analgesics, or general anesthetics. (2)
+ Streptomycin: The antiemetic agents should be used cautiously, since they may mask the presense of underlying organic abnormalities or the toxic effects of such drugs as Streptomycin. (2)

ANTIHEMORRHAGIC AGENTS + Fragicap-K: see Fragicap-K + Antihemorrhagic Agents

ANTIHISTAMINES: see listed under the following agents:

Adrenosem
Adrestat
Compazine
Coumadin
Eutonyl
Furoxone
Hedulin
Levoprome
Liquaemin Sodium
Matulane
Navane
Parnate
Permitil
Phenothiazine
Prolixin
Repoise
Serc
Temaril
Thorazine
Tindal
Trilafon
Tromexan
Vesprin

ANTIHISTAMINES

+ Alcohol: Patients should be cautioned against taking alcoholic beverages during antihistamine therapy, since the depressant action of these drugs is additive. (2)
+ Barbiturates: Patients should be cautioned against taking barbiturates during antihistamine therapy, since the depressant action of these drugs is additive. (2)

ANTIHYPERTENSIVE AGENTS: see listed under the following agents:

Aldactone
Aldomet
Anhydron
Aquatag
Bristuron
Compazine
Dartal
Diuril
Dopar
Dyazide
Dyrenium
Edecrin
Enduron
Esidrix
Eutonyl
Exna
Haldol
HydroDIURIL
Hydromox
Hygroton
Inversine
Isordil
Laradopa
Lasix

ANTIHYPERTENSIVE AGENTS (continued):

Levodopa	Proketazine
Levoprome	Prolixin
MAO Inhibitor	Regitine
Marplan	Renese
Mellaril	Repoise
Metahydrin	RoCYTE
Naqua	Saluron
Nardil	Singoserp
Naturetin	Singoserp-Esidrix
Navane	Stelazine
Niamid	Taractan
Nicalex	Tindal
Oretic	Trilafon
Pacatal	Unitensin
Parnate	Vesprin
Permitil	Vivactil
Phenothiazine	

ANTI-INFECTIVE AGENTS + Surgical Absorbable Hemostat: see Surgical Absorbable Hemostat + Anti-Infective Agents

ANTIMALARIAL DRUGS: see listed under the following agents:
- Quelicin Chloride Inj.
- Succinylcholine

ANTIMETABOLIC AGENTS
- + Cytotoxic Agents: see Antimetabolic Agents + Surgery
- + Radiation Therapy: see Antimetabolic Agents + Surgery
- + Surgery: The antimetabolic agents are contraindicated for patients in poor nutritional state or those with minimal leukocyte and thrombocyte counts because the bone marrow may become further depressed. Such conditions are likely to occur when the patient has had recent surgery, radiation therapy, or treatment with other cytotoxic drugs. (2)

ANTIMETABOLITES: see listed under the following agents:
- Attenuvax, Lyovac
- Biavax
- Cendevax
- Fungizone
- Lirugen
- Measles Virus Vaccine, Live, Attenuated

ANTIMETABOLITES (continued):

Measles Virus Vaccine, Live, Attenuated (Edministon B and Schwarz Strain)
Mumpsvax, Lyovac
M-Vac Measles Virus Vaccine, Live, Attenuated
Pfizer-Vax Measles-L
Rubeovax, Lyovac

ANTIMICROBIAL AGENTS: see listed under the following agents:

Staphcillin
Thio-Tepa

ANTI-NAUSEA SUPRETTES

+ Alcohol: see Anti-Nausea Suprettes + CNS Depressants
+ CNS Depressants: Patients should be warned against taking alcoholic beverages or other central nervous system depressants while taking Anti-Nausea Suprettes. (1)
+ Hypnotics: see Anti-Nausea Suprettes + Sedatives
+ Narcotics: see Anti-Nausea Suprettes + Sedatives
+ Sedatives: Use cautiously if employed with other sedative, hypnotic, or narcotic agents. (1)

ANTINEOPLASTIC AGENTS: see listed under the following agents:

Alkeran
Alkylating Agents
Cosmegen
Hydrea

ANTIPARKINSONISM AGENTS: see listed under the following agents:

Dopar
Eutonyl
Haldol
Laradopa
Levodopa
MAO Inhibitor
Parnate
Pertofrane
Tofranil

ANTIPYRINE: see Alkaline Agents + Acidic Media and Alkaline Agents + Acidifying Agents (Urinary)

ANTISEPTICS: see listed under the following agents:

Acne Mixed UBA
Attenuvax, Lyovac
Measles Virus Vaccine, Live, Attenuated
Mumpsvax, Lyovac
M-Vac Measles Virus Vaccine, Live, Attenuated
Pfizer-Vax Measles-L
Staphylococcus-Streptococcus UBA

ANTITHYROID DRUGS: see listed under the following agents:
Radiocaps-131
Sodium Iodide I-131

ANTURANE
+ Citrates: see Anturane + Salicylates
+ Gantrisin: see Anturane + Sulfadiazine
+ Insulin: see Anturane + Sulfonylureas, Hypoglycemic
+ Salicylates: Salicylates and citrates antagonize the action of Anturane and are contraindicated. (1, 2)
+ Sulfadiazine: Recent reports have indicated that Anturane potentiates the action of certain sulfonamides, such as sulfadiazine and Gantrisin (sulfisoxazole). In view of these observations, it is suggested that Anturane be used with caution in conjunction with sulfa drugs. (1) Mechanism: Anturane will displace sulfonamides from protein binding sites in blood plasma thus making increased quantities of free sulfonamides available. (116)
+ Sulfonamides: see Anturane + Sulfadiazine
+ Sulfonamides, Long-Acting: see Sulfonamides, Long-Acting + Anturane
+ Sulfonylureas, Hypoglycemic: Other pyrazole compounds (e.g., Butazolidin) have been observed to potentiate the hypoglycemic sulfonylurea agents, as well as insulin. In view of these observations, it is suggested that Anturane be used with caution in conjunction with the sulfonylurea hypoglycemic agents and insulin. (1, 2) Mechanism: It is postulated that the pyrazole compounds decrease the rate of metabolism of the hypoglycemic sulfonylureas. (18)

APODOL
+ Anesthetics: see Apodol + Narcotics
+ Narcotics: Special caution is indicated when Apodol is used in conjunction with other narcotics, with sedatives, or with anesthetics, since these agents may enhance respiratory depression. (1)
+ Sedatives: see Apodol + Narcotics

APPETITE SUPPRESSANTS: see listed under the following agents:
Hypertensin
Ismelin

APPETROL: see Miltown and Dextroamphetamine (Dexedrine)

APRESOLINE: see listed under the following agents:
Aquatag
Bristuron
Diuretics, Thiazide
Diuril
Esidrix
Exna

APRESOLINE (continued)

HydroDIURIL	Naturetin
Hydromox	Raudixin
Hygroton	Rauwiloid
Ismelin	Saluron
Metahydrin	Singoserp
Naqua	Unitensin

APRESOLINE

+ Alcohol: The narcotic effect of alcohol has been potentiated by Apresoline. (1)
+ Barbiturates: Although Apresoline alone does not have a sedative or hypnotic effect, extreme drowsiness has occurred in patients taking it with a barbiturate. (1)
+ Epinephrine: Apresoline may reduce the pressor response to epinephrine. (1)
+ Ganglionic Blocking Agents: When Apresoline is given in conjunction with ganglionic blocking agents, dosage must be reduced by at least 50%, and effects should be watched carefully. (1)

AQUACHLORAL SUPRETTES: see Chloral Hydrate

AQUACORT SUPRETTES

+ Corticosteroids: Concurrent therapy with oral or injectable corticosteroids or mercurials should be avoided. (1)
+ Mercurials: see Aquacort Suprettes + Corticosteroids

AQUALIN SUPRETTES + Theophylline: Do not administer other theophylline preparations concurrently. (1)

AQUAMEPHYTON: see Mephyton

AQUAMEPHYTON

+ Anticoagulants: When AquaMEPHYTON is used to correct excessive anticoagulant therapy induced hypoprothrombinemia, anticoagulant therapy still being indicated, the patient is again faced with the clotting hazards existing before starting the anticoagulant therapy. Temporary resistance to prothrombin-depressing anticoagulants may result, especially when larger doses of AquaMEPHYTON are used. (1)
+ Diluents: Diluents other than 0.9% Sodium Chloride Injection or 5% Dextrose and Sodium Chloride Injection should not be used with AquaMEPHYTON. (1)

AQUATAG: see Diuretics, Thiazide

AQUATAG (continued)

+ Antihypertensive Agents: Aquatag is a mildly antihypertensive agent in its own right and enhances the action of other antihypertensive drugs when used in combination. Since a lower blood pressure is frequently established under the potentiating effect of Aquatag, it is mandatory to reduce the dosage of therapeutic agents such as ganglionic blockers, Apresoline, Reserpine, or Ismelin by at least 50% immediately upon the addition of Aquatag to the therapeutic regimen in order to prevent excessive antihypertensive effects. (1, 2)
+ Apresoline: see Aquatag + Antihypertensive Agents
+ Corticosteroids: It should be noted that the concomitant administration of corticosteroids or corticotrophin may increase the loss of potassium. (1, 2)
+ Corticotrophin: see Aquatag + Corticosteroids
+ Diabetes: In patients with latent or overt diabetes mellitus, hypoglycemia and glycosuria may be precipitated or increased respectively by thiazide therapy. (1)
+ Digitalis: Hypokalemia, due to the thiazide, may result in increased sensitivity to the action of digitalis and the precipitation of digitalis toxicity. (1, 2)
+ Ganglionic Blockers: see Aquatag + Antihypertensive Agents
+ Ismelin: see Aquatag + Antihypertensive Agents
+ Norepinephrine: The thiazide diuretics may decrease arterial responsiveness to norepinephrine; therefore, caution should be observed in thiazide-treated patients undergoing surgery. (1, 2)
+ Reserpine: see Aquatag + Antihypertensive Agents
+ Trisoralen: see Trisoralen
+ Tubocurarine: The thiazides may increase the paralytic effects of tubocurarine. (1)

ARALEN: see also Alkaline Agents + Acidic Media and Alkaline Media + Acidifying Agents (Urinary)

ARALEN

+ Ammonium Chloride: If serious toxic symptoms occur, oral use of ammonium chloride (8 gm daily) in divided doses for three to four days a week for several months increases excretion of the drug. Acidification of the urine increases renal excretion of the 4-aminoquinolone compounds by 20 to 90%. (1)
+ Butazolidin: see Aralen + Drugs
+ Drugs: Aralen should not be used with drugs known to cause dermatitis and drug sensitization, such as Butazolidin and gold. (1)
+ Gold: see Aralen + Drugs

ARALEN (continued)
+ Hepatotoxic Drugs: Use Aralen with caution in conjunction with known hepatotoxic drugs, since Aralen is known to concentrate in the liver. (1)
+ Primaquine: see Primaquine + Aralen

ARALEN WITH PRIMAQUINE: see Aralen and Primaquine

ARALEN WITH PRIMAQUINE
+ Bone Marrow Depressants: Aralen with Primaquine is contraindicated in those patients receiving other potentially hemolytic drugs or depressants of myeloid elements of the bone marrow. (1)
+ Drugs, Hemolytic: see Aralen with Primaquine + Bone Marrow Depressants

ARAMINE: see listed under the following agents:
Aldomet
Eutonyl
Ismelin
Phenothiazine
Reserpine

ARAMINE
+ Adrenolytic Agents: The pressor effects of Aramine is decreased but not reversed by adrenolytic agents. (1)
+ Cyclopropane: Experience has demonstrated that Aramine is relatively free of the danger of causing cardiac arrhythmias when given as recommended. Nevertheless, it is advisable to avoid the use of this vasopressor with cyclopropane or halothane anesthesia unless clinical circumstances demand such use. (1) Mechanism: The retention of carbon dioxide, during anesthesia, causes an increase in sympathetic nervous activity and a liberation of norepinephrine within the myocardium and specialized conducting tissue. In addition, cyclopropane in some manner "sensitizes" the heart to the actions of norepinephrine and related catecholamines. (123)
+ Diabetes: Use Aramine with caution in diabetic patients because of its vasoconstrictor effect. (1)
+ Halothane: see Aramine + Cyclopropane
+ MAO Inhibitors: Monoamine oxidase inhibitors may potentiate the action of Aramine and other sympathomimetic agents. (1) Mechanism: see MAO Inhibitor + Sympathomimetics, Indirect-Acting.

ARFONAD + Penthrane: see Penthrane + Arfonad

ARISTOCORT + Diabetes: If the use of Aristocort is considered in this condition, the risks involved must be weighed against the gains involved. (1)

ARISTOCORT FORTE SUSPENSION
+ Diluents: The use of diluents containing preservatives, such as methylparaben, propylparaben, phenol, etc., must be avoided as these preparations tend to cause flocculation of the steroid. (1)
+ Methylparaben: see Aristocort Forte Suspension + Diluents
+ Phenol: see Aristocort Forte Suspension + Diluents
+ Propylparaben: see Aristocort Forte Suspension + Diluents

ARSENIC + Neo-Vagisol: see Neo-Vagisol + Arsenic

ARTANE: see listed under the following agents:
Dopar
Larodopa
Levodopa

ASBRON + Xanthine Derivatives: Do not administer within 12 hours after administration of, or concurrently with, other xanthine derivatives. (1)

ASCORBIC ACID: see listed under the following agents:
Hemastix
Premarin I.V.
Redisol

ASMOLIN + Anesthesia, Surgical: Ventricular arrhythmias and fibrillation may follow the use of epinephrine, particularly during surgical anesthesia. Use of Asmolin may be dangerous in such circumstances. (1)

ASPARTIC ACID + Velban: see Velban + Aspartic Acid

ASPIRIN
+ Anticoagulants: Aspirin will potentiate the hypoprothrombinemic effects of anticoagulants. (55)
+ Diuretics, Thiazide: Aspirin has been known to elevate serum uric acid; thiazide diuretics may also elevate serum uric acid. The patient who takes both runs a higher risk of precipitating acute gout in predisposed patients. (55)
+ Heparin: see Heparin + Aspirin
+ PAS: Aspirin in combination with para-amino salicylic acid (PAS) may cause an increased PAS toxicity. (55)

ASTRAFER + Iron: Oral iron should always be discontinued prior to instituting parenteral therapy because of the risk of exceeding the patient's

ASTRAFER + Iron (continued): unsaturated iron-binding capacity with subsequent potential signs of iron toxicity. (1)

ATABRINE: see Alkaline Agents + Acidifying Agents (Urinary)

ATABRINE: see listed under the following agents:
Pamaquine
Primaquine

ATABRINE
+ Ammonium Chloride: If serious toxic symptoms occur from overdosage or sensitivity, the oral administration of ammonium chloride (8 gm daily in divided doses) has been suggested for three or four days a week for several months after therapy has stopped, since acidification of urine increases renal excretion of most antimalarial compounds by 20 to 90%. (1)
+ Butazolidin: see Atabrine + Drugs
+ Drugs: Concomitant administration of medicaments such as Butazolidin, gold, and other drugs known to cause sensitization and dermatitis should be avoided. (1)
+ Gold: see Atabrine + Drugs

ATARAX
+ Alcohol: Patients should avoid the concomitant use of Atarax and alcohol, since the effects may be additive. (2)
+ Anticoagulants: It has been observed in some patients receiving anticoagulant medication concurrently with Atarax that the dosage of the anticoagulant will have to be decreased. (1, 112)
+ Barbiturates: see Atarax + Meperidine
+ CNS Depressants: The potentiating action of Atarax must be considered when the drug is used in conjunction with central nervous system depressants. Therefore, when central nervous system depressants are administered with Atarax, their dosage should be reduced up to 50%. (1)
+ MAO Inhibitors: see Atarax + Psychotropic Agents
+ Meperidine: Atarax may potentiate meperidine and barbiturates so their use in preanesthetic adjunctive therapy should be modified on an individual basis. See also Atarax + Narcotics. (1)
+ Narcotics: When Atarax is used preoperatively or prepartum, narcotic requirements may be reduced as much as 50%. (1, 2)
+ Phenothiazines: see Atarax + Psychotropic Agents
+ Psychotropic Agents: Other psychotropic agents, particularly phenothiazines or monoamine oxidase inhibitors, that are known to potentiate the action of other drugs should not be given with Atarax. (2)

ATROMID-S: see listed under the following agents:

Anticoagulants
Coumadin
Dicumarol
Panwarfin

ATROMID-S + Anticoagulants: Caution should be exercised when anticoagulants are given in conjunction with Atromid-S. The dosage of the anticoagulant should be reduced by one-half (depending on the individual case) to maintain the prothrombin time at the desired level to prevent bleeding complications. Frequent prothrombin determinations are advisable until it has been determined that the levels have been stabilized. (1, 13) Mechanism: see Anticoagulants + Atromid-S.

ATROPINE: see listed under the following agents:

Antidepressants, Tricyclic
Brevital
Cholinergic Drugs
Compazine
Cyclopropane
Dartal
Levoprome
MAO Inhibitor
Mellaril
Mestinon
Navane
Permitil
Phenothiazine
Prolixin
Prostigmin
Protopam
Repoise
Serentil
Thorazine
Tindal
Torecan
Trilafon
Vesprin

ATROPINE + Mestinon: Atropine may be used to abolish or obtund gastrointestinal side effects or other muscarinic reactions; but such use, by masking signs of overdosage, can lead to inadvertent induction of cholinergic crisis. (1)

ATROPINE-LIKE DRUGS + Benadryl: see Benadryl + Atropine-Like Drugs

ATTENUVAX, LYOVAC

+ Alkylating Agents: see Attenuvax, Lyovac + Corticosteroids
+ Antimetabolites: see Attenuvax, Lyovac + Corticosteroids
+ Antiseptics: see Attenuvax, Lyovac + Preservatives
+ Corticosteroids: Attenuvax, Lyovac is contraindicated in patients receiving therapy with corticosteroids, irradiation, alkylating agents, or antimetabolites. (1)
+ Detergents: see Attenuvax, Lyovac + Preservatives
+ Immune Serum Globulin (Human): see Attenuvax, Lyovac + Whole Blood (Human)

ATTENUVAX, LYOVAC (continued)

+ Immunization, Elective: In general, it is good practice to give Attenuvax one month before or after other elective immunization. (1)
+ Irradiation: see Attenuvax, Lyovac + Corticosteroids
+ Plasma, Human: see Attenuvax, Lyovac + Whole Blood (Human)
+ Preservatives: A new unused sterile disposable syringe with a 25 gauge 5/8 inch needle should be used for each injection of the vaccine because certain preservatives, antiseptics, and detergents will inactivate the live measles virus vaccine. (1)
+ Tuberculin Test: It has been reported that attenuated live-virus measles vaccine may temporarily depress tuberculin skin sensitivity. Therefore, if a tuberculin test is to be done, it should be scheduled before administering measles vaccine. This avoids the possibility of a false-negative response. (1)
+ Whole Blood (Human): Vaccination should be deferred for six weeks following receipt of whole blood (human), immune serum globulin (human), or human plasma. (1)

AUREOMYCIN

+ Antacids: see Aureomycin + Food
+ Coumadin: see Coumadin + Aureomycin
+ Dairy Products: see Aureomycin + Food
+ Food: Absorption of Aureomycin is impaired by the concomitant administration of high calcium content drugs such as antacid medications, foods, and some dairy products such as milk. Oral forms of Aureomycin should be given one hour before or two hours after meals. Mechanism: Tetracyclines are chelating agents that form insoluble chelates with multivalent cations such as calcium, magnesium, and aluminum. (1)
+ High Calcium Content Drugs: see Aureomycin + Food
+ Milk: see Aureomycin + Food
+ Trisoralen: see Trisoralen

AVENTYL: see also Antidepressants, Tricyclic

AVENTYL: see listed under the following agents:

MAO Inhibitors
Marplan
Nardil
Niamid
Parnate

AVENTYL

+ Alcohol: see Aventyl + CNS Drugs

AVENTYL (continued)

+ Anticholinergic Agents: Other agents that may potentiate the effects of Aventyl include anticholinergic agents, sympathomimetic compounds, barbiturates, alcohol, and thyroid preparations. (2)
+ Barbiturates: see Aventyl + Anticholinergic Agents
+ CNS Drugs: Concurrent administration of other central nervous system drugs or alcohol may potentiate the adverse effects of Aventyl. (1)
+ Electroconvulsive Therapy: Concomitant use of Aventyl and electroconvulsive therapy (with or without atropine, short acting barbiturates, and muscle relaxants) has not been thoroughly studied. In the event these treatments are used together, the physician should be aware of the possibility of added adverse effects. (1)
+ Hexobarbital-Methamphetamine: The use of Aventyl in office practice along with hexobarbital-methamphetamine in hypnonarcosis for severe anxiety has resulted in higher-than-usual incidence of stimulatory side effects. (1)
+ Hypotensive Agents: Special care should be taken when Aventyl is used with other agents that lower blood pressure (e.g., thiazide diuretics, phenothiazine compounds, vasodilators). (2)
+ Ismelin: Patients receiving a tricyclic antidepressant (e.g., Aventyl) may respond poorly to hypotensive agents such as Ismelin. (1) <u>Mechanism</u>: see Antidepressants, Tricyclic + Ismelin
+ Levophed: Aventyl may increase the pressor effect of Levophed. (2)
+ MAO Inhibitor: Concomitant use of Aventyl with a monoamine oxidase inhibitor is contraindicated. The potentiation of adverse effects may produce severe atropine-like reactions, tremors, hyperpyrexia, generalized clonic convulsions, delerium, and even death. It is advisable to discontinue the monoamine oxidase inhibitor for at least ten to twenty-one days before starting treatment with Aventyl. (1, 2) <u>Mechanism</u>: see MAO Inhibitors + Antidepressants, Tricyclic.
+ Phenethylamine: Aventyl decreases the pressor effect of phenethylamine. (2)
+ Phenothiazines: see Aventyl + Hypotensive Agents
+ Sympathomimetic Agents: see Aventyl + Anticholinergic Agents
+ Thiazide Diuretics: see Aventyl + Hypotensive Agents
+ Thyroid Preparations: see Aventyl + Anticholinergic Agents
+ Trisoralen: see Trisoralen
+ Vasodilators: see Aventyl + Hypotensive Agents

AZO GANTANOL + Dolonil: see Dolonil + Azo Gantanol

AZO GANTRISIN: see Gantrisin

AZO GANTRISIN + Dolonil: see Dolonil + Azo Gantrisin

AZOLATE + Dolonil: see Dolonil + Azolate

AZO-MANDELAMINE + Dolonil: see Dolonil + Azo-Mandelamine

AZOTREX + Dolonil: see Dolonil + Azotrex

B

BACITRACIN: see listed under the following agents:

- Anectine
- Magnesium Sulfate Inj.
- Succinylcholine

BACTERIOSTATIC AGENTS + Fungizone I.V.: see Fungizone I.V. + Bacteriostatic Agents

BACTERIOSTATS + Brevital Sodium: see Brevital Sodium + Bacteriostats

BAL + Iron: see Iron + BAL

BAMADEX: see Dextro-amphetamine and Meprobamate

BARBITURATES: see listed under the following agents:

- Alcohol
- Alvodine
- Amytal Sodium
- Analgesics
- Anhydron
- Anticholinesterase Agents
- Anticoagulants
- Antiemetics, Phenothiazine-type
- Antihistamines
- Apresoline
- Atarax
- Aventyl
- Benzodiazepines
- Bristuron
- Carbocaine
- Compazine
- Coumadin
- Dartal
- Diabinese
- Dicumarol
- Dilantin
- Diuretics, Thiazide
- Diuril
- Dymelor
- Elavil
- Esidrix
- Eutonyl
- Haldol
- Harmonyl
- Hedulin
- HydroDIURIL
- Hydromox
- Hygroton
- Hypoglycemic Drugs, Oral
- Ilopan
- Inapsine
- Ketaject
- Ketalar
- Largon
- Levoprome
- Lomotil
- Lorfan
- Magnesium Sulfate Inj.
- Marplan
- Matulane
- Medigesic
- Mepergan
- Methadone
- Metubine Iodide
- MAO Inhibitor

BARBITURATES (continued)

Modumate
Nalline
Narcotic Antagonist
Nardil
Navane
Niamid
Nisentil
Novocaine
Numorphan
Organophosphate Pesticides
Panwarfin
Parnate
Penthrane
Permitil
Pertofrane
Phenergan
Phenothiazines
Phenoxene
Placidyl
Prinadol
Probanthine with Dartal
Prolixin
Protopam
Quaalude
Repoise
Ritalin
Sparine
Sublimaze
Tacaryl
Talwin
Temaril
Theratuss
Thorazine
Tigan
Tindal
Trilafon
Tromexan
Valium
Vesprin
Vistaril

BASES, WEAK + Ammonium Chloride: see Ammonium Chloride + Weak Bases

BASIC COMPOUNDS + Paveril Phosphate: see Paveril Phosphate + Basic Compounds

BASIC SOLUTIONS + Keflin: see Keflin + Basic Solutions

BCG VACCINE + Corticosteroids: BCG Vaccine probably should not be given to individuals on chronic corticosteroid therapy. (1)

BELLADONNA DERIVATIVES: see listed under the following agents:

Mytelase
Tigan

BENADRYL

+ Anticholinergics: Benadryl has an atropine-like action which should be considered when prescribing Benadryl. (1)
+ Hypnotics: Hypnotics, sedatives, or tranquilizers, if used with Benadryl, should be prescribed with caution because of possible additive effect. (1)
+ Sedatives: see Benadryl + Hypnotics
+ Tranquilizers: see Benadryl + Hypnotics

BENASEPT VAGINAL JELLY

+ Anionic Compounds: Douches containing anionic compounds or soap should be avoided. (1)
+ Soap: see Benasept Vaginal Jelly + Anionic Compounds

BENEDICT'S SOLUTION: see listed under the following agents:

Benemid
Keflin
NegGram
Skelaxin

BENEMID: see listed under the following agents:

Diabinese
Dymelor
Edecrin
Geopen
Hypoglycemic Drugs, Oral
Keflin
Orinase
Penbritin
Polycillin
Pyopen
Tolinase

BENEMID

+ Aminosalicylic Acid (PAS): Benemid increases the plasma levels of aminosalicylic acid. (2)
+ p-Aminohippuric Acid (PAH): Benemid delays elimination of PAH. (1)
+ Benedict's Solution: A reducing substance may appear in the urine of patients treated with Benemid. Although this disappears with discontinuance of therapy, a false diagnosis of glycosuria may be made because of false-positive Benedict's test. (1, 2)
+ Bromsulphalein (Sodium Sulfobromophthalein): There is some evidence that Benemid inhibits the urinary and hepatic excretion of Bromsulphalein. (2)
+ Erythromycin: In amimals, tubular reabsorption of erythromycin is inhibited. (1)
+ Neo-Iopax: Benemid delays elimination of Neo-Iopax. (1)
+ Pantothenic Acid: Benemid delays elimination of Pantothenic Acid. (1)
+ Penicillin: Benemid will cause an elevation and prolongation of plasma levels of penicillin by whatever route the antibiotic is given. Elevation of penillemia approximating twofold to fourfold has been demonstrated for various forms of penicillin (e.g., sodium and potassium penicillin G, benzathine penicillin G, penicillin O, procaine penicillin G, and phenoxymethyl penicillin (penicillin V)). (1, 2) Mechanism: The delay in excretion of penicillin is due to a direct tubular effect of the Benemid. (50)

BENEMID (continued)

+ Phenolsulfonphthalein: The renal clearance of phenolsulfonphthalein is reduced to about one-fifth of the normal rate when the dosage of Benemid is adequate. (1, 2)
+ Potassium Citrate: see Benemid + Sodium Bicarbonate
+ Salicylates: Salicylates should not be given with Benemid, since co-administration results in inhibition of the uricosuric activity of the Benemid. (1, 2) Mechanism: Salicylates may interfere with the uricosuric effects of Benemid by antagonistic action on the tubular sites of uric acid clearance but not by actually altering clearance of either drug. (50)
+ Sodium Bicarbonate: As urates tend to crystallize out of an acid urine, a liberal fluid intake is recommended, as well as sufficient sodium bicarbonate (3 to 7.5 gm daily) or potassium citrate (7.5 gm daily) to maintain an alkaline urine. The acid-base balance of the patient should be watched when this amount of alkali is given. (1)
+ Sulfa Drugs: Benemid raises the plasma level of conjugated sulfa drugs; therefore plasma concentrations should be determined from time to time when a sulfa drug and Benemid are co-administered for prolonged periods of time. (1, 2)
+ Urokon: In animals, renal excretion of Urokon Sodium is decreased. (1)

BENZEDRINE + MAO Inhibitors: Do not use Benzedrine in patients taking MAO Inhibitors. (1) Mechanism: see MAO Inhibitors + Sympathomimetics, Indirect-Acting

BENZODIAZEPINES: see Librium, Valium, and Serax

BENZODIAZEPINES

+ Alcohol: The benzodiazepines may potentiate the sedative and/or atropine-like effects of alcohol, barbiturates, phenothiazines, and tricyclic antidepressants. (50, 55)
+ Anticoagulants, Oral: The benzodiazepines may alter the effects of oral anticoagulants. (50)
+ Antidepressants, Tricyclic: see Benzodiazepines + Alcohol
+ Barbiturates: see Benzodiazepines + Alcohol
+ MAO Inhibitors: Monoamine oxidase inhibitors enhance the sedative effects of benzodiazepines and severe atropine-like reactions may occur. (55, 59)
+ Phenothiazines: see Benzodiazepines + Alcohol (59)

BIAVAX

+ ACTH: Use of Biavax is contraindicated in patients receiving therapy with ACTH, corticosteroids, irradiation, alkylating agents, or

BIAVAX (continued)

+ ACTH (continued): antimetabolites because these agents depress the normal defense mechanisms. (1)
+ Alkylating Agents: see Biavax + ACTH
+ Antimetabolites: see Biavax + ACTH
+ Corticosteroids: see Biavax + ACTH
+ Irradiation: see Biavax + ACTH
+ Tuberculin Test: It has been reported that attenuated live virus mumps vaccine may temporarily depress tuberculin skin sensitivity for four weeks or longer. Therefore, if a tuberculin test is to be done, it should be scheduled before administering mumps vaccine. This avoids the possibility of a false-negative response. (1)

BIPHETAMINE-T: This agent contains methqualone. See Quaalude or Sopor

BLOOD: see listed under the following agents:

Hypertensin
Lipomul I.V.
Travert

BLOOD

+ Calcium: Isotonic saline solutions containing physiologic amounts of other cations in addition to sodium may be used, provided that materials containing calcium are avoided. Calcium will clot the usual citrated bank blood. (52)
+ Dextrose 5%: Dextrose 5% in distilled water should not be added to blood transfusions. Its low ionic strength results in aggregation of red blood cells. Prolonged incubation of red blood cells in dextrose solution leads to impaired viability and even hemolysis. In situations where there is need to minimize sodium infusion, one-quarter strength saline in dextrose may safely be used. (52)

BLOOD CROSSMATCHING + Dextran: see Dextran + Blood Crossmatching

BLOOD DERIVATIVES: see listed under the following agents:

Diuril
Edecrin

BLOOD GROUPING + Dextran: see Dextran + Blood Grouping

BLOOD TRANSFUSION: see listed under the following agents:

Cendevax
Measles Virus Vaccine, Live, Attenuated
Pfizer-Vax Measles-L
Rubeovax, Lyovac

BLOOD TYPING + Dextran: see Dextran + Blood Typing

BLOOD, WHOLE: see listed under the following agents:

Attenuvax, Lyovac
Diuril
Edecrin
Hypertensin
Levophed
Liquamar
Lirugen
Measles Virus Vaccine, Live, Attenuated
Merthiolate
Miradon
M-Vac Measles Virus Vaccine, Live, Attenuated
Sintrom

BONE MARROW DEPRESSANTS: see listed under the following agents:

Aralen with Primaquine
Chloromycetin
Cytosar
Fluorouracil
Leukeran
Matulane
Mithracin
Mychel
Permitil
Phenothiazines
Repoise
Stelazine
Thorazine
Tindal
Trilafon

BORIC ACID + Stoxil: see Stoxil + Boric Acid

BREVITAL Sodium

+ Acidic Solutions: Solutions of Brevital Sodium should not be mixed with acidic solutions, such as Atropine Sulfate, Metubine Iodide, and Succinylcholine. The soluble sodium salts of barbiturates are the forms used for intravenous administration. Solubility is maintained only at a relatively high (basic) pH. (1)
+ Antabuse: see Brevital Sodium + Hypotensive Agents
+ Atropine Sulfate: see Brevital Sodium + Acidic Solutions
+ Bacteriostats: When Brevital Sodium is prepared for intravenous use, sterile water for injection is the preferred diluent. Bacteriostats are incompatible. (1)
+ Hypotensive Agents: Patients who have received hypotensive agents, steroids, or Antabuse must be considered increased anesthetic risks. (1)
+ Metubine Iodide: see Brevital Sodium + Acidic Solutions and Metubine Iodide + Brevital Sodium
+ Steroids: see Brevital Sodium + Hypotensive Agents
+ Succinylcholine: see Brevital Sodium + Acidic Solutions. Succinylcholine is rapidly hydrolyzed by alkaline solutions and therefore loses potency rapidly. Such mixtures, if used at all, must be used within a few minutes of preparation; however, separate injections are preferable.

BRISTURON: see Diuretics, Thiazide

BRISTURON

+ Alcohol: Orthostatic hypotension caused by Bristuron may be potentiated when combined with either alcohol, barbiturates, or narcotics. (1)
+ Anesthesia, General: see Bristuron + Tubocurarine and Bristuron + Pressor Amines
+ Antihypertensive Agents: Since Bristuron is capable of enhancing the effects of other antihypertensive agents (e.g., ganglionic blocking agents, Apresoline, Ismelin, veratrum derivatives, and rauwolfia alkaloids), combination therapy with these drugs may necessitate a reduction in the dose of the latter. The dose of ganglionic blocking agents should be reduced by at least one-half. (1, 2)
+ Apresoline: see Bristuron + Antihypertensive Agents
+ Barbiturates: see Bristuron + Alcohol
+ Corticotrophin: see Bristuron + Steroids
+ Curare and Derivatives: see Bristuron + Tubocurarine
+ Diabetes: Hyperglycemia and glycosuria have occurred in patients receiving thiazides. Patients who have diabetes mellitus or who are suspected of being prediabetic should be kept under close observation if treated with this agent. Adjustment of diabetic control is frequently indicated. (1, 2)
+ Digitalis: Patients taking digitalis should be observed for signs of digitalis toxicity, since depletion of the body stores of potassium may lessen digitalis requirements. Potentiation of digitalis effects upon the heart will cause atrial or ventricular arrhythmias. (1, 2)
+ Ganglionic Blocking Agents: see Bristuron + Antihypertensive Agents
+ Ismelin: see Bristuron + Antihypertensive Agents
+ Levophed: see Bristuron + Pressor Amines
+ Narcotics: see Bristuron + Alcohol
+ Pressor Amines: Since thiazide compounds decrease the response of the arterial system to pressor amines (e.g., Levophed), caution should be exercised in administering this drug to patients who undergo general anesthesia and surgery. It is advisable to discontinue therapy for one week or so prior to elective surgery. (1, 2)
+ Rauwolfia Alkaloids: see Bristuron + Antihypertensive Agents
+ Steroids: Decreased serum potassium is more likely to occur in cases associated with concomitant steroid or corticotrophin administration. (1, 2)
+ Trisoralen: see Trisoralen
+ Tubocurarine: Since thiazide compounds enhance the effects of tubocurarine, great caution should be exercised in administering this

BRISTURON (continued)

+ Tubocurarine (continued): drug to patients who undergo general anesthesia and surgery or who are treated with curare or its derivatives. It is advisable to discontinue therapy for one week or so prior to elective surgery. (1,2)
+ Veratrum Derivatives: see Bristuron + Antihypertensive Agents

BROMIDES + Dilantin: see Dilantin + Bromides

BROMSULPHALEIN + Benemid: see Benemid + Bromsulphalein

BRONDECON: see Choledyl

BRONDECON + Xanthine Preparations: Brondecon should not be administered within 12 hours following oral, parenteral, or rectal use of other xanthine-containing preparations. Simultaneous use of additional xanthine by any route should be avoided in all patients, especially children. (1)

BRONKOMETER

+ Epinephrine: Because of the possibility of excessive tachycardia, Bronkometer should not be administered concomitantly with epinephrine or other sympathomimetic amines. It may be alternated with these agents. (1)
+ Sympathomimetic Amines: see Bronkometer + Epinephrine

BRONKOSOL

+ Epinephrine: Because of the possibility of excessive tachycardia, Bronkosol should not be administered concimitantly with epinephrine or other sympathomimetic amines. It may be alternated with these agents. (1)
+ Sympathomimetic Amines: see Bronkosol + Epinephrine

BRUCELLERGEN + Adrenocortical Hormones: Negative or doubtful reactions may be the result of suppression of dermal sensitivity by adrenocortical hormones. (1)

BUFFERING AGENTS + Surital: see Surital + Buffering Agents

BUREN + Dolonil: see Dolonil + Buren

BUTAZOLIDIN: see also Acidic Agents + Alkaline Media

BUTAZOLIDIN: see listed under the following agents:

Antacids	Atabrine
Anticoagulants	Coumadin
Aralen	Diabinese

BUTAZOLIDIN (continued)

Dicumarol
Dymelor
Hedulin
Hypoglycemic Drugs, Oral
Indocin
Orinase
Panwarfin
Plaquenil
Ponstel
Questran
Sintrom
Sulfonamides, Long-Acting
Tolinase
Tromexan

BUTAZOLIDIN

+ Aminopyrine: Butazolidin increases the metabolism of aminopyrine thereby decreasing the action of aminopyrine. (120)

+ Antacids: Antacids inhibit the effect of Butazolidin by interfering with the absorption, since Butazolidin is a weak acid. (55)

+ Anticoagulants, Coumarin-Type: Coumarin-type anticoagulants depress prothrombin activity. This is accentuated in some cases when Butazolidin is simultaneously employed in treatment; occasional severe bleeding has been reported. For this reason, patients receiving coumarin-type anticoagulant therapy should be very carefully followed for evidence of excessive increase of prothrombin time (increased hypoprothrombinemia) when Butazolidin is added to this regimen. Anticoagulant therapy can then be properly adjusted, if necessary. When prescribed alone, Butazolidin has not been shown to influence prothrombin activity. (1) Mechanism: Butazolidin displaces coumarin-type anticoagulants from plasma protein-binding sites, thus causing an increase in anticoagulant action with resultant increased hypoprothrombinemia. (50, 55, 118)

+ Chemotherapeutic Agents: The use of Butazolidin in conjunction with other potent chemotherapeutic agents may greatly increase the possibility of toxic reactions, and this practice is therefore inadvisable. (1)

+ Diabinese: see Butazolidin + Sulfonylurea Agents

+ Dymelor: see Butazolidin + Sulfonylurea Agents. Hypoglycemia occurred after Butazolidin was added to a regimen using Dymelor, clearance of the active metabolite hydroxyhexamide appeared diminished. Further studies are needed to document this as the sole factor contributing to the increased hypoglycemia. (61)

+ Insulin: see Butazolidin + Sulfonylurea Agents

+ Orinase: see Butazolidin + Sulfonylurea Agents. Mechanism: see Orinase + Butazolidin.

+ Penicillin: Butazolidin increases the antibacterial activity of acidic antibiotics such as penicillin. Mechanism: Displacement of the

BUTAZOLIDIN (continued)

+ Penicillin (continued): penicillin from protein-binding sites by the Butazolidin, thereby increasing the concentration of free unbound penicillin. (118)
+ Sulfonamide-Type Agents: see Butazolidin + Sulfonylurea Agents. Mechanism: Displacement of the sulfonamide from protein-binding sites by the Butazolidin, thereby increasing the concentration of free unbound sulfonamide. (118)
+ Sulfonylurea Agents: Recent studies have indicated that pyrazole compounds, such as Butazolidin, may potentiate the pharmacologic action of sulfonylurea (e.g., Dymelor, Orinase, Tolinase, Diabinese), sulfonamide-type agents, and insulin. Patients receiving such concomitant therapy, therefore, should be carefully observed for this effect. (1, 32, 55)
+ Sul-Spantab: Butazolidin increases the proportion of unbound sulfonamide if used concomitantly, thereby enhancing the antibacterial activity. Mechanism: Displacement of the sulfonamide from protein binding sites by the Butazolidin, thereby increasing the concentration of free unbound sulfonamides. (118)
+ Tolinase: see Butazolidin + Sulfonylurea Agents. Mechanism: see Tolinase + Butazolidin

C

CAFFEINE: see listed under the following agents:
- Eutonyl
- MAO Inhibitors

CAFFEINE with SODIUM BENZOATE + Darvon: see Darvon + Caffeine with Sodium Benzoate

CALCIUM: see listed under the following agents:
- Blood
- Cuemid

CALCIUM, PARENTERAL: see listed under the following agents:
- Crystodigin
- Davoxin
- Digitalis
- Myodigin
- Pil-Digis

CALCIUM CHLORIDE SOL. + Digitalis: Because of the danger involved in the simultaneous use of calcium salts and drugs of the digitalis group, a digitalized patient should not receive an intravenous injection of a calcium compound unless the indications are clearly defined. (1)

CALCIUM-CONTAINING DRUGS: see listed under the following agents:
- Achromycin
- Aureomycin
- Declomycin

CALCIUM-CONTAINING SOLUTIONS: see listed under the following agents:
- Achromycin I.V.
- Tetrachel I.V.

CALCIUM GLUCONATE-GLUCOHEPTANATE + Digitalis: Do not give concomitantly. Parenteral calcium may provoke arrhythmias in digitalized patients. (1)

CALPHOSAN + Digitalis: As there is a similarity in the actions of calcium and digitalis on the contractility and excitability of the heart muscle, Calphosan is contraindicated in fully digitalized patients. (1)

CALSCORBATE + Digitalis: Calscorbate is contraindicated in patients receiving digitalis since cardiac arrhythmias may be precipitated. (1)

CANCER CHEMOTHERAPEUTIC AGENTS + Orimune Poliovirus Vaccine, Live, Oral, Trivalent: see Orimune Poliovirus Vaccine, Live, Oral, Trivalent + Cancer Chemotherapeutic Agents

CARBOCAINE

+ Antiemetic Agents: see Carbocaine + Sedatives
+ Barbiturates: For the control of convulsions due to the local anesthetic, ultrashort acting barbiturates may be used with caution. Due to the depressant effect, barbiturates may do more harm than good. (1)
+ Chloroform: see Carbocaine + Cyclopropane. Mechanism: see Chloroform + Epinephrine
+ Cyclopropane: Local anesthetics containing epinephrine or other vasoconstrictors should not be used during the first 15 minutes following administration of cyclopropane, chloroform, Fluothane, or other related drugs, since such concomitant use may produce arrhythmias. (1) Mechanism: see Cyclopropane + Epinephrine.
+ Fluothane: see Carbocaine + Cyclopropane. Mechanism: see Fluothane + Epinephrine
+ Oxytocic Drugs: In obstetrics, if vasopressor drugs are used during caudal analgesia either to correct hypotension or to add to the local anesthetic solution, the obstetrician should be cautioned that if an oxytocic drug (oxytocin, ergonovine maleate, or methylergonovine maleate) is employed during the postpartum period, severe persistent hypertension and even rupture of a cerebral blood vessel may occur. Hence, oxytocic drugs should not be used. (1)
+ Sedatives: Excessive premedication with sedatives, tranquilizers, and antiemetic agents prior to local anesthesia should be avoided, especially in infants, small children, and elderly adults. (1)
+ Tranquilizers: see Carbocaine + Sedatives

CARBONIC ANHYDRASE INHIBITOR: see listed under the following agents:
Humorsol
Phospholine Iodide

CARCHOLIN + Anesthesia, Topical: Caution should be observed in administering Carcholin after any procedure that reduces epithelial barrier of the cornea and conjunctiva, such as instrumental tonometry or topical anesthesia. (1)

CARDIAC GLYCOSIDES + Regitine: see Regitine + Cardiac Glycosides

CARDIO-GREEN

+ Heparin: Heparin preparations containing sodium bisulfite reduce the absorption peak of Cardio-Green and therefore should be avoided. (1)
+ Sodium Bisulfite: see Cardio-Green + Heparin
+ Water for Injection: Only Hynson, Wescott and Dunning Aqueous Solvent (which is especially prepared sterile Water for Injection) should be

CARDIO-GREEN (continued)
+ Water for Injection (continued): used to dissolve Cardio-Green because there have been reports of incompatibility with some commercially available Water for Injection. (1)

CARDIOQUIN: see Quinidine

CAROTENE + Neomycin: see Neomycin + Carotene

CARTRAX: see Atarax

CATECHOLAMINE DEPLETING DRUGS + Inderal: see Inderal + Catecholamine Depleting Drugs

CATECHOLAMINE INJECTION + Choloxin: see Choloxin + Catecholamine Injection

CATECHOLAMINES: see listed under the following drugs:
Euthroid
Thyrolar

CATRON + Demerol: see Demerol + Catron

CELESTONE + Diabetes: Although hyperglycemia, glycosuria, and increased insulin requirements usually do not occur with Celestone in the controlled diabetic patient, close observation of such patients should be maintained during therapy. (1)

CELESTONE SOLUSPAN + Corticosteroids: A portion of the administered dose of Celestone Soluspan Injection is absorbed systemically following intra-articular injection. In patients being treated concomitantly with peroral or parenteral corticosteroids, especially those receiving large doses, the systemic absorption of the drug should be considered in determining intra-articular dosage. (1)

CENDEVAX
+ Alkylating Agents: see Cendevax + Steroids
+ Antimetabolites: see Cendevax + Steroids
+ Blood Transfusion: see Cendevax + Gamma Globulin
+ Gamma Globulin: Vaccination should be delayed until at least six weeks after gamma globulin therapy or blood transfusion because of the possible suppressive effect of passive antibodies. (1)
+ Irradiation: see Cendevax + Steroids
+ Steroids: Attenuated rubella virus infection might be potentiated when resistance has been lowered by therapy with steroids, alkylating drugs, antimetabolites, or irradiation. Vaccination of patients under these conditions is contraindicated. (1)

CENDEVAX (continued)

+ Vaccines, Live Virus: Simultaneous administration of live rubella virus vaccine and other live virus vaccines should be avoided until the safety and efficacy of such use has been established. Until then, it is recommended that rubella vaccination be given at least one month before or after administration of other live virus vaccines. (1)

CENTRAL NERVOUS SYSTEM DEPRESSANTS: see listed under the following agents:

Alcohol
Amytal Sodium Sterile
Antiemetics, Phenothiazine-type
Anti-Nausea Suprettes
Atarax
Aventyl
Barbiturates
Compazine
Coplexen
Dalmane
Dartal
Disophrol
Doriden
Elavil
Eutonyl
Haldol
Inapsine
Largon
Levoprome
Librium
MAO Inhibitor
Matulane
Medomin
Mellaril
Nardil
Navane
Noludar
Pacatal
Parest
Parnate
Periactin
Permitil
Phenergan
Phenothiazine
Placidyl
Prinadol
Proketazine
Prolixin
Quaalude
Quide
Repoise
Serentil
Solacen
Somnofac
Sparine
Stelazine
Sublimaze
Taractan
Tegopen
Temaril
Thorazine
Tindal
Torecan
Trilafon
Tybatran
ULO
Ultran
Valium
Valmid
Vesprin
Vistaril
Xylocaine

CENTRAL NERVOUS SYSTEM DRUGS: see listed under the following agents:

Pertofrane
Tigan

CENTRAL NERVOUS SYSTEM STIMULANTS: see listed under the following agents:

- Compazine
- Deprol
- Iphyllin
- Lufyllin
- Preludin
- Symmetrel
- ULO

CHEMOTHERAPEUTIC AGENTS: see listed under the following agents:

- Butazolidin
- Cytoxan
- Mustargen

CHLORAL HYDRATE: see listed under the following agents

- Anticoagulants
- Coumadin
- Dicumarol
- Eutonyl
- MAO Inhibitors
- Panwarfin

CHLOROFORM: see listed under the following agents:

- Carbocaine
- Inderal
- Suprarenin
- Vasoconstrictors
- Vasopressors

CHLOROFORM

\+ Epinephrine: Chloroform sensitizes the myocardium to the action of epinephrine and Levophed; therefore it should not be used when injection of these amines is contemplated, since ventricular tachycardia or fibrillation may be induced. (2, 126)

\+ Levophed: see Chloroform + Epinephrine

CHLOROMYCETIN: see listed under the following agents:

- Coumadin
- Penicillin

CHLOROMYCETIN + Bone Marrow Depressants: Concurrent therapy with other drugs that may cause bone marrow depression should be avoided. (1)

CHLOROMYCETIN I.M. + Diluent: The suspension is prepared by using Water for Injection that does not contain benzyl alcohol. (1)

CHLOROTHIAZIDE: see listed under the following agents:

- Nardil
- Ostensin
- Questran

CHOLECYSTOGRAPHIC MEDIA I.V. + Oragrafin: see Oragrafin + Cholecystographic Media I.V.

CHOLEDYL + Xanthine Preparations: Choledyl should not be administered within 12 hours following oral, parenteral, or rectal use of other xanthine-containing preparations. Simultaneous use of additional xanthines by any route should be avoided in all patients, especially children. (1)

CHOLINE: see listed under the following agents:
Coumadin
Sintrom
Tromexan

CHOLINERGIC AGENTS: see listed under the following agents:
Cozyme
Mylaxen
Mytelase
Succinylcholine

CHOLINERGIC AGENTS + Atropine: The use of atropine or belladonna preparations may be necessary to prevent or counteract signs of mild overdosage with cholinergic agents. However, there is danger that the use of these anticholinergic agents may prevent the early recognition of an impending cholinergic crisis. (2)

CHOLINESTERASE INHIBITORS: see listed under the following agents:
Quelicin Chloride Inj.
Succinylcholine
Sucostrin
Tubocurarine

CHOLINESTERASE INHIBITORS + Skeletal Muscle Relaxants "Depolarizers": Cholinesterase inhibitors (e.g., Prostigmin, Physostigmine, Tensilon) potentiate the muscle paralyzing action of the "depolarizing" drugs (e.g., Succinylcholine, Syncurine) and their use in the treatment of apnea caused by these drugs is therefore contraindicated. (2)
Mechanism: Cholinesterase inhibitors, by blocking the action of pseudocholinesterase in the plasma, may be associated with prolonged paralysis of respiration following the use of depolarizing skeletal muscle relaxants. The cholinesterase inhibitors may prolong the block and the resulting apnea by preventing enzymatic hydrolysis of the depolarizing skeletal muscle relaxants at the motor end plates. Cholinesterase inhibitors, by also preventing the destruction of acetylcholine which acts as a depolarizing agent at the motor end plates, augment the depolarizing action of the depolarizing skeletal muscle relaxants. (127)

CHOLOGRAFIN + Oragrafin: see Oragrafin + Chulografin

CHOLOXIN: see listed under the following agents:

Anticoagulants
Dicumarol
Panwarfin

CHOLOXIN

+ Anticoagulants: Several studies indicate that Choloxin potentiates the effects of anticoagulants, such as warfarin or Dicumarol, on prothrombin time. Reductions of warfarin drug dosage by as much as 30% have been required in some patients. Consequently, the dosage of such anticoagulants should be reduced by one-third upon initiation of Choloxin therapy and the dosage subsequently readjusted on the basis of laboratory findings. The prothrombin time of patients receiving anticoagulant therapy concomitantly with Choloxin therapy should be observed as frequently as necessary, but at least weekly, during the first few weeks of treatment. Choloxin has also been shown apparently to decrease the concentration of Factors VII, VIII, and IX and platelet activity in some patients. Therefore, during anticoagulant therapy, attention should be paid to other clotting factors besides the one-stage prothrombin time. Spontaneous bleeding has been observed in a Choloxin-treated patient on warfarin who had a prothrombin time within the therapeutic range but whose prothrombin and Factor VII concentrations were greatly decreased. In surgical patients, it is wise to consider withdrawal of the drug two weeks prior to surgery if the use of anticoagulants during surgery is contemplated. (1) Mechanism: see Anticoagulants + Choloxin.

+ Catecholamine Injections: Epinephrine injection is patients with coronary artery disease may precipitate an episode of coronary insufficiency. This condition may be enhanced in patients receiving thyroid analogues. This phenomenon should be kept in mind when catecholamine injections are required in Choloxin-treated hypothyroid patients with coronary artery disease. (1)

+ Diabetes: There are reports that Choloxin in diabetic patients is capable of increasing blood sugar levels with a resultant increase in requirements of insulin or oral hypoglycemic agents. Special attention should be paid to parameters necessary for good control of the diabetic state in Choloxin-treated diabetic patients and to dosage requirements of insulin or other antidiabetic drugs. If Choloxin is later withdrawn from patients who had required an increase of insulin (or oral hypoglycemic agents) dosage during its administration, the dosage of antidiabetic drug should be reduced and adjusted to

CHOLOXIN (continued)

+ Diabetes (continued): maintain good control of the diabetic patient. (1, 33)
+ Dicumarol: see Choloxin + Anticoagulants
+ Epinephrine: see Choloxin + Catecholamine Injection
+ Surgery: Since the possibility of precipitating cardiac arrhythmias during surgery may be greater in patients treated with thyroid hormones, it may be wise to discontinue Choloxin in euthyroid patients at least two weeks prior to elective surgery. During emergency operations in euthyroid patients, and in surgery in hypothyroid patients in whom it may not be advisable to withdraw thyroid therapy, the patients should be carefully observed. (1)
+ Thyroid Medication: Special consideration must be given to the dosage of other thyroid medications used concomitantly with Choloxin. As with all thyroactive drugs, hypothyroid patients are more sensitive to a given dose of Choloxin than euthyroid patients. (1)
+ Warfarin: see Choloxin + Anticoagulants

CHYMOLASE + Anticoagulants: Although not contraindicated, Chymolase should be used with caution in patients who exhibit abnormalities of the blood-clotting mechanism. (1)

CHYMORAL + Anticoagulants: see Anticoagulants + Chymoral. There is no evidence that Chymoral or Chymoral-100 in the recommended dosage affects the blood-clotting mechanism. However, the usual laboratory tests and precautions should be observed carefully when anticoagulants are used concomitantly. (1, 2)

CINCHONA ALKALOIDS: see listed under the following agents:

Succinylcholine
Tubocurarine
Tubocurarine Chloride Inj.

CITRATES + Anturane: see Anturane + Citrates

CLEANSERS, ABRASIVE: see listed under the following agents:

Loroxide Lotion
Vanoxide Lotion

CLEOCIN + Erythromycin: Antagonism has been demonstrated between Cleocin and erythromycin in vitro. Because of possible clinical significance, these two drugs should not be administered simultaneously. (1)

CLINITEST TABLETS: see listed under the following agents:
Keflin
NegGram

COCAINE: see listed under the following agents:
MAO Inhibitors
Marplan
Nardil
Nembutal Sodium Inj.
Niamid

CODEINE: see listed under the following agents:
MAO Inhibitors
Quaalude

COGENTIN: see listed under the following agents:
Dopar
Larodopa
Levodopa

COGENTIN
+ Phenothiazines: Permanent extrapyramidal symptoms have been described with prolonged phenothiazine therapy. The masking action of Cogentin on the possible development of these symptoms has not been investigated.
In the treatment of side effects of phenothiazine derivatives and reserpine in patients with a mental disorder, occasionally there may be intensification of mental symptoms. In such cases, at times, increased doses of antiparkinsonian drugs should be kept under careful observation, especially at the beginning of treatment or if dosage is increased. (1)
+ Reserpine: see Cogentin + Phenothiazines

COLBENEMID: see Benemid

COLD REMEDIES: see listed under the following agents:
Eutonyl
Marplan
MAO Inhibitors

COLY-MYCIN M INJECTABLE: see listed under the following agents:
Anectine
Garamycin
Succinylcholine

COLY-MYCIN M INJECTABLE (continued)

+ Antibiotics: Since other antibiotics (Kanmycin, Streptomycin, Dihydrostreptomycin, Polymixin, Neomycin) also have varying neurotoxic or nephrotoxic potential, they should be given concomitantly with Coly-Mycin M injection with the greatest caution. Permanent nerve damage, such as deafness or vestibular damage, probably does not occur with Coly-Mycin M Injectable. Reports of such episodes generally reveal that another polypeptide antibiotic with a potential for permanent neurotoxicity has been given concomitantly. (1)
+ Antibiotics, Nephrotoxic: see Coly-Mycin M Injectable + Antibiotics
+ Antibiotics, Neurotoxic: see Coly-Mycin M Injectable + Antibiotics
+ Dihydrostreptomycin: see Coly-Mycin M Injectable + Antibiotics
+ Drugs, Curare-Like: Exercise caution when used concomitantly with curare-like drugs. Probably all polypeptide antibiotics have some potential for blocking nerve transmission at the neuromuscular junction (curariform action). (1)
+ Kantrex: see Coly-Mycin M Injectable + Antibiotics
+ Muscle Relaxants, Curariform: Curariform muscle relaxants should be used with extreme caution in patients receiving Coly-Mycin M injection, since their pharmacologic action also interferes with nerve transmission at neuromuscular junction, and apnea--requiring assisted respiration--can result. (1)
+ Neomycin: see Coly-Mycin M Injectable + Antibiotics
+ Polymixin: see Coly-Mycin M Injectable + Antibiotics
+ Streptomycin: see Coly-Mycin M Injectable + Antibiotics

COMBID: see Compazine

COMPAZINE: see Phenothiazines

COMPAZINE

+ Alcohol: see Compazine + CNS Depressants
+ Analgesics: see Compazine + CNS Depressants
+ Antihistamines: see Compazine + CNS Depressants
+ Antihypertensive Agents: Compazine may potentiate the action of antihypertensive agents, and care should be exercised when used concomitantly. (2)
+ Atropine: Phenothiazines will potentiate atropine. (1)
+ Barbiturates: see Compazine + CNS Depressants
+ CNS Depressants: Compazine is contraindicated in cases of greatly depressed states due to central nervous system depressants. Although Compazine has little or no clinically significant potentiating activities, phenothiazines may potentiate depressant drugs (opiates,

COMPAZINE (continued)

+ CNS Depressants (continued): analgesics, antihistamines, barbiturates, alcohol). (1, 2)
+ CNS Stimulants: If a stimulant is desired, as in central nervous system depression, amphetamine, dextroamphetamine, caffeine, or sodium benzoate is recommended. Stimulants that may cause convulsions (e.g., Picrotoxin or Metrazol) should be avoided. (1)
+ Drug Overdosage: Compazine, due to its antiemetic action, may mask signs of overdosage of toxic drugs. (1)
+ Epinephrine: see Compazine + Pressor Agents
+ Metrazol: see Compazine + CNS Stimulants
+ Opiates: see Compazine + CNS Depressants
+ Phosphorous Insecticides: Phenothiazines have been reported to potentiate phosphorous insecticides. (1)
+ Picrotoxin: see Compazine + CNS Stimulants
+ Pressor Agents: If hypotension occurs with Compazine and a vasoconstrictor is needed, Levophed and Neosynephrine are the most suitable. Other pressor agents, including epinephrine, are not recommended, since phenothiazine derivatives may reverse the usual elevating action of these drugs and cause further lowering of blood pressure. (1, 2)
+ Vasoconstrictors: see Compazine + Pressor Agents

COMPAZINE INJECTION + Drugs: It is recommended that Compazine not be mixed with other agents in the syringe. (1)

COMPOUNDS, HIGH MOLECULAR WEIGHT + Keflin: see Keflin + Compounds, High Molecular Weight

CONTRACEPTIVES, ORAL

+ Anticoagulants: see Anticoagulants + Contraceptives, Oral
+ Diabetes: A decrease in glucose tolerance occurs occasionally and for this reason the status of patients with overt diabetes who are receiving oral contraceptives should be reviewed periodically. (2)

CONTRAST MEDIA I.V.P. + Dyclone: see Dyclone + Contrast Media I.V.P.

COPAVIN

+ Ephedrine: see Copavin + Picrotoxin
+ Picrotoxin: Stimulants such as Picrotoxin and ephedrine are contraindicated in Copavin overdosage. (1)

COPLEXEN: see Temaril and Propadrine

COPLEXEN + CNS Depressants: If administered with central nervous

COPLEXEN + CNS Depressants (continued): system depressants, dosage should be adjusted carefully to avoid possible excessive sedation. (1)

CORONARY VASODILATORS (NITRATES) + Alcohol: Sharp falls in blood pressure have occurred after the ingestion of alcoholic beverages by patients who have also been taking nitrates. (2)

CORROSIVE SUBSTANCES + Ipechar Poison Control Kit: see Ipechar Poison Control Kit + Corrosive Substances

CORTEF: see Adrenocorticosteroids

CORTEF + Diabetes: Diabetes mellitus may be aggravated if Cortef is used in a diabetic patient. (1)

CORTENEMA: see Hydrocortisone and Adrenocorticosteroids

CORTICOSTEROIDS: see Adrenocorticosteroids

CORTICOSTEROIDS: see listed under the following agents:

Anesthetics, General
Aquacort Suprettes
Aquatag
Attenuvax, Lyovac
BCG Vaccine
Biavax
Brusturon
Celestone Soluspan
Coumadin
Desitin HC Oint. & Suppos.
Diuril
Exna
Fungizone I.V.
Hedulin
Indocin
Mumpsvax, Lyovac
Periactin
Ponstel
Protef Rectal Suppos.
Rubeovax, Lyovac
Sintrom
Smallpox Vaccine Dried
Trinalis Suprettes
Tromexan

CORTICOSTEROIDS + Diabetes: Diabetic patients frequently need an increase in insulin dosage if treated with corticosteroids. (2)

CORTICOTROPHIN: see listed under the following agents:

Anhydron
Aquatag
Bristuron
Diuretics, Thiazide
Diuril
Enduron
Esidrix
Exna
HydroDIURIL
Hydromox
Metahydrin
Naqua
Renese
Saluron
Smallpox Vaccine

CORTIPHATE + Diabetes: The side effects most commonly associated with prolonged systemic steroid therapy include alteration of glucose metabolism with aggravation of diabetes mellitus. (1)

CORTISONE ACETATE

+ Diabetes: Glucocorticoid steroids may aggravate diabetes mellitus so that higher insulin dosage may become necessary or manifestations of latent diabetes mellitus may be precipitated. Hyperglycemia and glycosuria, due in part to increase in gluconeogenesis, can be controlled by reducing the daily dose of cortisone. In diabetic patients, the insulin requirements may need to be increased temporarily when cortisone is being used in high doses. (1)

+ Diuretics, Thiazide: Concomitant use may cause a dangerous loss of potassium. (2)

CORTISPORIN OTIC DROPS + Heat: Excessive heating of the drops should be avoided. (1)

CORTONE

+ Diabetes: Cortisone causes gluconeogenesis; therefore hyperglycemia and glycosuria may occur, glucose tolerance may be altered, and diabetes mellitus may be aggravated. These effects usually are reversible when therapy is discontinued or sometimes when dosage is decreased. (1)

+ Diuretics: Diuretics if used concomitantly with Cortone may provoke a further dangerous loss of potassium (1)

CORTONE ACETATE SALINE SUSPENSION

+ Diluents: No attempt should be made to alter the suspension. Diluting or mixing them with other substances may affect the state of suspension or change the rate of absorption and reduce their effectiveness. Refrigeration is not desirable as agglomerates may be formed if stored at low temperature. (1)

+ Drugs: see Cortone Acetate Saline Suspension + Diluents

CORTRIL

+ Diabetes: Increased gluconeogenesis with hyperglycemia can occur quite frequently, and insulin requirements are increased in diabetics on hydrocortisone therapy. (1)

+ Diuretics: If there is an indication of hypochloremic, hypopotassemic alkalosis, the dosage of hydrocortisone should be reduced and 1 to 4 gm of potassium chloride administered daily. Diuretics, if needed, must be administered with care. (1)

COSMEGEN

+ Drugs, Antineoplastic: The danger of toxic reactions may be increased

COSMEGEN (continued)

+ Drugs, Antineoplastic (continued): when Cosmegen is used with other antineoplastic agents. The usual doses are decreased if concomitant therapy is used. (1, 2)
+ X-Ray Therapy: There is much evidence that Cosmegen potentiates the effects of x-ray therapy. The converse also appears likely; i.e., Cosmegen may be more effective when radiation therapy also is given. Erythema from previous x-ray therapy may be reactivated by Cosmegen alone, especially when the interval between the two forms of therapy is brief. Severe reactions may ensue if high doses of both Cosmegen and radiation therapy are used or if the patient is particularly sensitive to such combined therapy. (1, 2)

COUMADIN: see Anticoagulants

COUMADIN: The following agents may be responsible for increased prothrombin time response (a further increase of the hypoprothrombinemic state): (1)

Alcohol (response unpredictable)
Anabolic Steroids
Analexin
Anesthetics
Antibiotics
Atromid-S (85)
Butazolidin (84)
Dietary deficiencies (Vitamin C, Protein, Choline, Cystine)
Diphenylhydantoin (84)
Drugs affecting Blood Elements
Hepatotoxic Drugs
Narcotics, Prolonged
Prolonged Hot Weather
Propylthiouracil
Quinidine (23)
Quinine (23)
Salicylates (in excess 1 gm/day)
Sulfonamides
Tandearil

COUMADIN: The following agents may be responsible for decreased prothrombin time response (a decrease of the hypoprothrombinemic state): (1)

Alcohol (response unpredictable)
Antihistamines
Barbiturates (48, 83, 84, 85)

COUMADIN (continued, agents that may be responsible for decreased prothrombin time response):

Chloral Hydrate
Corticosteroids
Doriden (48, 84, 85)
Foods rich in Vitamin K (alfalfa, cabbage, carrots, cauliflower, corn, fish, fish oils, liver, mushrooms, oats, peas, pork, potatoes, soybeans, spinach, strawberry, tomati (green), tomato (ripe), wheat, wheat bran, wheat germ
Griseofulvin (84)
Meprobamate (84, 85)
Mineral Oil
Placidyl (84)
Vitamin K

COUMARINS: see acidic agents + Alkaline media

COZYME

+ Cholinergic Agents: Neostigmine or other cholinergic agents should not be used with Cozyme solution. Their use together may cause overstimulation of the intestine. (1)
+ Neostigmine: see Cozyme + Cholinergic Agents
+ Parasympathomimetic Drugs: Theoretical considerations suggest that if parasympathetic drugs have been used, Cozyme should not be administered for 12 hours after these agents have been discontinued. (2)
+ Succinylcholine: Respiratory embarrassment was observed in one patient when Cozyme was administered shortly after cessation of succinylcholine therapy; thus, it may be advisable to wait one hour after discontinuing succinylcholine before using Cozyme. (2)

C-QUENS + Diabetes: Because a decreased glucose tolerance has been observed in some patients taking oral contraceptives, diabetic patients taking C-Quens should be closely followed. (1, 2)

CREMOMYCIN: see Sulfasuxidine and Neomycin

CREMOMYCIN + Anesthesia: Since a curare-like neuromuscular block has been reported after parenteral neomycin, the possibility of an interaction between this neomycin effect and other agents used by an anesthesiologist should be considered if a patient has received large doses of neomycin preoperatively. (1)

CRESATIN

+ Acetate Fabrics: Avoid contact of Cresatin with acetate fabrics. (1)

CRESATIN (continued)

+ Plastics: Avoid contact of Cresatin with plastics (i.e., ear pieces of certain hearing aids). Certain plastics can be dissolved in Cresatin. (1)

CRESOL + Furadantin Sodium Sterile: see Furadantin Sodium Sterile + Cresol

CRYSTODIGIN

+ Calcium, Parenteral: see Crystodigin + Digitalis
+ Digitalis: Patients taking digitalis preparations must not be given the rapid digitalizing dose of Crystodigin or parenteral calcium. (1)

CRYSTODIGIN INJECTION + Calcium I.V.: Patients under the influence of digitalis should not receive calcium intravenously, because elevation of the serum calcium concentration increases the effect of digitalis, and deaths have been reported from intoxication induced in this way. (1)

CUEMID: see Questran

CUEMID

+ Anticoagulants: Chronic use of Cuemid may be associated with increased bleeding tendency due to hypoprothrombinemia associated with vitamin K deficiency. (1)
+ Calcium: Cuemid may increase the absorption of calcium, although data on this effect are not conclusive. (2)
+ Drugs, Oral: To reduce the risk of adsorption of other oral medication, it is important that Cuemid be given at least four hours after a dosage of other drugs. (1, 2)
+ Vitamins, Fat-Soluble: Cuemid can interfere with the adsorption of fat soluble vitamins, and deficiencies of such vitamins can also occur independently of therapy in patients with primary biliary cirrhosis. Accordingly, it has been suggested that patients undergoing treatment be given vitamins A, D, and K by injection. (1, 2)

CUPRIMINE + Iron: Iron deficiency may develop, especially in menstruating women and in children. In Wilson's disease this may be a result of adding a low copper diet, which is probably also low in iron, sulfurated potash or carboxylic cation exchange resin, and Cuprimine to the effects of blood loss or growth. If necessary, iron may be given in short courses without the sulfurated potash or the resin. (1)

CURARE + Ether: see Ether + Curare

CURARE and DERIVATIVES: see listed under the following agents:

- Bristuron
- Lasix
- Saluron
- Tensilon

CURARE-LIKE DRUGS: see listed under the following agents:

- Coly-Mycin M Inj.
- Neomycin
- Seconal Sodium

CURARE-TYPE MUSCLE RELAXANTS: see listed under the following agents:

- Succinylcholine
- Sucostrin

CURARIFORM DRUGS + Fluothane: see Fluothane + Curariform Drugs

CYCLEX: see HydroDIURIL and Meprobamate

CYCLOPROPANE: see listed under the following agents:

- Aramine
- Carbocaine
- Dopram
- Epinephrine
- Levophed
- Methedrine
- Nalline
- Pressonex
- Pressorol
- Prostigmin
- Quelicin Chloride Inj.
- Succinylcholine
- Suprarenin
- Vasoconstrictors
- Vasopressors

CYCLOPROPANE

+ Atropine: The administration of intravenous atropine during Cyclopropane anesthesia may be followed by severe and dangerous arrhythmias, particularly when atropine has not been given as premedication. These are less frequent and less severe when the concentration of Cyclopropane is low, and the incidence is also diminished by slow injection and by administration of small incremental doses until the desired effect is obtained. (110, 128)

+ Epinephrine: The administration of epinephrine during anesthesia with cyclopropane may be dangerous, for this drug combination tends to increase the likelihood of ventricular tachycardia or fibrillation. (2, 14) Mechanism: Cyclopropane sensitizes the heart to the actions of norepinephrine and related catecholamines. (123)

+ Flaxedil: Flaxedil, which has a vagolytic action, has been shown to precipitate arrhythmias when used with Cyclopropane. (110, 128)

CYCLOPROPANE (continued)

+ Levophed: The administration of Levophed during anesthesia with Cyclopropane may be dangerous, for this drug combination tends to increase the likelihood of ventricular tachycardia or fibrillation. (2) Mechanism: Cyclopropane sensitizes the heart to the actions of norepinephrine and related catecholamines. (123)
+ Methedrine: Methedrine is generally contraindicated during Cyclopropane anesthesia. (1)
+ Pressonex: Use Pressonex with caution during Cyclopropane anesthesia. (1)

CYSTINE: see listed under the following agents:

Coumadin
Sintrom
Tromexan

CYTOMEL

+ Thyroglobulin (Proloid): see Cytomel + Thyroid
+ Thyroid: When switching a patient to Cytomel from thyroid, L-thyroxine, or thyroglobulin, discontinue the other medication, initiate Cytomel in a low dosage and increase gradually according to the patient's response. When selecting a starting dosage, bear in mind that this drug has a rapid onset of action and that residual effect of other thyroid preparation may persist for the first several weeks of therapy. (1)
+ L-Thyroxine (Letter, Levoid, Synthroid, Titroid): see Cytomel + Thyroid

CYTOSAR + Bone-Marrow Depressants: Therapy with Cytosar should not be started in patients with pre-existing drug-induced bone marrow suppression unless the clinician feels that such management offers the most hopeful alternative to the patient. (1)

CYTOXAN

+ Chemotherapeutic Agents: see Cytoxan + X-Ray Therapy
+ Cytotoxic Drugs: see Cytoxan + X-Ray Therapy
+ X-Ray Therapy: Prior treatment with x-ray, cytotoxic drugs, or prior chemotherapeutic agents frequently causes an earlier or exaggerated leukopenia or thrombocytopenia after Cytoxan medication. Reduction in dose and caution are advisable in the treatment of patients who have been given x-ray, cytotoxic drugs, or prior chemotherapeutic agents. (1, 2)

CYTOTOXIC AGENTS: see listed under the following agents:

Antimetabolic Agents
Cytoxan
Hydrea
Uracil Mustard

CYTOTOXIC CHEMOTHERAPY: see listed under the following agents:

Thioguanine
Vercyte

D

DAIRY PRODUCTS: see listed under the following agents:
Aureomycin
Declomycin

DALMANE
+ Alcohol: see Dalmane + CNS Depressants
+ CNS Depressants: Patients receiving Dalmane should be cautioned about possible combined effects with alcohol and other central nervous system depressants. If Dalmane is to be combined with other drugs with known hypnotic properties or central nervous system depressant effects, due consideration should be given to potential additive effects. (1)
+ Hypnotics: see Dalmane + CNS Depressants

DANILONE: see Hedulin

DANILONE: see listed under the following agents:
Analexin
Haldol

DARAPRIM + Sulfonamides: There is a synergistic effect between Daraprim and sulfonamides. (1)

DARBID + Trisoralen: see Trisoralen

DARBID (overdosage) + Pilocarpine: Although pilocarpine or similar drugs are sometimes recommended for the relief of dry mouth, many authorities feel that these drugs are not indicated, since they relieve the minor peripheral effect but do not influence the more serious central effects and thus may merely mask signs of drug activity. (1)

DARCIL + Food: Absorption of Darcil is greater if the drug is given when the stomach is empty. Darcil should be given not less than one-half hour before or two hours after meals for optimal absorption. (1, 2)

DARTAL: Since Dartal belongs to the class of phenothiazines, side effects characteristic of this group of compounds may theoretically occur. (1) See Phenothiazines.

DARTAL
+ Alcohol: see Dartal + CNS Depressants
+ Anesthetics: see Dartal + CNS Depressants

DARTAL (continued)

+ Anticonvulsants: Caution should be exercised if Dartal is given to epileptic patients or to those in a depressed state. In some of these patients receiving Dartal the dose of concomitant medication may require adjustment, and attention should be paid to such possibilities. (1)
+ Antihypertensives: Dartal may potentiate the action of antihypertensive agents, and care should be exercised when used concomitantly. (2)
+ Atropine: Since Dartal belongs to the class of phenothiazines, side effects characteristic of this group of compounds may theoretically occur. Among these side effects, none of which has been known to occur with Dartal, are the following: potentiation of the effects of atropine, heat, and phosphorous insecticides. (1)
+ Barbiturates: see Dartal + CNS Depressants
+ CNS Depressants: Dartal may potentiate the effects of central nervous system depressants (e.g., general anesthetics, hypnotics, barbiturates, narcotics, alcohol), and care should be exercised when used concomitantly. (1, 2)
+ Epinephrine: Epinephrine should not be used concomitantly, for Dartal may reverse its action and cause profound hypotension. Vasopressors such as Levophed may be used. (1, 2)
+ Heat: see Dartal + Atropine
+ Hypnotics: see Dartal + CNS Depressants
+ Narcotics: see Dartal + CNS Depressants
+ Phosphorous Insecticides: see Dartal + Atropine

DARVON: see listed under the following agents:
Disipal
Norflex

DARVON

+ Amphetamine: see Darvon + Analeptic Drugs
+ Analeptic Drugs: If an overdosage of Darvon is accidentally or intentionally ingested, analeptic drugs (caffeine with sodium benzoate, picrotoxin, amphetamine) should not be used because fatal convulsions may be produced. (1, 2)
+ Caffeine with Sodium Benzoate: see Darvon + Analeptic Drugs
+ Norflex: Since mental confusion, anxiety, and tremors have been reported in patients receiving orphenadrine and Darvon concurrently, it is not recommended that Norflex be given in combination with Darvon. (1)
+ Picrotoxin: see Darvon + Analeptic Drugs

DARVO-TRAN: see Darvon and Ultran

DAVOXIN: see Digitalis

DAVOXIN
+ Calcium: The administration of calcium intravenously may induce digitalis intoxication. (1)
+ Diuretic Agents: Potassium depletion, usually caused by the use of diuretic agents or electrolyte manipulations, sensitizes the heart to digitalis and may produce arrhythmias even with recommended doses. In this event reduction of dosage may be necessary and administration of potassium may be indicated. (1)

DBI: see listed under the following agents:
Orinase
Tolinase

DECADRON + Diabetes: Since Decadron is a glucocorticoid, it may aggravate diabetes mellitus so that higher insulin dosage may become necessary or it may precipitate manifestations of latent diabetes mellitus. Alterations of glucose metabolism with aggravation of diabetes mellitus including hyperglycemia and glycosuria. (1, 2)

DECLOMYCIN
+ Antacids: see Declomycin + Food
+ Dairy Products: see Declomycin + Food
+ Drugs, High Calcium Content: see Declomycin + Food
+ Food: Absorption is impaired by the concomitant administration of high calcium content drugs such as antacid medications, foods, and some dairy products such as milk. Oral forms of Declomycin should be given one hour before or two hours after meals. (1)
+ Milk: see Declomycin + Food
+ Trisoralen: see Trisoralen

DELADUMONE + Diabetes: Because of a possible decrease in glucose tolerance, diabetic patients should be followed closely. (1)

DELENAR: Delenar contains orphenadrine. See Darvon + Norflex.

DELFEN VAGINAL CREAM + Douche: If a douche is desired for cleansing purposes, it should be deferred for at least six hours after intercourse. (1)

DELFEN VAGINAL FOAM + Douche: If a douche is taken, always wait at least six to eight hours after intercourse. (1)

DELTA-CORTEF + Diabetes: Glucocorticoid steroids may aggravate diabetes mellitus so that higher insulin dosage may become necessary or manifestations of latent diabetes mellitus may be precipitated. (1)

DELTASONE + Diabetes: Glucocorticoid steroids may aggravate diabetes mellitus so that higher insulin dosage may become necessary or manifestations of latent diabetes mellitus may be precipitated. (1)

DEMEROL: see Meperidine

DEMEROL: see listed under the following agents:
- Eutonyl
- Marplan
- Nardil
- Niamid
- Parnate

DEMEROL
- \+ Catron: see Demerol + MAO Inhibitors
- \+ Isoniazid and Derivatives: see Demerol + MAO Inhibitors
- \+ MAO Inhibitors: The activity of many drugs including Demerol is greatly, and even dangerously, potentiated in patients receiving MAO inhibitors, isoniazid or its derivatives, Proketazine, Marplan, Niamid, Nardil, Catron, and Parnate. Serious and even fatal reactions (with coma, hypotension, and peripheral vascular collapse) have been reported in some instances in which a narcotic was administered to a patient who has taken a MAO inhibitor. The latter compounds retard the metabolism and detoxification of many drugs. In the treatment of reactions the intravenous use of hydrocortisone and vasopressors has been recommended. (1, 55)
- \+ Nardil: see Demerol + MAO Inhibitors
- \+ Niamid: see Demerol + MAO Inhibitors
- \+ Parnate: see Demerol + MAO Inhibitors
- \+ Proketazine: see Demerol + MAO Inhibitors

DEPO-HEPARIN + Prothrombinopenic Drugs: It has been shown that high concentrations of heparin in the blood will cause an apparent lowering of prothrombin level; however, appreciable lowering does not occur unless the coagulation time (Lee-White) is more than 40 minutes. To avoid false prothrombin levels during the period of combined heparin-prothrombinopenic therapy, the daily test should be performed before the administration of the prothrombinopenic drug and at a time when the coagulation time is less than 40 minutes, namely, 12 to 24 hours after injection of the repository sodium heparin. (1)

DEPO-MEDROL: see Medrol

DEPO-MEDROL + Solutions: Depo-Medrol should not be mixed or diluted with other solutions. (1)

DEPO-PROVERA + Diabetes: A decreased glucose tolerance has been observed in a small percentage of patients on estrogen-progesterone combinations. The mechanism of this decrease is obscure. For this reason diabetic patients should be carefully observed while receiving Depo-Provera therapy. (1)

DEPROL: see Miltown

DEPROL
+ Alcohol: It is possible that the effects of excessive concumption of alcohol may be increased by meprobamate. (1)
+ Anticholinergics: Use with caution in combination with anticholinergic preparations because of the anticholinergic properties of benactyzine hydrochloride. (1)
+ CNS Stimulants: In cases of Deprol overdosage CNS stimulants and pressor amines should be cautiously administered. (1)
+ Pressor Amines: see Deprol + CNS Stimulants

DERONIL + Diabetes: Deronil should be used with great caution in patients with diabetes mellitus, since it causes disturbances in glucose metabolism. (2)

DESBUTAL + MAO Inhibitors: Desbutal is contraindicated in patients receiving monoamine oxidase inhibitors. (1) Mechanism: see Mao Inhibitor + Sympathomimetics, Indirect-Acting.

DESITIN HC OINTMENT + Corticosteroids: Use concomitantly cautiously in patients receiving oral or parenteral corticosteroid therapy, since 20 to 40% of rectally administered hydrocortisone may be absorbed. (1)

DESITIN HC SUPPOSITORIES + Corticosteroids: Use concomitantly cautiously in patients receiving oral or parenteral corticosteroid therapy, since 20 to 40% of rectally administered hydrocortisone may be absorbed. (1)

DESOXYN + MAO Inhibitor: Desoxyn is contraindicated in patients receiving a monoamine oxidase inhibitor. (1) Mechanism: see MAO Inhibitor + Sympathomimetics, Indirect-Acting.

DETERGENTS: see listed under the following agents:
Attenuvax, Lyovac
Measles Virus Vaccine, Live, Attenuated

DETERGENTS (continued)
M-Vac Measles Virus Vaccine, Live, Attenuated
Pfizer-Vax Measles-L

DEXAMETH + Diabetes: Since dexamethasone is a glucocorticoid, it has an inherent potential for affecting carbohydrate metabolism. When used in low or average dosages, hyperglycemia and glycosuria have been reported infrequently. Alteration of glucose metabolism, associated with adrenocortical hormone therapy, will cause aggravation of diabetes mellitus. (1, 2)

DEXAMYL: see Dexedrine

DEXEDRINE + MAO Inhibitors: Do not use Dexedrine in patients taking MAO inhibitors. (1) Mechanism: see MAO Inhibitor + Sympathomimetics, Indirect-Acting.

DEXTRAN
+ Blood Crossmatching: see Dextran + Blood Grouping
+ Blood Grouping: Although Dextran does not alter a patient's blood group or type, Dextran may interfere with blood grouping, typing, and crossmatching. Whenever possible, blood for typing and crossmatching should be drawn before administration of Dextran. If difficulties occur with blood drawn afterward, the cells may be washed with isotonic sodium chloride or other techniques may be employed. Note: Roleau formation (false agglutination) may occur in normal blood as well as in blood containing Dextran. (1)
+ Blood Typing: see Dextran + Blood Grouping

DEXTROAMPHETAMINE + MAO Inhibitor: see MAO Inhibitor + Sympathomimetics, Indirect-Acting

DEXTROSE INJECTION: see listed under the following agents:
Blood
Fungizone I. V.

DEXTROMETHORPHAN + Sublimaze: see Sublimaze + Dextromethorphan

DEXTRO-TUSSIN + Digitalis: Dextro-Tussin should be used with caution in patients receiving digitalis. (1)

DIABETES: see listed under the following agents:
HP Acthar Gel
Adrenal Corticosteroids
Adrenergic Agents
Adrenocorticosteroids
Aldactazide
Alphadrol
Anhydron
Antabuse
Aquatag
Aramine

DIABETES (continued)

- Aristocort
- Bristuron
- Celestone
- Choloxin
- Contraceptives, Oral
- Cortef
- Corticosteroids
- Cortiphate
- Cortisone Acetate
- Cortone
- Cortril
- C-Quens
- Decadron
- Deladumone
- Delta-Cortef
- Deltasone
- Depo-Provera
- Deronil
- Dexemeth
- Diuretics, Thiazide
- Diuril
- Dopar
- Dyazide
- Edecrin
- Enduron
- Enovid
- Esidrix
- Euthroid
- Eutonyl
- Exna
- Furoxone
- Gammacorten
- Haldrone
- Hexadrol
- Hydeltra-TBA
- Hydeltrasol Inj.
- Hydrocortone
- HydroDIURIL
- Hydromox
- Hygroton
- Inderal
- Indocin
- Kenacort
- Kenacort, Parenteral
- Laradopa
- Lasix
- Letter
- Maltsuprex
- Medrol
- Metahydrin
- Meticortelone
- Meticorten
- Mintezol
- Naqua
- Naturetin
- Neocylone
- Nicalex
- Norinyl
- Norlestrin
- Norquen
- Oracon
- Oretic
- Ortho-Novum
- Ovral
- Ovulen
- Predisal
- Prednis
- Prednisolone
- Prednisone
- Predsem
- Preludin
- Provest
- Pyrazinamide
- Renese
- Repoise
- Salcort
- Saluron
- Skelaxin
- Somnacort
- Stemex
- Sterane
- Sterapred

DIABETES (continued)

Sterasal-K
Synthroid
Thorazine
Thyrolar
Tofranil
Tromexan
X-Prep Liquid

DIABINESE: see listed under the following agents:

Tolinase
Trisoralen

DIABINESE

+ Alcohol: Diabinese may decrease tolerance to alcohol. In some patients a disulfiram (Antabuse)-like reaction may be produced after the ingestion of alcohol. (1, 2, 76)
+ Barbiturates: Since animal studies suggest that the action of barbiturates and other sedatives and hypnotics may be prolonged by therapy with Diabinese, they should be employed cautiously. (1, 2)
+ Benemid: see Diabinese + Butazolidin
+ Butazolidin: Caution should be exercised when Butazolidin, Tandearil, antibacterial sulfonamides, salicylates, Benemid, Dicumarol, or MAO inhibitors are administered concomitantly with Diabinese, as hypoglycemia resulting from either potentiation or accumulation of sulfonylureas has been reported. (1, 2)
+ Dicumarol: see Diabinese + Butazolidin
+ Diuretics, Thiazide: The thiazide diuretics should be given cautiously with Diabinese, since the thiazides have been reported to aggravate the diabetic state and to result in increased requirements of Diabinese, temporary loss of control, and even secondary failure. (2)
+ Hypnotics: see Diabinese + Barbiturates
+ MAO Inhibitors: see Diabinese + Butazolidin
+ Phenothiazines: The possibility that phenothiazine-type tranquilizers may cause jaundice or abnormal results in liver function tests should be borne in mind when these agents are used with Diabinese. (2)
+ Salicylates: see Diabinese + Butazolidin
+ Sedatives: see Diabinese + Barbiturates
+ Sulfonamides, Antibacterial: see Diabinese + Butazolidin
+ Tandearil: see Diabinese + Butazolidin

DIAMOX

+ Miotics: A complementary effect has been noted when Diamox has been used in conjunction with miotics or mydriatics as the case demanded. (1)
+ Mydriatics: see Diamox + Miotics

DIANABOL

+ Anticoagulants, Coumarin: Dianabol potentiates the effects of coumarin anticoagulants and inandione anticoagulants which could result in hemorrhage. Patients receiving anticoagulants and Dianabol should be followed closely with prothrombin time determinations and appropriate adjustment in anticoagulant dosage. (1, 11, 60)
+ Anticoagulants, Inandione: see Dianabol + Anticoagulants, Coumarin
+ Tandearil: The anabolic steroid Dianabol causes an elevation in Tandearil plasma levels in man. The elevation is not due to a slowed rate of metabolism but to an altered distribution between plasma and tissues. (121) Mechanism: Dianabol apparently inhibits glucuronyl transferase, an enzyme needed for the conjugation of Tandearil with glucuronic acid, thereby slowing metabolism of Tandearil. (120)

DIBENZYLINE

+ Epinephrine: see Dibenzyline + Pressor Agents
+ Levophed: see Levophed + Dibenzyline
+ Pressor Agents: Usual pressor agents are not effective in the treatment of Dibenzyline overdosage. Epinephrine is contraindicated because it may cause a further fall in blood pressure (epinephrine reversal). Intravenous infusion of Levophed may be used to combat severe hypotension reactions because the pressor effect is not reversed by Dibenzyline and may not be completely blocked. (1)

DICUMAROL: see Anticoagulants

DICUMAROL: see listed under the following agents:

Analexin
Choloxin
Diabinese
Diphenylhydantoin
Euthroid
Orinase
Phenobarbital
Sulfonamides, Long-Acting
Thyrolar

DICUMAROL: The following agents may inhibit the effects of Dicumarol: (1)

Chloral Hydrate
Griseofulvin
Phenobarbital and other Barbiturates

DICUMAROL: The following agents may potentiate the effects of Dicumarol: (1)

ACTH
Adrenocorticosteroids
Alcohol
Antibiotics, Broad Spectrum
Atromid-S
Butazolidin
Choloxin
Dilantin (diphenylhydantoin)
Indocin
Methylthiouracil
Nilevar
Quinidine

DICUMAROL (continued, agents that may potentiate Dicumarol):

Quinine
Radioactive Compounds
Tandearil
Salicylates

DICUMAROL

+ Diphenylhydantoin: A two-way interaction between Dicumarol and diphenylhydantoin has been reported. Diphenylhydantoin apparently increases the anticoagulant effect by displacing Dicumarol from plasma binding sites. On the other hand, Dicumarol elevates the serum hydantoin concentration probably by inhibiting its enzymatic degradation in the liver which has led to drug intoxication due to diphenylhydantoin. (1, 40, 55, 73)

+ Heparin Sodium: When sodium heparin is given with Dicumarol, the one-stage prothrombin time determination should be delayed for a period of three to four hours following heparin administration. (1)

+ Orinase: The coumarin anticoagulants are known to potentiate the hypoglycemic effect of Orinase by inhibiting its degradation in the liver. (1, 56)

+ X-Ray: Exposure to x-ray may lead to prolonged prothrombin time. (1)

DIET +Anticoagulants: see Anticoagulants + Diet

DIETHYLSTILBESTEROL in ETHYL OLEATE + Plastics: Ethyl Oleate, the vehicle used in Ampoules No. 549, is incompatible with plastics. A glass syringe should be used for administration. (1)

DIGITALIS: see listed under the following agents:

Adrenocorticosteroids
Anhydron
Antabuse
Aquatag
Atarax
Bristuron
Calcium Chloride Sol.
Calcium Gluconate-Glucoheptanate Sol.
Calphosan
Calscorbate Ampuls
Dextro-Tussin
Diuretics, Thiazide
Diuril
Diutensin
Diutensin-R
Edecrin
Enduron
Ephed-Organidin
Esidrix
Exna
HydroDIURIL
Hydromox
Hygroton
Ismelin
Isuprel
Kayexalate
KIE Tab. and Syrup
Lanoxin
Lasix
Liquaemin Sodium
Metahydrin
Naqua
Naturetin

DIGITALIS (continued)

Oretic
Perithiazide SA
Persantine
Pertofrane
Pronestyl
Protalba
Provell Maleate
Quelicin Chloride Inj.
Quinaglute Dura-Tabs.
Quinidex
Quinidine Sulfate
Quinora
Raudixin
Rauwiloid
Rauwiloid plus Veriloid
Regitine
Renese
Saluron
Salutensin
Sedorzyl
Serpasil
Singoserp
SKF Special Resin No. 648
Succinylcholine
Unitensin
Veriloid

DIGITALIS

+ Adrenocorticosteroids: Since adrenocorticosteroids may cause potassium loss, caution must be observed when used concomitantly so as not to precipitate digitalis intoxication and arrhythmias. (26)
+ Calcium, Parenteral: Parenteral calcium may provoke arrhythmias in digitalized patients. (26, 55)
+ Diuretics, Thiazide: The thiazide diuretics lower total body potassium stores. Potassium depletion in turn enhances the toxicity of digitalis, so that even a fraction of a normal dose can cause an arrhythmia ranging anywhere from ventricular premature contractions to ventricular tachycardia or ventricular fibrillation. (26, 50, 55)
+ Reserpine: There is clinical and experimental evidence that simultaneous administration of these drugs can result in arterial arrhythmias, ectopic ventricular activity, and varying degrees of heart block. (25, 74)

DIGITOXIN + Lanoxin: see Lanoxin + Digitoxin

DIGOXIN + Lanoxin: see Lanoxin + Digoxin

DIHYDROSTREPTOMYCIN: see listed under the following agents:

Anectine
Coly-Mycin M Inj.
Kantrex Inj.
Succinylcholine
Sucostrin

DILANTIN: see Diphenylhydantoin

DILANTIN

+ Anticonvulsants: see Dilantin + Phenobarbital
+ Barbiturates: see Dilantin + Morphine
+ Bromides: see Dilantin + Phenobarbital
+ Morphine: When patients in status epilepticus are sedated with morphine or barbiturates, administration of Dilantin should be delayed until the effects of the sedative injections are diminished. (1)
+ Phenobarbital: If the patient has been receiving other medication, such as phenobarbital or bromides, sudden withdrawal may precipitate a series of attacks before Dilantin has been given in sufficient amounts to exercise control. This may be avoided by gradually replacing with Dilantin the anticonvulsant previously used. (1) See Dilantin + Morphine.

DILOR

+ Ephedrine: Do not administer Dilor simultaneously with ephedrine to children. (1)
+ Sympathomimetic Drugs: Dilor should not be combined with sympathomimetic drugs. (1)

DILUENTS: see listed under the following agents:

AquaMEPHYTON
Aristocort Forte Suspension
Chloromycetin Intramuscular
Cortone Acetate Saline Suspension
Ilotycin Gluceptate I. V.
M-Vac Measles Virus Vaccine, Live, Attenuated
Orimune Poliovirus Vaccine, Live, Oral, Trivalent
Pfizer-Vax Measles-L
Surital

DILUENTS, ACID + Surital: see Surital + Acid Diluents

DIMOCILLIN-RT + Neutrapen: Since methicillin is relatively unaffected by penicillinase, it is unlikely that the use of B. cereus penicillinase (Neutrapen) would be of value in the management of allergic reactions. (1)

DIPHENYLHYDANTOIN: see listed under the following agents:

Alcohol
Anticoagulants
Dicumarol
Panwarfin
Phenobarbital
Ritalin

DIPHENYLHYDANTOIN

+ Analexin: Phenyramidol (Analexin) inhibits the metabolism of diphenylhydantoin when the two drugs are given concomitantly. The half-life of diphenylhydantoin is doubled and plasma levels are significantly increased. Concurrent administration of these two drugs should be avoided, since high plasma concentrations of diphenylhydantoin have been associated with such manifestations as nystagmus, ataxia, and changes of mentation. Mechanism: Studies suggest that the hydroxylation of diphenylhydantoin by a hepatic microsomal enzyme may be inhibited by Analexin: decrease in metabolism of diphenylhydantoin. (29, 55)
+ Dicumarol: Dicumarol increases the biologic half-life of diphenylhydantoin in man, presumably by inhibition of p-hydroxylation of diphenylhydantoin (decrease in the metabolism of diphenylhydantoin). High plasma concentrations of diphenylhydantoin have been associated with such symptoms as nystagmus, ataxia, and changes of mentation. (55, 129) Diphenylhydantoin apparently increases the anticoagulant effect of Dicumarol by displacing Dicumarol from plasma binding sites. (1)
+ Isoniazid and Para-aminosalicylic Acid: Simultaneous administration of isoniazid and para-aminosalicylic acid also impairs the metabolism of diphenylhydantoin. (30, 33)
+ Phenobarbital: Phenobarbital (a known inducer of microsomal enzyme activity) increases the rate of metabolism of diphenylhydantoin and diminishes its pharmacologic activity. The increase in diphenylhydantoin metabolism is not a problem in the management of epileptic patients, since phenobarbital also possesses anticonvulsant activity. (33, 55, 129)

DIPYRONE

+ Anticoagulants: Dipyrone aggravates prothrombin deficiency. (1)
+ Thorazine: Dipyrone should not be used concomitantly with Thorazine because of resulting severe hypothermia. (1)

DISIPAL + Darvon: Since mental confusion, anxiety, and tremors have been reported in patients receiving orphenadrine and Darvon concurrently, it is recommended that Disipal not be given in combination with Darvon. (1)

DISOPHROL

+ Alcohol: see Disophrol + CNS Depressants
+ CNS Depressants: Patients receiving Disophrol Chronotab tablets should be cautioned about possible additive effects with alcohol and

DISOPHROL
+ CNS Depressants (continued): other central nervous system depressants (hypnotics, sedatives, tranquilizers).
+ Hypnotics: see Disophrol + CNS Depressants
+ Sedatives: see Disophrol + CNS Depressants
+ Tranquilizers: see Disophrol + CNS Depressants

DIUPRES: see Diuril and Reserpine

DIURETICS: see listed under the following agents:

Aldactone
Cortone
Cortril
Davoxin
Dyrenium
Edecrin
Eskalith
Eutonyl
Gitalgin
Hydrocortone
Inversine
Lanoxin
Lithane
Lithium Carbonate
Lithonate
MAO Inhibitor
Mio-Pressin
Modumate
Myodigin
Ostensin
Parnate
Pil-Digis
Prednisolone
Prednisone
Protalba

DIURETICS, CARBONIC ANHYDRASE INHIBITING: see listed under the following agents:

Edecrin
Medrol

DIURETICS, MERCURIAL: see listed under the following agents:

Aldactone
Analgesics
Medrol
Methadone
Naturetin

DIURETICS, THIAZIDE: see also under individual trade name preparations.

DIURETICS, THIAZIDE: see listed under the following agents:

Adrenocorticosteroids
Aldactone
Aldomet
Aspirin
Aventyl
Cortisone
Diabetes
Diabinese
Digitalis
Dymelor
Dyrenium
Flaxedil

DIURETICS, THIAZIDE (continued)

Hedulin
Hypoglycemic Drugs, Oral
Ismelin
MAO Inhibitors
Niamid
Norpramin
Orinase
Parnate
Pertofrane
Phenothiazines
Tolinase
d-Tubocurarine
Vivactil

DIURETICS, THIAZIDE

+ Adrenocortical Steroids: see Diuretics, Thiazide + Corticotrophin
+ Alcohol: Orthostatic hypotension may occur which may be potentiated when thiazide diuretics is combined with either alcohol, barbiturates, or narcotics.
+ Ammonium Chloride: Each product may cause an elevation of serum ammonia and precipitate onset of coma in patients with liver disease. (2)
+ Anesthesia: Hypotensive episodes under anesthesia have been observed in some patients receiving thiazides. (2)
+ Apresoline: see Diuretics, Thiazide + Hypotensive Agents, and Ismelin
+ Barbiturates: see Diuretics, Thiazide + Alcohol
+ Corticotrophin: Patients in whom the excretion of potassium is likely to be excessive (those who are receiving digitalis, adrenocortical steroids, or corticotrophin (ACTH, ACTHAR) should receive supplementary potassium as a salt or preferably food high in potassium content (e.g., orange juice). (2)
+ Diabetes: Development of mild hyperglycemia in patients with latent diabetes and aggravation of pre-existing diabetes has been noted occasionally with all of these agents. These effects are usually reversible upon the discontinuation of the drug. The hyperglycemic action of these drugs should not preclude their use in diabetic patients, but regular blood sugar determinations should be made during their prolonged administration. Increased doses of hypoglycemic agents may be needed to control the elevated blood sugar concentrations. Insulin requirements may be increased, decreased, or remain unchanged. (2,17) Thiazide diuretics may induce carbohydrate intolerance in diabetic patients by (1) direct or indirect toxic effect on islet cells of the pancreas, (2) decreased sensitivity to insulin, (3) decreased secretion of insulin, (4) increased binding of insulin, or (5) a secondary effect resulting from other biochemical alterations that often accompany thiazide treatment, such as hyperuricemia and hypokalemia. (33) See Diuretics, Thiazide + Sulfonylureas, Hypoglycemic.
+ Digitalis: Hypokalemia due to the thiazides may result in increased sensitivity to the action of digitalis; changes in the dosage require-

DIURETICS, THIAZIDE (continued)

+ Digitalis (continued): ments of both drugs must be anticipated. (2) See also Diuretics, Thiazide + Corticotrophin.
+ Diuretics, Mercurial: Thiazides may increase the response to mercurials when both agents are given together.
+ Ecolid: In elderly patients with advanced arteriosclerosis, the use of these drugs may precipitate coronary or cerebral thrombosis due to severe orthostatic hypotension, especially if the patient is receiving a ganglionic blocking agent (e.g., Ecolid, Inversine). (2) See Diuretics, Thiazide + Hypotensive Agents, and Ismelin.
+ Eutonyl: see Diuretics, Thiazide + Hypotensive Agents
+ Ganglionic Blocking Agents: see Diuretics, Thiazide + Hypotensive Agents and Diuretics, Thiazide + Ecolid
+ Hypotensive Agents: Because of a potentiating effect when oral diuretics are used with other hypotensive agents, the dosage of these other agents (ganglionic blocking agents, veratrum alkaloids, rauwolfia and its alkaloids, Apresoline, Eutonyl) should be reduced markedly. (2, 55) See Diuretics, Thiazide + Ismelin, Ecolid, and MAO Inhibitors.
+ Inversine: see Diuretics, Thiazide + Ismelin
+ Ismelin: When these diuretics are given with other potent hypotensive agents (Ismelin, Inversine, Apresoline), the dose of the latter should be reduced by at least 50% to prevent excessive hypotensive effects, especially in the upright position. (2) See Diuretics, Thiazide + Hypotensive Agents, and Ecolid.
+ Levophed: see Diuretics, Thiazide + Pressor Amines
+ MAO Inhibitors: An increased hypotensive effect may occur when these agents are used concomitantly due to an additive effect. (55)
+ Narcotics: see Diuretics, Thiazide + Alcohol
+ Pressor Amines: Recent reports indicate that thiazides may decrease arterial responsiveness to pressor amines (e.g., Levophed). (2)
+ Rauwolfia and its Alkaloids: see Diuretics, Thiazide + Hypotensive Agents
+ Sulfonylureas, Hypoglycemic: It is believed that thiazide diuretics elevate blood glucose by suppression of insulin secretion. The mechanism of action of the sulfonylurea oral hypoglycemic agents (Orinase, Tolinase, Diabinese, Dymelor) is the stimulation of release of pancreatic insulin. It is apparent that the pancreatic effects of the two classes of drugs are opposed. (55, 62)
+ Trisoralen: see Trisoralen
+ Tubocurarine: Recent reports indicate that thiazides may augment the paralyzing action of tubocurarine in surgical patients. (2)

DIURETICS, THIAZIDE (continued)

+ Uricosuric Agents: Since these oral diuretics decrease the excretion of uric acid, they may aggravate gout or interfere with its treatment with uricosuric agents. (2)
+ Veratrum Alkaloids: see Diuretics, Thiazide + Hypotensive Agents

DIURIL: see Diuretics, Thiazide

DIURIL: see listed under the following agents:

MAO Inhibitors
Niamid

DIURIL

+ ACTH: see Diuril + Steroids
+ Alcohol: Orthostatic hypotension may occur with Diuril and be potentiated by alcohol, barbiturates, or narcotics. (1)
+ Antihypertensive Agents: Diuril potentiates the action of other antihypertensive agents. Therefore the dosage of these agents, especially the ganglion blockers, must be reduced by at least 50% as soon as the diuretic is added to the regimen. This must be done to avoid an excessive drop in pressure and other unwanted effects. (1, 2)
+ Apresoline: Diuril potentiates the hypotensive effect of Apresoline; therefore the dosage of the Apresoline must be reduced markedly. (2)
+ Barbiturates: see Diuril + Alcohol
+ Diabetes: Insulin requirements in diabetic patients may be increased, decreased, or unchanged. In latent diabetics Diuril in common with other benzothiadiazines may cause hyperglycemia and glycosuria. (1, 2)
+ Digitalis: Interference with adequate electrolyte intake will contribute to hypokalemia. Digitalis therapy may exaggerate metabolic effects of hypokalemia especially with reference to myocardial activity. (1, 2)
+ Ganglionic Blocking Agents: see Diuril + Antihypertensive Agents
+ Narcotics: see Diuril + Alcohol
+ Norepinephrine: Diuril decreases arterial responsiveness to norepinephrine, as do the other thiazides, necessitating due care in surgical patients. It is recommended that the thiazides be discontinued 48 hours before elective surgery. (1, 2)
+ Steroids: Hypokalemia may develop with Diuril as with other potent diuretics, especially during concomitant steroid or ACTH administration. (1, 2)
+ Trisoralen: see Trisoralen

DIURIL (continued)

+ Tubocurarine: Thiazide drugs may increase the responsiveness to tubocurarine. (1, 2)

DIURIL LYOVAC

+ Blood, Whole: Avoid simultaneous administration of solutions of Diuril with whole blood or its derivatives. (1)
+ Blood Derivatives: see Diuril Lyovac + Blood, Whole

DIUTENSIN: see Unitensin and Enduron

DIUTENSIN

+ Ammonium Chloride: In liver disease with concomitant thiazide therapy, ammonium chloride should not be administered since each may cause an elevation of serum ammonia and precipitate onset of coma. (1)
+ Digitalis: The concomitant use of digitalis and the Diutensin products may increase the risk of digitalis intoxication (this is due chiefly to the saluretic effects of methychlothiazide and may experience a change in electrolyte balance with subsequent hypokalemia and perhaps to the vagotonic effects of Unitensin and/or Reserpine). (1)

DIUTENSIN-R: see Unitensin, Enduron, and Reserpine

DIUTENSIN-R + Digitalis: The concomitant use of digitalis and the Diutensin products may increase the possibility of digitalis-like intoxication (due chiefly to the saluretic effects of methyclothiazide and perhaps to the vagotonic effects of Unitensin and/or Reserpine). (1)

DOLONIL + Pyridium: Since Dolonil contains Pyridium (phenazopyridine hydrochloride) in pharmacologically adequate amounts, other preparations containing this compound should not be administered concomitantly. (1) The following preparations contain phenazopyridine:

Azo Gantanol
Azo Gantrisin
Azolate
Azo-Mandelamine
Azotrex
Buren
Donnasep
Mandaley
Renasul A
Suladyne
Thiosulfil Duo-Pak
Thiosulfil-A Forte
Uremide
Uriplex
Urobiotic
Uropeutic

DONNASEP + Dolonil: see Dolonil + Donnasep

DOPAMINE: see listed under the following agents:

Nardil
Parnate

DOPAR: see Levodopa

DOPAR

+ Aldomet: see Dopar + Antihypertensive Agents
+ Antihypertensive Agents: Careful monitoring is required in patients receiving antihypertensive drugs, particularly Aldomet, reserpine, or ganglionic blocking agents. (1)
+ Antiparkinsonism Drugs: Dopar may be used concomitantly with other antiparkinson drugs, such as Cogentin, Artane, or Kemadrin, in which cases the usual dose of each may need to be reduced. (1)
+ Artane: see Dopar + Antiparkinsonism Drugs
+ Cogentin: see Dopar + Antiparkinsonism Drugs
+ Diabetes: The control of the diabetic patient may be adversely affected by treatment with levodopa. Therefore use of Dopar in diabetics should be attended with frequent monitoring of the patient and proper adjustment of the antidiabetic regimen. (1)
+ Ganglionic Blocking Agents: see Dopar + Antihypertensive Agents
+ Kemadrin: see Dopar + Antiparkinsonism Drugs
+ MAO Inhibitor: Monoamine oxidase inhibitors and Dopar should not be given concomitantly, and these inhibitors must be discontinued two weeks prior to initiating therapy with Dopar. (1)
+ Psychoactive Drugs: If the concomitant administration of psychoactive drugs is deemed necessary, such drugs should be administered with caution and patients should be carefully observed for unusual side effects. (1)
+ Pyridoxine Hydrochloride: It has been reported that pyridoxine hydrochloride in oral doses of 10 to 25 mg rapidly reverses the antiparkinson effects of Dopar. This should be considered before recommending multivitamin preparations containing pyridoxine hydrochloride (vitamin B_6). (1)
+ Reserpine: see Dopar + Antihypertensive Agents

DOPRAM

+ Alkaline Solutions: Since Dopram has an acid pH, it should not be mised with alkaline solutions such as Pentothal Sodium. (1)
+ Anesthetics: Since an increase in epinephrine release has been noted with Dopram, it is recommended that initiation of therapy with this respiratory stimulant be delayed for at least 10 minutes following the discontinuance of anesthetics known to sensitize the myocardium to catecholamines, such as Fluothane and cyclopropane. (1)
+ Cyclopropane: see Dopram + Anesthetics
+ Fluothane: see Dopram + Anesthetics
+ MAO Inhibitors: see Dopram + Sympathomimetic Drugs

DOPRAM (continued)

+ Muscle Relaxant Drugs: In patients who have been lightly curarized, Dopram may temporarily mask the residual effects of muscle relaxant drugs. (1)
+ Pentothal Sodium: see Dopram + Alkaline Solutions
+ Sympathomimetic Drugs: Dopram should be administered cautiously to patients receiving sympathomimetic or monoamine oxidase inhibiting drugs, since a synergistic pressor effect may occur. (1, 2)

DORIDEN: see listed under the following agents:

Anticoagulants
Coumadin
Dicumarol
Panwarfin
Quaalude

DORIDEN

+ Alcohol: see Doriden + CNS Depressants
+ Anticoagulants: Doriden may depress the response to coumarin anticoagulants by increasing the synthesis of drug metabolizing enzymes in the liver leading to more rapid inactivation of the anticoagulant. Dosage of coumarin anticoagulants may require adjustments during and on cessation of Doriden therapy. (1, 9, 55)
+ Barbiturates: see Doriden + CNS Depressants
+ CNS Depressants: Patients should be cautioned about the possible combined effects with alcohol or other CNS depressants. The effects of Doriden are exaggerated by concomitant ingestion of other sedatives such as alcohol, barbiturates, etc., and suicidal efforts commonly involve multiple drugs of the sedative-tranquilizer types. (1)

DOUCHE: see listed under the following agents:

Delfen Vaginal Cream
Delfen Vaginal Foam
Lorophyn Suppos. and Jelly
Mycostatin Vaginal Tablets
Preceptin Vaginal Gel

DPT VACCINE + Adrenocorticosteroids: Administration of adrenocorticosteroids concomitantly with immunizing agents should be avoided because the steroid may interfere with antibody response. (1)

DRAMAMINE + Streptomycin: Dramamine is capable of masking the symptoms of streptomycin toxicity and should be used with caution in these circumstances. (1)

DRINALFA + MAO Inhibitors: Drinalfa in contraindicated in patients receiving monoamine oxidase inhibitors, since there is evidence that the action of sympathomimetic amines may be profoundly potentiated following the use of monoamine oxidase inhibitors. (1) Mechanism: see MAO Inhibitors + Sympathomimetics, Indirect-Acting.

DRONACTIN: see Decadron and Periactic

DRUG OVERDOSAGE: see listed under the following agents:
Compazine
Repoise
Stelazine
Thorazine
Tigan

DRUGS: see listed under the following agents:
Aldomet
Aminosol
Aralen
Atabrine
Compazine
Cortone Acetate Saline Suspension
Cuemid
Lipomul I.V.
Paradione
Plaquenil
Quaalude
Questran
Quide
Stoxil
Tandearil
Tegetrol
Trecator
Tridione
Trilafon
Vivactil
Xylocaine

DRUGS, ANTITHYROID + Oriodide-131: see Oriodide-131 + Drugs, Antithyroid

DRUGS, ANTIBUBERCULOSIS: see listed under the following agents:
Diphenylhydantoin
Seromycin

DRUGS, CARDIAC-DEPRESSING + Phenothiazines: see Phenothiazines + Drugs, Cardiac-Depressing

DRUGS, CARDIOVASCULAR-DEPRESSING + Xylocaine: see Xylocaine + Drugs, Cardiovascular-Depressing

DRUGS, CURARE-LIKE + Sucostrin: see Sucostrin + Drugs, Curare-Like

DRUGS, CYTOTOXIC: see listed under the following agents:
Leukeran
Vercyte

DRUGS, DEPRESSANT + Quelicin Chloride Inj.: see Quelicin Chloride Inj. + Drugs, Depressant

DRUGS, ELECTROLYTE-DEPLETING: see listed under the following agents:
Tubocurarine Chloride Inj.
Viocin Sulfate

DRUGS, HEMATOPOIETIC-DEPRESSANTS + Peganone: see Peganone + Drugs, Hematopoietic-Depressants

DRUGS, HEMOLYTIC + Aralen with Primaquine: see Aralen with Primaquine + Drugs, Hemolytic

DRUGS, HEPATOTOXIC: see listed under the following agents:
Anticoagulants
Aralen
Coumadin
Plaquenil
Sintrom
Tromexan

DRUGS, IMMUNOSUPPRESSIVE + Smallpox Vaccine, Dried: see Smallpox Vaccine, Dried + Drugs, Immunosuppresive

DRUGS, IMPAIRING RENAL FUNCTION: see listed under the following agents:
Furadantin
Macrodantin

DRUGS, NEPHROTOXIC: see listed under the following agents:
Garamycin
Neomycin

DRUGS, NON-NARCOTIC: see listed under the following agents:
Lorfan
Nalline

DRUGS, ONCOLYTIC + Velban: see Velban + Drugs, Oncolytic

DRUGS, OTOTOXIC: see listed under the following agents:
Garamycin
Kantrex Inj.
Neomycin
Lasix

DRUGS, OVER-THE-COUNTER, COLD: see listed under the following agents:
Nardil
Parnate

DRUGS, OVER-THE-COUNTER, HAY FEVER: see listed under the following agents:

Nardil
Parnate

DRUGS-OVER-THE-COUNTER, REDUCING: see listed under the following agents:

Nardil
Parnate

DRUGS, POTASSIUM-DEPLETING: see listed under the following agents:

Flaxedil
Tubocurarine

DRUGS, PROPRIETARY: see listed under the following agents:

MAO Inhibitor
Marplan
Nardil
Niamid
Parnate

DRUGS, TOXIC + Permitil: see Permitil + Drugs, Toxic

DULCOLAX

+ Antacids: Dulcolax tablets must be swallowed whole, not chewed or crushed, and should not be taken within one hour of antacids or milk. (1)
+ Milk: see Dulcolax + Antacids

DUO-MEDIHALER: see Medihaler-Epi and Isoproterenol

DYAZIDE: see Dyrenium and Hydrochlorothiazide

DYAZIDE

+ Aldactone: If Dyazide is used concomitantly with Aldactone, the physician should frequently determine serum potassium levels, since both the Dyrenium component in Dyazide and Aldactone can cause potassium retention. There have been reports of two deaths in patients receiving concomitant Dyrenium or Dyazide and Aldactone. (1)
+ Antihypertensive Agents: Since Dyazide has an antihypertensive effect, its use with another antihypertensive drug requires reduced dosage of the latter agents. When Dyazide is added to another antihypertensive drug already being used as therapy, the dose of the other antihypertensive drug should be reduced by half, particularly if it is a ganglionic blocking agent. Subsequent adjustment of dosage should be made as required. (1)

DYAZIDE (continued)

+ Diabetes: Dyazide, because of its hydrochlorothiazide component, may cause hyperglycemia and glycosuria in patients predisposed to diabetes, and the insulin requirements may be altered. (1)
+ Ganglionic Blocking Agents: see Dyazide + Antihypertensive Agents
+ Potassium: Because of the potassium-conserving effect of Dyrenium, hypokalemia is an uncommon occurrence with the use of Dyazide. Patients should not be placed on dietary potassium supplements or potassium salts in conjunction with Dyazide unless hypokalemia or their dietary intake of potassium is markedly impaired. (1)

DYCLONE + Contrast Media I.V.P.: Dyclone is contraindicated in cystoscopic procedures following intravenous pyelography because an iodine precipitate occurs which interferes with visualization. (1)

DYMELOR: see also Hypoglycemic Drugs, Oral

DYMELOR

+ Alcohol: Very rarely, patients receiving any of the sulfonylureas may experience peculiar symptoms referred to as the "disulfiram reaction" following the ingestion of alcohol. (1, 2, 76) Mechanism: see Alcohol + Sulfonylurea, Hypoglycemic.
+ Barbiturates: The effects of barbiturates and other sedatives and hypnotics may be prolonged when Dymelor is used. (2)
+ Benemid: Because of the sulfonylurea-potentiating effects reported with certain drugs, hypoglycemia may occur when such agents as the antimicrobial sulfas, Butazolidin, or Benemid are a part of the treatment program of patients receiving Dymelor. (1, 2)
+ Butazolidin: see Dymelor + Benemid. Mechanism: This study demonstrated that Butazolidin can potentiate the hypoglycemic effects of Dymelor by interfering with the renal excretion of the active metabolite hydroxyhexamide. (32)
+ Diuretics, Thiazide: Caution should be exercised in the administration of thiazide diuretics to diabetic patients on sulfonylurea therapy, since the thiazides have been reported to aggravate the diabetic state and to result in increased requirements of the sulfonylurea, temporary loss of control, or even secondary failure. (1, 2)
+ Hypnotics: see Dymelor + Barbiturates
+ Insulin: In the changeover from Insulin to Dymelor, hypoglycemia can occur at the time both drugs are given simultaneously. (1, 76)
+ Phenothiazines: The possibility that phenothiazine-type tranquilizers may cause jaundice or abnormal results in liver function tests should be borne in mind when these agents are used with Dymelor. (2)

DYMELOR (continued)
+ Sedatives: see Dymelor + Barbiturates
+ Sulfas, Antimicrobial: see Dymelor + Benemid
+ Trisoralen: see Trisoralen. (1)

DYNAPEN + Meals: Studies indicate that this material is best absorbed when taken on an empty stomach, preferably one or two hours before meals. (1)

DYRENIUM
+ Aldactone: If Dyrenium is used concomitantly with Aldactone, the physician should frequently determine serum potassium levels, since both of these agents can cause potassium retention and in some cases hyperkalemia. (1)
+ Antihypertensive Agents: Although Dyrenium has not proved to be a consistent antihypertensive agent, the physician should be aware of a possible lowering of blood pressure. Concomitant use with antihypertensive drugs may result in an antihypertensive effect. (1, 2)
+ Diuretics: When required, Dyrenium may be given concomitantly with other diuretics. It has been used most frequently with thiazides. When it is combined with another diuretic, total dosage of each agent should usually be lowered initially and then adjusted to the patient's need. (1, 2)
+ Diuretics, Thiazide: The naturetic effects of both agents are potentiated when they are given concomitantly. (2) See also Dyrenium + Diuretics.
+ Lasix: see Lasix + Dyrenium
+ Potassium: Because of the potassium conserving effects of Dyrenium, potassium supplements, either as medication or as a potassium rich diet, should not be used with Dyrenium. (1, 2)

E

ECOLID: see Ganglionic Blocking Agents

ECOLID
+ Anesthesia: Because of potentiation by other drugs, the dosage of Ecolid must be adjusted prior to surgery involving anesthesia. (2)
+ Diuretics, Thiazide: see Diuretics, Thiazide + Ecolid
+ Trisoralen: see Trisoralen (2)

EDECRIN
+ Alcohol: Because studies in dogs have shown a transient increase in blood ethanol levels with parenteral Edecrin, the possibility that this drug may augment the effects of alcohol ingestion should be considered. (1)
+ Aldactone: There is a synergistic effect when Edecrin is used concomitantly with Aldactone. Advantage may be taken of this observation not only to increase the diuretic effect in patients who are extremely resistant to therapy but also to prevent an especially serious consequence, such as potassium loss. (37)
+ Antihypertensive Agents: The safety and efficacy of Edecrin in hypertension have not been established. However, the dosage of co-administered antihypertensive agents may require adjustment. Orthostatic hypotension may occur in patients receiving other antihypertensive agents when given Edecrin. (1)
+ Benemid: Benemid can delay the excretion of Edecrin and prolong the duration of action. (37)
+ Blood, Whole: Avoid simultaneous administration of parenteral Edecrin with whole blood or its derivatives; it is incompatible. (1)
+ Blood Derivatives: see Edecrin + Blood, Whole
+ Diabetes: The effects of Edecrin on carbohydrate metabolism have not yet been fully delineated. However, patients who have developed hyperglycemia following thiazide administration have not done so with Edecrin. Nevertheless, hyperglycemia has occurred in a few patients, most of whom had decompensated cirrhosis of the liver. (1, 37)
+ Digitalis: Too vigorous or excessive diuresis in patients who are digitalized; the development of hypokalemia may precipitate cardiac arrhythmias. (1)
+ Diuretics: Edecrin has additive effects when used with other diuretics. (1, 37)

EDECRIN (continued)

+ Diuretics, Carbonic Anhydrase Inhibiting: Edecrin may potentiate the action of carbonic anhydrase inhibitors with augmentation of natriuresis and kaliuresis. Therefore, when adding Edecrin, the initial dose and changes of dose should be in 25 mg increments to avoid electrolyte depletion. (1)
+ Dyrenium: There is a synergistic effect when Edecrin is used concomitantly with Dyrenium. Advantage may be taken of this observation not only to increase the diuretic effect in patients who are extremely resistant to therapy but also to prevent an especially serious consequence, such as potassium loss. (37)

EDRISAL + MAO Inhibitors: Do not use Edrisal in patients taking monoamine oxidase inhibitors. (1) Mechanism: see MAO Inhibitors + Sympathomimetics, Indirect-Acting

ELAVIL: see Antidepressants, Tricyclic

ELAVIL: see listed under the following agents:

Eutonyl	Nardil
MAO Inhibitors	Parnate
Matulane	Placidyl

ELAVIL

+ Alcohol: Patients should be cautioned that the response to alcohol may be potentiated when taking Elavil. (1)
+ Antidepressants: The potency of Elavil is such that addition of other antidepressant drugs generally does not result in any additional therapeutic benefit. Untoward reactions have been reported after the combined use of antidepressant agents having varying modes of activity. Accordingly, combined use with other antidepressant drugs should be undertaken only with due recognition of the possibility of potentiation. (1)
+ Barbiturates: It has been demonstrated that Elavil potentiates the effects of barbiturates in mice. Although no data have been presented to demonstrate this effect in man, care should be taken when barbiturates or other central nervous system depressants are administered concomitantly with Elavil. (2)
+ CNS Depressants: see Elavil + Barbiturates
+ MAO Inhibitors: Monoamine oxidase inhibitor drugs may potentiate other drug effects and such potentiation may even cause death. Accordingly, Elavil should not be given to patients who have been receiving MAO inhibitors for at least two weeks after stopping the MAO inhibitor. In such patients it is recommended that therapy

ELAVIL (continued)

+ MAO Inhibitors (continued): with Elavil be initiated cautiously with gradual increase in the dosage required to obtain a satisfactory response. (1, 2) Mechanism: see MAO Inhibitor + Antidepressants, Tricyclic.

ELECTROCONVULSIVE THERAPY + Aventyl: see Aventyl + Electroconvulsive Therapy

ELECTROLYTE FLUIDS + Lytren Powder: see Lytren Powder + Electrolyte Fluids

ELECTROLYTE IMBALANCE: see listed under the following agents:

Flaxedil
Succinylcholine
Sucostrin

ELECTROSHOCK THERAPY: see listed under the following agents:

Harmonyl
Rauwiloid
Reserpine
Serpasil

ELIXOPHYLLINE + Theophylline Preparations: Do not administer other theophylline preparations concurrently. (1)

EMIVAN

+ Anesthesia: Too rapid intravenous injection of Emivan to the anesthetized or comatose patient should be avoided to minimize coughing, sneezing, laryngospasm, or pruritic manifestations. (1)
+ MAO Inhibitors: There is an apparent additive stimulant effect between Emivan and the monoamine oxidase inhibitors in experimental animals; therefore care in adjustment of dosage of either or both drugs should be observed when one is used in the presence of the other. (1, 2)

ENARAX: see Atarax

EN-CHLOR: see Chloral Hydrate

EN-CHLOR + Alcohol: En-Chlor should not be given to patients who are taking or having recently taken alcohol. (1)

ENDURON: see Diuretics, Thiazide

ENDURON

+ Adrenocortical Steroids: To avoid the possibility of hypokalemia developing, precautions are particularly indicated in those patients receiving corticotrophin or adrenocortical steroids. (1)
+ Antihypertensive Agents: Like other benzothiadiazines, Enduron also has antihypertensive properties and may be used to enhance the antihypertensive action of other drugs. When another antihyperten-

ENDURON (continued)

+ Antihypertensive Agents (continued): sive agent is to be added to the regimen, this should be accomplished gradually. Ganglionic blocking agents should be given at one-half the usual dose, since their effect is potentiated by pretreatment with Enduron. (1)
+ Corticotrophin: see Enduron + Adrenocortical Steroids
+ Diabetes: Alteration of carbohydrate metabolism in newborn infants has been suggested as a possible occurrence when expectant mothers show decreased glucose tolerance while on thiazide drugs. Elevation of blood sugar has occurred with the use of thiazide drugs. (1, 2)
+ Digitalis: Hypokalemia may occur during therapy with Enduron. Potassium depletion can be hazardous in patients taking digitalis. Myocardial sensitivity to digitalis is increased in the presence of reduced serum potassium, and signs of digitalis intoxication may be produced by formerly tolerated doses of digitalis. (1, 2)
+ Ganglionic Blocking Agents: see Enduron + Antihypertensive Agents
+ Harmonyl: An enhanced response frequently follows the concurrent administration of Enduron and Harmonyl. (1) See also Enduron + Antihypertensive Agents.
+ Norepinephrine: see Enduron + Vasopressor Agents
+ Trisoralen: see Trisoralen
+ Tubocurarine: Thiazide drugs may increase the responsiveness to tubocurarine. (1, 2)
+ Vasopressor Agents: It has been observed that thiazide drugs may reduce arterial responsiveness to levarterenol (norepinephrine). Accordingly, the dosage of vasopressor agents may need to be modified in surgical patients who have been receiving thiazide drugs. (1, 2)

ENDURONYL: see Enduron and Harmonyl

ENDURONYL

+ Anesthesia: Enduronyl should be withdrawn or dosage reduced during the two weeks prior to elective surgery. An unexpected degree of hypotension and bradycardia has been reported during anesthesia in patients under treatment with rauwolfia alkaloids. When emergency surgical procedures are necessary, parenteral administration of vagal blocking agents to prevent or reverse hypotension and/or bradycardia may be considered. (1) Mechanism: see Reserpine + Anesthesia.
+ Anticonvulsants: In epileptic patients therapy may necessitate dosage adjustment of anticonvulsant medication. Rauwolfia alkaloids lower the convulsive threshold and shorten seizure latency. (1)

ENEMAS: see listed under the following agents:
- Sulfasuxidine
- Sulfathalidine

ENOVID + Diabetes: There is no direct evidence that Enovid alters the diabetic state. However, in a few instances some degree of difficulty in the management of diabetic patients has been reported in connection with Enovid therapy. It is possible that a change in insulin dosage may be required. (1, 2)

EPHED-ORGANIDIN + Digitalis: Cardiac arrythmias may occur if given to digitalized patients. (1)

EPHEDRINE: see listed under the following agents:
- Aminet Suppositories
- Copavin
- Dilor
- Eutonyl
- Furoxone
- Ismelin
- MAO Inhibitors
- Norisodrine

EPIDOL + Trisoralen: see Trisoralen

EPINEPHRINE: see listed under the following agents:
- Apresoline
- Bronkometer
- Bronkosol
- Chloroform
- Choloxin
- Compazine
- Cyclopropane
- Dartal
- Dibenzyline
- Ethyl Chloride
- Euthroid
- Fluoromar
- Fluothane
- Haldol
- Inapsine
- Inderal
- Inversine
- Isoproterenol
- Isuprel
- Largon
- Levoprome
- Medihaler-Iso
- Mellaril
- Metycaine
- Navane
- Nebair
- Nephenalin
- Norisodrine
- Novocain
- Pacatal
- Penthrane
- Permitil
- Phenothiazines
- Priscoline
- Proketazine
- Prolixin
- Proterenol
- Quide
- Regitine
- Repoise
- Ritalin
- Sparine

EPINEPHRINE (continued):

Stelazine	Tindal
Taractan	Torecan
Temaril	Trilafon
Thorazine	Vapo-N-Iso Metermatic
Thyrolar	Vesprin

EPINEPHRINE
+ Chloroform: Sensitization of the myocardium to the effects of epinephrine occurs with clinical concentrations of chloroform, Trilene, cyclopropane, and Fluothane but not with the ethers or thiopental, causing arrhythmias (ventricular fibrillation). (110)
+ Cyclopropane: see Epinephrine + Chloroform
+ Fluothane: see Epinephrine + Chloroform
+ Trilene: see Epinephrine + Chloroform

EQUAGESIC: see Equinal and Zactirin

EQUANIL: see Meprobamate

EQUANIL + Alcohol: Special care should be taken to warn patients taking Equanil that their tolerance to alcohol may be lowered with resultant slowing of reaction time and impairment of judgment and coordination. (1)

ERGOT ALKALOIDS: see listed under the following agents:
Methedrine
Vasoxyl

ERGOT DERIVATIVES + Fluothane: see Fluothane + Ergot Derivatives

ERGOTRATE MALEATE Inj.
+ Anesthesia, Caudal: see Ergotrate Maleate Inj. + Anesthesia, Regional
+ Anesthesia, Regional: Elevations of blood pressure (sometimes extreme) may occur in a small percentage of patients, most frequently in association with regional anesthesia (caudal or spinal), previous administration of a vasoconstrictor, and the intravenous route of administration of the oxytocic. The mechanism of such hypertension is obscure, since it may occur in the absence of anesthesia, vasoconstrictors, and oxytocics. (1)
+ Anesthesia, Spinal: see Ergotrate Maleate Inj. + Anesthesia, Regional
+ Oxytocics: see Ergotrate Maleate Inj. + Anesthesia, Regional
+ Vasoconstrictors: see Ergotrate Maleate Inj. + Anesthesia, Regional

ERYTHROCIN LACTOBIONATE + Mucomyst: see Mucomyst + Erythromycin Lactobionate

ERYTHROMYCIN: see listed under the following agents:
Benemid
Cleocin
Penicillin

ESCHAROTIC AGENTS + Oxycel: see Oxycel + Escharotic Agents

ESIDRIX: see Diuretics, Thiazide

ESIDRIX
+ ACTH: see Esidrix + Steroias
+ Alcohol: Orthostatic hypotension may occur with thiazides and may be potentiated by alcohol, barbiturates, or narcotics. (1)
+ Aldomet: see Esidrix + Antihypertensive Agents
+ Anesthesia: Hypotensive episodes under anesthesia have been observed in patients receiving thiazides. If emergency surgery is indicated, preanesthetic and anesthetic agents should be administered in reduced dosage. (1)
+ Antihypertensive Agents: Esidrix has the marked ability to potentiate other antihypertensive agents, including reserpine and other rauwolfia derivatives, Apresoline, Ismelin, veratrum alkaloids, and Aldomet. Dosage of ganglionic blockers should be halved. (1, 2)
+ Apresoline: see Esidrix + Antihypertensive Agents
+ Barbiturates: see Esidrix + Alcohol
+ Diabetes: Thiazide derivatives, particularly in large doses, may decrease glucose tolerance; therefore Esidrix should be used cautiously in diabetics. Insulin requirements in diabetic patients may be increased, decreased, or unchanged. Latent diabetes may become manifest during thiazide administration. (1, 2)
+ Digitalis: Interference with adequate oral intake of electrolytes will contribute to hypokalemia. Digitalis therapy may exaggerate metabolic effects of hypokalemia especially with reference to myocardial activity. (Signs of digitalis intoxication may be produced by formerly tolerated doses of digitalis.) (1, 2)
+ Ganglionic Blocking Agents: see Esidrix + Antihypertensive Agents
+ Ismelin: see Esidrix + Antihypertensive Agents
+ Mercurials, Parenteral: Esidrix and parenterally administered mercurials may be given concomitantly; the effects are additive. (1)
+ Narcotics: see Esidrix + Alcohol
+ Norepinephrine: Thiazides may decrease arterial responsiveness to norepinephrine. If possible, withdraw therapy two weeks prior to surgery. (1, 2)
+ Rauwolfia Derivatives: see Esidrix + Antihypertensive Agents
+ Reserpine: see Esidrix + Antihypertensive Agents

ESIDRIX (continued)

+ Steroids: As with other potent diuretics, hypokalemia may develop with thiazides especially during brisk diuresis or during concomitant administration of steroids or ACTH. (1, 2)
+ Trisoralen: see Trisoralen
+ Tubocurarine: Thiazides may increase responsiveness to tubocurarine. (1, 2)
+ Veratrum Alkaloids: see Esidrix + Antihypertensive Agents

ESIMIL: see Esidrix and Ismelin

ESIMIL

+ Anesthesia: If possible, discontinue Esimil two weeks before surgery to eliminate the possibility of vascular collapse during anesthesia. Furthermore, depletion of catecholamines may increase the hazard of cardiac arrest during anesthesia. If emergency surgical procedures are necessary, give anesthesia cautiously. Have atropine, oxygen, and vasopressor solutions ready for immediate use. Hydrochlorothiazide decreases responsiveness to exogenously administered norepinephrine while Ismelin increases responsiveness to this agent. Therefore the administration of norepinephrine to patients on Esimil may be attended by a greater propensity for the production of cardiac arrhythmias. Use caution in administering vasopressor agents. Hypotensive episodes under anesthesia have been observed in patients receiving thiazides. (1)
+ Norepinephrine: see Esimil + Anesthesia
+ Vasopressor Agents: see Esimil + Anesthesia

ESKADIAZINE + Sulfonylureas, Hypoglycemic: Sulfonamides may potentiate the hypoglycemic action of sulfonylureas. (1) Mechanism: see Sulfonylureas, Hypoglycemic + Sulfonamides.

ESKALITH: see Lithium Carbonate

ESKALITH + Diuretics: Toxicity of lithium has been shown to be potentiated when sodium intake is restricted. Therefore it is essential for the patient to maintain a normal diet including salt, and an adequate fluid intake (2500 to 3000 cc) at least during the initial stabilizing period. Decreased tolerance to lithium has been reported to ensue from protracted sweating or diarrhea and, if such occur, supplemental fluid and salt should be administered. Diuretics should not be used concomitantly with Eskalith. (1)

ESKASERP: see Reserpine

ESKASERP (continued)

+ Anesthetic Agents: If possible, reserpine therapy should be discontinued for two weeks before elective surgery, since anesthetic agents potentiate the hypotensive action of reserpine. (1) Mechanism: see Reserpine + Anesthesia.

ESKATROL: see Dexedrine and Compazine

ESKAY'S NEURO PHOSPHATES + Morphine: If spasms or other signs of excitability are present due to overdosage of Eskay's Neuro Phosphates, do not give morphine. (1)

ESKAY'S THERANATES: see Eskay's Neuro Phosphates

ESTOMUL: This preparation contains orphenadrine hydrochloride. See Disipal + Darvon.

ESTROGENS: see listed under the following agents:

Anticoagulants
Phenobarbital
Provera

ETHAMIDE + Miotics: It is recommended that Ethamide be employed concomitantly with miotic therapy since evidence suggests that the effect is additive. (1)

ETHER: see listed under the following agents:

Antibiotics
Flaxedil
Inderal
MAO Inhibitor
Marplan
Mecostrin
Mylaxen
Nalline
Nardil
Niamid
Seconal Sodium
Succinylcholine
Tubocurarine
Tubocurarine Chloride Inj.

ETHER

+ Curare: Ether appreciably potentiates the action of curarizing agents by virtue of its own muscle relaxing properties which may cause irreversible hypotension or apnea. This fact must be given due consideration in determining dosage when curare preparations and ether are used concomitantly. (1, 2)

+ Levophed: Ether diminishes the pressor response to the infusion of Levophed. The reduction in total peripheral resistance which it causes may be due, at least in part, to interference with the ability of blood vessels to constrict in response to the catecholamines. (110)

+ Tubocurarine: Although much of the relaxing effect of Ether on muscle is due to depression of reflex activity on the spinal cord, it has a

ETHER (continued)

+ Tubocurarine (continued): peripheral neuromuscular blocking effect, being synergistic with tubocurarine. (110) See Ether + Curare.

ETHERS + Antibiotics: see Antibiotics + Ethers

ETHYL CHLORIDE + Epinephrine: In the induction period the myocardium is sensitized to epinephrine and sympathetic nerve impulses so that ventricular arrhythmias may develop. (1)

ETRAFON: see Trilafon and Elavil

EUTHROID

+ Anticoagulants: The institution of thyroid replacement therapy may potentiate the anticoagulant effects with agents such as warfarin or Dicumarol. Dosage of anticoagulants should be reduced by one-third upon initiation of Euthroid therapy. Subsequent anticoagulant adjustment should be made on the basis of frequent prothrombin determinations. (1)
+ Catecholamines: see Euthroid + Epinephrine
+ Diabetes: Initiation of thyroid replacement therapy in patients with diabetes must be carefully monitored because of potential fluctuation in daily insulin or oral hypoglycemic requirements. (1)
+ Dicumarol: see Euthroid + Anticoagulants
+ Epinephrine: Injection of epinephrine in patients with coronary artery disease may precipitate an episode of coronary insufficiency. This may be enhanced in patients receiving thyroid preparations. Careful observation is required if catecholamines are administered to patients in this category. (1)
+ Surgery: Euthroid treated patients with coronary artery disease should be carefully observed during surgery, since the possibility of precipitating cardiac arrhythmias may be greater in patients treated with thyroid hormones. (1)
+ Warfarin: see Euthroid + Anticoagulants

EUTONYL: see MAO Inhibitors

EUTONYL: see listed under the following agents:

Antidepressants, Tricyclic
Diuretics
Parnate

EUTONYL

+ Alcohol: see Eutonyl + CNS Depressants (55)
+ Aldomet: Aldomet should not be used with Eutonyl, since the combination may cause hyperexcitability. (1, 2, 55)

EUTONYL (continued)

+ Amphetamine: see Eutonyl + Sympathomimetic Amines
+ Analgesics: see Eutonyl + Narcotics
+ Anesthetics: see Eutonyl + CNS Depressants. Eutonyl may augment the hypotensive effects of anesthetic agents and surgery. For this reason the drug should be discontinued from ten days to two weeks prior to surgery. In the event of emergency surgery, the dose of the anesthetic agents should be adjusted to the response of the patient. If severe hypotension should occur, this can be controlled by small doses of a vasopressor agent such as Levophed. (1, 2, 55)
+ Anesthetics, Local: It has been reported clinically that certain individuals receiving Eutonyl for a prolonged period of time are refractory to the nerve-blocking effects of local anesthetics, e.g., Xylocaine. (1)
+ Anorexiants: see Eutonyl + Sympathomimetic Amines
+ Antidepressants: Tofranil, Elavil, Norpramin, Pertofrane, and Aventyl or similar antidepressants should not be used with Eutonyl. The use of these drugs with a monoamine oxidase inhibitor has been reported to cause vascular collapse and hyperthermia which may be fatal. A drug free interval (about two weeks) should separate therapy with Eutonyl and the use of these agents. (1, 2, 55)
+ Antihistamines: see Eutonyl + CNS Depressants
+ Antihypertensive Agents: When Eutonyl is used in combination with other antihypertensive agents, the initial dose of this agent should not exceed 25 mg daily. Eutonyl will potentiate the effects of antihypertensive agents. (2)
+ Antiparkinsonism Agents: Eutonyl should be used with caution in patients with Parkinsonism as it may increase symptoms. In addition, great care is required if Eutonyl is administered in conjunction with antiparkinsonism agents. (1, 55)
+ Aramine: see Eutonyl + Sympathomimetics, Indirect-Acting
+ Aventyl: see Eutonyl + Antidepressants
+ Barbiturates: see Eutonyl + CNS Depressants (55)
+ Caffeine: The therapeutic response of caffeine may be changed or exaggerated in patients receiving a monoamine oxidase inhibitor, such as Eutonyl. Caffeine should be used cautiously and in reduced dosage in patients taking Eutonyl. (1)
+ Chloral Hydrate: Eutonyl will potentiate the effects of chloral hydrate.(55) Chloral hydrate should be used cautiously and at reduced dosage in patients taking Eutonyl. (1) See Eutonyl + CNS Depressants.
+ CNS Depressants: An exaggerated response to central nervous system depressants may be manifested by acute hypotension and increased

EUTONYL (continued)

+ CNS Depressants (continued): sedative effects. The therapeutic response to a variety of drugs may be changed or exaggerated in patients receiving a monoamine oxidase inhibitor, such as Eutonyl. Alcohol, narcotics (meperidine should be avoided), antihistamines, anesthetics, barbiturates, and other hypnotics, sedatives or tranquilizers should be used cautiously and at reduced dosage in patients receiving Eutonyl. (1, 2)
+ Cold Remedies: see Eutonyl + Sympathomimetic Amines
+ Diabetes: In some patients taking monoamine oxidase inhibitors, a reduction of blood sugar has been reported. (1, 2, 55)
+ Diuretics: The addition of an oral diuretic to the dosage regimen of Eutonyl enhances the antihypertensive response and diminishes the dosage requirement of Eutonyl. (1, 55)
+ Elavil: see Eutonyl + Antidepressants
+ Ephedrine: see Eutonyl + Sympathomimetic Amines
+ Foods: Aged cheese, broad beans, or other foods which require the action of bacteria or molds for their preparation or preservation because of the presence of pressor substances such as tyramine are contraindicated in patients receiving Eutonyl. In some patients receiving Eutonyl, the above foods may precipitate an abrupt rise in blood pressure accompanied by some or all of the following: severe headache, chest pain, profuse sweating, alteration of heart rate, and intracranial bleeding. A phenothiazine or Regitine may be administered parenterally for treatment of such an acute hypertensive reaction. (1, 2, 34, 55)
+ Ganglionic Blocking Agents: Eutonyl may potentiate the action of ganglionic blocking agents. (2)
+ Hypnotics: see Eutonyl + CNS Depressants
+ Insulin: Eutonyl may potentiate the hypoglycemic effects of Insulin. (55)
+ Ismelin, Parenteral: Parenteral forms of Ismelin are contraindicated during and for one week following treatment with Eutonyl, since a hypertensive reaction may occur from a sudden release of catecholamines due to the Ismelin. (1)
+ MAO Inhibitors: Other monoamine oxidase inhibitors should not be used with Eutonyl, since they may augment the effects of Eutonyl. (1)
+ Meperidine: see Eutonyl + Narcotics and Demerol + MAO Inhibitors (55)
+ Narcotics: see Eutonyl + CNS Depressants: In the event of emergency surgery, smaller than usual doses (one-fourth to one-fifth) of narcotics, analgesics, and other premedications should be used. Parenteral administration of these drugs should be avoided if possible.

EUTONYL (continued)

+ Narcotics (continued): It is advisable to avoid the use of meperidine. (1, 2)
+ Nasal Decongestants: see Eutonyl + Sympathomimetic Amines
+ Neosynephrine: see Eutonyl + Sympathomimetics, Indirect-Acting
+ Norpramin: see Eutonyl + Antidepressants
+ Parnate: If a patient on Eutonyl were to receive Parnate, he might suffer a hypertensive crisis. (55)
+ Pertofrane: see Eutonyl + Antidepressants
+ Phenothiazines: A phenothiazine, such as Thorazine, Stelazine, or Trilafon, may be potentiated by Eutonyl resulting in hypotension as well as in the extrapyramidal symptoms that sometimes accompany administration of these tranquilizers. (55)
+ Preludin: see Eutonyl + Ritalin
+ Propadrine: see Eutonyl + Sympathomimetics, Indirect-Acting
+ Reserpine, Parenteral: Parenteral forms of reserpine is contraindicated during and for one week following treatment with Eutonyl, since a hypertensive reaction may occur from a sudden release of catecholamines due to the reserpine. (1, 55)
+ Ritalin: Ritalin and Preludin used concomitantly with Eutonyl have produced hypertensive reactions presumably by the same action as the amphetamines. (34) See also Eutonyl + Sympathomimetic Amines.
+ Sedatives: see Eutonyl + CNS Depressants
+ Sympathomimetic Amines: Centrally acting sympathomimetic amines, such as amphetamine and its derivatives (also found in anoretic preparations), are contraindicated in patients receiving Eutonyl. Eutonyl may potentiate the action of sympathomimetic amines. In some patients receiving Eutonyl, the centrally acting sympathomimetics may precipitate an abrupt rise in blood pressure accompanied by some or all of the following: severe headache, chest pain, profuse sweating, tachycardia or bradycardia, palpitation, photophobia, and intracranial bleeding (which may be fatal).
  Peripherally acting sympathomimetic drugs, such as ephedrine and its derivatives (also found in nasal decongestants and cold remedies), are contraindicated in patients receiving Eutonyl, as a dangerously high increase in blood pressure may occur. (1, 2, 34, 55) Mechanism: see MAO Inhibitors + Sympathomimetics, Indirect-Acting.
+ Sympathomimetics, Indirect-Acting: Concomitant use of indirect-acting sympathomimetics (Wyamine, Aramine, Neosynephrine, Propadrine, etc.) is likely to result in excessively high blood pressure and should be used with great caution. Mechanism: see MAO Inhibitors + Sympathomimetics, Indirect-Acting. (34, 55)

EUTONYL (continued)
+ Tofranil: see Eutonyl + Antidepressants
+ Tranquilizers: see Eutonyl + CNS Depressants (55)
+ Wyamine: see Eutonyl + Sympathomimetics, Indirect-Acting
+ Xylocaine: see Eutonyl + Anesthetics, Local

EUTRON: see Eutonyl and Enduron

EXERCISE: see listed under the following agents:
Ansolysen
Inversine

EXNA: see Diuretics, Thiazide

EXNA
+ Anesthesia: Since thiazides may decrease arterial responsiveness to norepinephrine, Exna should be withdrawn two weeks before elective surgery. If emergency surgery is indicated, pre-anesthetic and anesthetic agents should be administered in reduced dosage. (1)
+ Antihypertensive Agents: Exna enhances the activity of antihypertensive agents when used concomitantly. (1)
+ Apresoline: see Exna + Ganglionic Blocking Agents
+ Corticosteroids: Since corticosteroids and corticotrophin may increase the loss of potassium, they should be used with caution in patients taking thiazide diuretics. (1, 2)
+ Corticotrophin: see Exna + Corticosteroids
+ Diabetes: Latent diabetes may become frank or frank diabetes may be worsened in patients receiving thiazide diuretics. Therefore caution should be observed in patients with tendencies toward diabetes. (1, 2)
+ Digitalis: Because potassium deficiency can be a complication of diuretic therapy, patients must be observed regularly for early signs of hypokalemia. Digitalis toxicity may be precipitated by hypokalemia. (1, 2)
+ Ganglionic Blocking Agents: It is mandatory to reduce the dosage of therapeutic agents, such as ganglionic blocking agents or Apresoline, by at least 50% immediately upon the addition of Exna to the therapeutic regimen to prevent excessive hypertensive effects. (1, 2)
+ Ismelin: see Exna + Reserpine
+ Mercurials: Exna enhances the activity of mercurials when given concomitantly. (1)
+ Norepinephrine: see Exna + Anesthesia, and Pressor Amines
+ Pressor Amines: Results of recent reports indicate that thiazides may decrease the arterial responsiveness to pressor amines (e.g., norepinephrine). (2)

EXNA (continued)

+ Reserpine: Exna enhances the antihypertensive effect of reserpine and Ismelin. (2)
+ Trisoralen: see Trisoralen
+ Tubocurarine: Results of recent reports indicate that thiazides may augment the paralyzing action of tubocurarine in surgical patients. (2)

EXNA-R: see Exna and Reserpine

F

FEHLING'S SOLUTION: see listed under the following agents:

Keflin
NegGram

FEOSOL ELIXIR

+ Milk: Feosol Elixir should not be given in milk because of the lowering of iron utilization. (1)
+ Vehicles: Feosol Elixir may form a precipitate with various pharmaceutical specialties having a wine vehicle. (1)
+ Vitamin C: Feosol Elixir is compatible with most water-soluble vitamins, except vitamin C. (1)

FLAGYL + Alcohol: Episodes of gastrointestinal disturbances have been reported in patients receiving Flagyl, characterized by abdominal cramps with or without nausea or vomiting, and sometimes accompanied by flushing or headache, especially in patients who take alcoholic beverages while on medication with Flagyl. A modification of the taste of alcoholic beverages has also been noted. Patients who are known to take moderate or excessive amounts of alcohol in any form, therefore, should be advised of these possibilities. (1)

FLAXEDIL: see listed under the following agents:

Cyclopropane
Mylaxen
Penthrane
Tensilon

FLAXEDIL

+ Antibiotics: Flaxedil acts by polarizing the motor end plates and are potentiated by certain antibiotics with similar activity (e.g., Neomycin, Streptomycin, Polymixin) and by drugs that cause potassium depletion, such as thiazide diuretics. The effects of Flaxedil (polarizing muscle relaxants) can be reversed by cholinesterase inhibitors; this therapy is contraindicated during Syncurine (depolarizing) paralysis. (1, 55)
+ Cyclopropane: Ventricular arrhythmias may arise after injection of Flaxedil during cyclopropane anesthesia. (14)
+ Diuretics, Thiazide: see Flaxedil + Antibiotics
+ Drugs, Electrolyte-Depleting: The action of Flaxedil may be altered by changes in body temperature, dehydration, and electrolyte imbalance. (1, 55) See Flaxedil + Antibiotics.

FLAXEDIL (continued)

+ Drugs, Potassium-Depleting: see Flaxedil + Antibiotics and Drugs, Electrolyte-Depleting
+ Ether: Flaxedil may be potentiated by several anesthetic agents, particularly ether, Penthrane, and Fluroxene. (1)
+ Fluroxene: see Flaxedil + Ether
+ Neomycin: see Flaxedil + Antibiotics
+ Neostigmine: Neostigmine is an effective antidote for Flaxedil, but it is advisable to administer atropine prior to or simultaneously with Neostigmine to counteract the muscarinic action of Neostigmine. Neostigmine has a prolonged action which may be cumulative. Therefore extreme caution is needed if Neostigmine is to be used as an antidote for Flaxedil. (1)
+ Penthrane: see Flaxedil + Ether
+ Polymixin: see Flaxedil + Antibiotics
+ Streptomycin: see Flaxedil + Antibiotics

FLOROPRYL: see listed under the following agents:

Succinylcholine
Sucostrin

FLOROPRYL

+ Insecticides, Organophosphate: Persons receiving cholinesterase inhibitors who are exposed to organophosphate-type insecticides and pesticides, such as gardners, organophosphate manufacturing or warehouse workers, farmers, residents of communities which are undergoing insecticide or pesticide spraying or dusting, should be warned of the added systemic effects possible from absorption through the respiratory tract or skin. Wearing of respiratory masks, frequent washing, and clothing changes may be advisable. (1)
+ Pilocarpine: Pilocarpine will interfere with the intensity of action of Floropryl by competing with the acetylcholine that is freed through the inactivation of cholinesterase by Floropryl. If pilocarpine is used, it should be continued along with Floropryl for a few days, then gradually reduced, and finally discontinued. (1)
+ Succinylcholine: Succinylcholine should not be administered before or during general anesthesia to patients receiving cholinesterase inhibitors because of possible respiratory and cardiovascular collapse. (1)
+ Water: Floropryl hydrolyzes in the presence of water to form hydrofluoric acid. To prevent absorption of moisture and loss of potency, the vial of opthalmic solution and the ointment tube should be kept

FLOROPRYL (continued)

+ Water (continued): tightly closed. The dropper or the tip of the tube should not be washed or allowed to touch the eyelid or other moist surface. Opthalmic solution Floropryl in anhydrous peanut oil must not be diluted. (1)

FLUIDS + Zyloprim: see Zyloprim + Fluids

FLUIDS, INTRAVENOUS + Plasmanate: see Plasmanate + Fluids, Intravenous

FLUOROMAR

+ Epinephrine: Although it has been reported that myocardial irritability does not occur during concurrent use of epinephrine and Fluoromar in man, there is some experimental evidence that the administration of catecholamines (epinephrine, Levophed) during Fluoromar anesthesia may produce potentially serious cardiac arrhythmias. (2)
+ Flaxedil: see Flaxedil + Fluoromar (Fluroxene)
+ Levophed: see Fluoromar + Epinephrine
+ Tubocurarine: Although much of the relaxing effect of Fluoromar on muscle is due to the depression of reflex activity on the spinal cord, Fluoromar has a peripheral neuromuscular blocking effect being synergistic with tubocurarine. (11)

FLUOROURACIL

+ Alkylating Agents: see Fluorouracil + Surgery
+ Bone Marrow Depressants: Any form of therapy which adds to the stress of the patient, interferes with nutrition, or depresses bone marrow function will increase the toxicity of Fluorouracil. Fluorouracil is contraindicated for those patients with bone marrow previously depressed by other therapy. (1)
+ Radiation Therapy: see Fluorouracil + Surgery
+ Surgery: Fluorouracil is contraindicated for patients in a poor nutritional state, those with bone marrow depression, or those with azotemia. Such conditions are likely to occur when the patient has undergone major surgery, received high doses of radiation, or recent use of alkylating agents. (1, 2)
+ Trisoralen: see Trisoralen

FLUOTHANE: see listed under the following agents:

Aramine
Carbocaine
Dopram
Levophed
Vasopressors

FLUOTHANE (continued)

+ Epinephrine: Fluothane sensitizes the myocardium to the action of epinephrine and Levophed; therefore it should not be used when injection of these amines is contemplated, since ventricular tachycardia or fibrillation may be induced. However, small doses of these agents have been given subcutaneously to minimize bleeding in the presence of Fluothane anesthesia without apparent ill effect. Other vasopressors (e.g., Vasoxyl, methamphetamine) that are less likely to provoke arrhythmias when given intravenously should be used to combat hypotension. (1, 2)
+ Ergot Derivatives: Fluothane is primarily a uterine muscle relaxant. The uterine relaxation obtained, unless carefully controlled, may fail to respond to ergot derivatives and oxytocic posterior pituitary extract. (1)
+ Curariform Drugs: D-tubocurarine or its chemically related compounds may be employed with Fluothane, but caution should be observed. The use of curariform drugs should be one-third to one-half that usually applied during other anesthetic techniques. (1)
+ Levophed: see Fluothane + Epinephrine
+ Methium: Fluothane should be used with caution in patients who are receiving certain drugs (e.g., Methium, reserpine, Thorazine), since this anesthetic may increase the sensitivity of patients to the hypotensive effect of these drugs. (2)
+ Oxytocic Posterior Pituitary Extract: see Fluothane + Ergot Derivatives
+ Reserpine: see Fluothane + Methium
+ Thorazine: see Fluothane + Methium
+ d-Tubocurarine: see Fluothane + Curariform Drugs. Fluothane increases the magnitude and duration of action of d-tubocurarine. (36)

FLUROTHYL

+ Phenothiazines: see Flurothyl + Tranquilizers
+ Tranquilizers: The concomitant use of tranquilizers, particularly phenothiazine derivatives, may enhance the possibility of vascular collapse. (2, 87)

FOLIC ACID ANTAGONIST + Leucovorin Calcium Inj.: see Leucovorin Calcium Inj.

FOOD: see listed under the following agents:

Achromycin
Ansolysen
Aureomycin
Coumadin
Darcil
Declomycin

FOOD (continued)

Eutonyl
Hedulin
Kesso-Tetra
MAO Inhibitors
Marplan
Maxipen
Nardil
Niamid
Ostensin
Parnate
Penicillin
Resistopen
Semopen
Sintrom
Steclin
Syncillin
Tegopen
Tetramax
Triethylene Melamine (TEM)
Tromexan

FOOD

\+ Orbenin: see Food + Prostaphlin

\+ Prostaphlin: Oxacillin (Prostaphlin, Resistopen), cloxacillin (Tegopen, Orbenin), and nafcillin (Unipen) are acid resistant. Despite their acid resistance, these penicillins are poorly absorbed unless administered to patients in the postabsorptive state. Thus, it is important that they be given at least three hours after or one hour before meals. (38)

\+ Resistopen: see Food + Prostaphlin

\+ Tegopen: see Food + Prostaphlin

\+ Unipen: see Food + Prostaphlin

FOOD, CALCIUM-CONTAINING: see listed under the following agents:

Mysteclin F
Panmycin
Retet
Rondomycin
Signemycin
Steclin
Sumycin
Terramycin
Tetrachel
Tetracyn
Tetramax
Tetrex

FOODS, TYRAMINE-CONTAINING: see listed under the following agents: (For foods that contain either tyramine or DOPA see MAO Inhibitor + Food)

Furoxone
Marplan
Matulane
Parnate

FORHISTAL

\+ Hypnotics: Because antihistamines often produce drowsiness, give hypnotics cautiously to patients receiving Forhistal. (1)

\+ Sedatives: Because antihistamines often produce drowsiness, give sedatives cautiously to patients receiving Forhistal. (1)

FRAGICAP-K
+ Anticoagulants: Use with caution when anticoagulant and antihemorrhagic agents are administered concurrently. (1)
+ Antihemorrhagic Agents: see Fragicap-K + Anticoagulants

FRUIT JUICE: see listed under the following agents:
Ilotycin Ethyl Carbonate Drops
Lytren Powder

FUADIN + Iron: Iron salts should be given after the completion of treatment with Fuadin and not concurrently. (1)

FULVICIN: see Griseofulvin

FULVICIN + Trisoralen: see Trisoralen

FUNGIZONE Intravenous
+ Antibiotics: Since deep fungal infections sometimes emerge in patients undergoing therapy with antibiotics, nitrogen mustard, or antimetabolites, these agents should not be used concomitantly with Fungizone if avoidable. (1, 2)
+ Antimetabolites: see Fungizone Intravenous + Antibiotics
+ Bacteriostatic Agents: The presence of a bacteriostatic agent (e.g., benzyl alcohol) in the diluent may cause precipitation of the antibiotic. Do not use the infusion solution if there is any evidence of precipitation or foreign matter. (1, 2)
+ Corticosteroids: Since deep fungal infections sometimes emerge in patients undergoing therapy with corticosteroids, the concomitant administration should be avoided unless necessary to control drug reactions or underlying disease (e.g., Addison's disease). (1, 2)
+ Dextrose Inj. 5%: The pH of the Dextrose Inj. 5% U.S.P. must have a pH above 4.2, otherwise it must be buffered with the following composition: Dibasic Sodium Phosphate (anhydrous) 1.59 gm, Monobasic Sodium Phosphate (anhydrous) 0.96 gm, Water for Injection U.S.P. qs. 100.0 cc. The buffer should be sterilized before it is added to the Dextrose Injection. If the pH of the Dextrose Injection is below 4.2, then one or two cubic centimeters of the buffer is added before it is used to dilute the concentrated solution of Fungizone. (1)
+ Nitrogen Mustard: see Fungizone Intravenous + Antibiotics
+ pH: Only 5% Dextrose Injection U.S.P. with a pH above 4.2 for further dilution of Fungizone Intravenous is recommended. (1) See Fungizone Intravenous + Dextrose Inj.

FUNGIZONE Intravenous (continued)
+ Sodium Chloride Sol. 5%: The use of Sodium Chloride Sol. 5% as a diluent for Fungizone Intravenous will cause precipitation of the antibiotic. (1)

FURADANTIN: see Acidic Agents + Alkaline Media

FURADANTIN: see Acidic Agents + Alkanizing Agents (Urinary)

FURADANTIN
+ Antacids: see Antacids + Furadantin
+ Drugs Impairing Renal Function: Furadantin should not be given along with drugs which may impair renal function. (1)

FURADANTIN Sodium Sterile
+ Cresol: see Furadantin Sodium Sterile + Methylparaben
+ Methylparaben: Do not mix Furadantin Sodium Sterile with solutions containing methyl- and propyl-parabens, phenol, or cresol as preservatives, since these compounds cause Furadantin to be precipitated out of solution. (1)
+ Phenol: see Furadantin Sodium Sterile + Methylparaben
+ Propylparaben: see Furadantin Sodium Sterile + Methylparaben

FUROXONE: Furoxone was found to be a potent MAO Inhibitor (63). See MAO Inhibitors.

FUROXONE
+ Alcohol: Rarely, disulfiram-like reactions to alcohol consumption during Furoxone therapy have been reported and consist of flushing, slight temperature elevation, dyspnea, and in some instances a sense of constriction within the chest. Therefore alcohol in any form is contraindicated during therapy and for four days thereafter. (1, 55)
+ Amphetamines: see Furoxone + Sympathomimetic Amines, Indirect-Acting
+ Anorectics: see Furoxone + Sympathomimetic Amines, Indirect-Acting
+ Antihistamines: Antihistamines should be used in reduced dosage and with caution in patients taking Furoxone. (1)
+ Diabetes: Furoxone may potentiate or prolong the hypoglycemic action of insulin. (1, 55)
+ Ephedrine: see Furoxone + Sympathomimetic Amines, Indirect-Acting
+ Foods, Tyramine-Containing: Tyramine-containing foods, such as broad beans, yeast extracts, strong unpasteurized cheeses, beer, wine, pickled herring, chicken livers, and fermented products, are contraindicated. (1, 55)

FUROXONE (continued)

\+ MAO Inhibitors: In general monoamine oxidase inhibitors are contraindicated or should be used with caution and at reduced dosage in patients receiving Furoxone.
Effective inhibition of monoamine oxidase by Furoxone has been demonstrated experimentally in man. If Furoxone is administered in doses larger than recommended or in excess of five days, the indications must be weighed against the possible hazards of hypertensive crisis related to the accumulation of monoamine oxidase inhibition. (1, 63)

\+ Narcotics: Narcotics should be used in reduced dosage and with caution in patients taking Furoxone. (1, 55)

\+ Nasal Decongestants: see Furoxone + Sympathomimetic Amines, Indirect Acting

\+ Phenylephrine: see Furoxone + Sympathomimetic Amines, Indirect-Acting

\+ Sedatives: Sedatives should be used in reduced dosage and with caution in patients taking Furoxone. (1, 55)

\+ Sympathomimetic Amines, Indirect-Acting: Indirectly acting sympathomimetic amines, such as those found in nasal decongestants (phenylephrine, ephedrine), and anorectics (amphetamines), are contraindicated when Furoxone is being administered to patients. (1, 55)

\+ Tranquilizers: Tranquilizers should be used in reduced dosage and with caution in patients taking Furoxone. (1, 55)

G

GAMASTAN + Live Measles Virus Vaccine: see Immune Serum Globulin (Human)

GAMMACORTEN + Diabetes: Gammacorten, like other glucocorticoids, may aggravate diabetes mellitus so that higher insulin dosage may become necessary or manifestations of latent diabetes mellitus may be precipitated. (1, 2)

GAMMAGEE + Live Measles Vaccine: For modification of the symptoms induced by concomitant administration of attenuated live measles virus vaccine, Gammagee is administered in a dose of 0.01 cc per pound of body weight. This should be injected immediately after the vaccine into the deltoid muscle of the opposite arm. Gammagee must not be administered intravenously. The administration of the vaccine before Gammagee permits the virus to infect cells of the injection site prior to distribution of measles antibody from the immune serum globulin. (1)

GAMMA GLOBULIN: see listed under the following agents:

- Cendevax
- Measles Virus Vaccine, Live, Attenuated
- M-Vac Measles Virus Vaccine, Live, Attenuated
- Rubeovax, Lyovac

GAMULIN + Live Measles Virus Vaccine: see Immune Serum Globulin (Human)

GANGLIONIC BLOCKING AGENTS: see listed under the following agents:

- Aldactone
- Aldomet
- Anhydron
- Apresoline
- Aquatag
- Bristuron
- Diuretics, Thiazide
- Diuril
- Dopar
- Dyazide
- Enduron
- Esidrix
- Eutonyl
- Exna
- HydroDIURIL
- Hydromox
- Hygroton
- Ismelin
- Laradopa
- Levodopa
- Metahydrin
- Naqua
- Naturetin
- Oretic
- Penthrane
- Perithiazide SA
- Raudixin
- Rauwiloid

GALGLIONIC BLOCKING AGENTS: see listed under the following agents:

Renese
Reserpine
Saluron
Singoserp-Esidrix
Unitensin
Urecholine

GANTRISIN + Anturane: see Anturane + Gantrisin

GANTRISIN OPTHALMIC Sol. & Oint. + Silver Preparations: Do not use Gantrisin Opthalmic preparations with silver preparations. (1)

GARAMYCIN
+ Coly-Mycin M: see Garamycin + Ototoxic Drugs
+ Kantrex: see Garamycin + Ototoxic Drugs
+ Nephrotoxic Drugs: see Garamycin + Ototoxic Drugs
+ Ototoxic Drugs: Garamycin is potentially ototoxic and nephrotoxic. Concurrent administration of potentially ototoxic drugs such as streptomycin and Kanmycin or of potentially nephrotoxic drugs such as polymixin, Coly-Mycin M, and Kanmycin (Kantrex) with Garamycin has not been shown to afford any clinical advantages and, moreover, may result in additive toxicity. Monitoring of vestibular, cochlear, and renal function will provide guidance for therapy in such cases. (1)
+ pH: Garamycin is 8 to 32 times more active at pH 7.5 than at pH 5.5 against several common urinary tract pathogens. Results of in vitro studies indicate that alkinization of the urine may be a useful therapeutic adjunct. (1)
+ Polymixin: see Garamycin + Ototoxic Drugs
+ Streptomycin: see Garamycin + Ototoxic Drugs

GEMONIL + Anticonvulsant Drugs: When Gemonil is added to an established regimen to replace or supplement other anticonvulsant therapy, dosage of other medication should be gradually reduced while increasing that of Gemonil so as to avoid or minimize recurrence of seizures. (1)

GEOPEN + Benemid: The administration of Benemid results in somewhat higher and more prolonged serum levels. (1)

GERILETS FILMTAB + MAO Inhibitors: Because of its methamphetamine content, Gerilets should be used with caution in patients taking a monoamine oxidase inhibitor. Mechanism: see MAO Inhibitors + Sympathomimetics, Indirect-Acting. (1)

GERILID + Coronary Vasodilator: Caution is advised when there is concomitant administration of a coronary vasodilator. (1)

GERILIQUID + Coronary Vasodilator: Caution is advised when there is concomitant administration of a coronary vasodilator. (1)

GITALGIN + Diuretics: The use of certain diuretic agents can cause potassium deficiency which renders the myocardium more sensitive to digitalis intoxication so that arrhythmias are produced without actual overdosage of digitalis. (1)

GLUCOCORTICOIDS: see listed under the following agents:
Aldactone
Phenobarbital

GLUCOSE + Neomycin: see Neomycin + Glucose

GLYCOSIDES, CARDIAC + Mylaxen: see Mylaxen + Glycosides, Cardiac

GOLD: see listed under the following agents:
Aralen
Atabrine
Para-amino Benzoic Acid
Plaquenil

GRIFULVIN: see Griseofulvin

GRIFULVIN + Trisoralen: see Trisoralen

GRISACTIN: see Griseofulvin

GRISACTIN + Trisoralen: see Trisoralen

GRISEOFULVIN: see listed under the following agents:
Anticoagulants
Coumadin
Dicumarol
Panwarfin
Phenobarbital
Trisoralen

GRISEOFULVIN + Phenobarbital: Phenobarbital may decrease the blood level of Griseofulvin. (33, 55)

H

HAIR PREPARATIONS + Sebucare: see Sebucare + Hair Preparations

HALDOL

+ Alcohol: see Haldol + CNS Depressants
+ Analgesics: see Haldol + CNS Depressants. An increase in depression may be observed in newborn infants when Haldol is used concomitantly with an analgesic agent during labor. (86)
+ Anesthetics, General: see Haldol + CNS Depressants
+ Anticoagulants: Haldol should be used with caution in patients receiving anticoagulants, since interference with the effects of one anticoagulant (phenindione) has been reported in a single patient. (1, 86, 109)
+ Anticonvulsants: When instituting therapy with Haldol in patients receiving anticonvulsant medication, the dose of the anticonvulsant should not be altered. Anticonvulsant dosage may require subsequent adjustment, particularly when Haldol brings under control those psychotic symptoms that were responsible for precipitating convulsive seizures. Haldol does not potentiate anticonvulsants. (1, 86)
+ Antihypertensive Agents: see Haldol + CNS Depressants
+ Antiparkinsonism Drugs: If antiparkinsonism drugs should be used to control Parkinson-like symptoms concomitantly with Haldol, they may have to be continued after Haldol is stopped because tissue levels of Haldol persist for some time whereas the antiparkinson drugs are excreted more rapidly. Extrapyramidal symptoms may occur if both drugs are discontinued simultaneously. (1)
+ Barbiturates: Haldol potentiates the central nervous system depressant action of barbiturates. It does not potentiate the anticonvulsant action of barbiturates. (1, 86)
+ CNS Depressants: Care should be exercised when antihypertensive agents, general anesthetics, hypnotics, alcohol, analgesics, and other central nervous system depressants are used concomitantly with Haldol, since it may potentiate their action. (1, 86)
+ Danilone: see Haldol + Anticoagulants
+ Epinephrine: Epinephrine should not be used in cases of hypotension with Haldol, since Haldol may block the vasoconstrictor effects of this drug and reverse its action causing profound hypotension. Vasopressors such as Levophed can be used. (1, 86)
+ Hedulin: see Haldol + Anticoagulants
+ Hypnotics: see Haldol + CNS Depressants

HALDOL (continued)

+ Kemadrin: A toxic psychosis developed in one patient while receiving Haldol and Kemadrin concomitantly. (86)
+ Trisoralen: see Trisoralen

HALDRONE

+ Antibiotics: Great care should be taken if Haldrone is used in overwhelming infections even though appropriate antibiotic therapy is being used. There is still dispute over the safety and effectiveness of steroids as an adjunct to antibiotics and other specific therapy in overwhelming infections. (1)
+ Diabetes: Haldrone should be used with great caution in patients with diabetes mellitus, since it may cause disturbances in glucose metabolism. (1, 2)

HALOTHANE: see Fluothane

HALOTHANE: see listed under the following agents:

Epinephrine
Neosynephrine Parenteral
Pressonex
Quelicin Chloride Inj.
Succinylcholine
Vasoconstrictors

HALOTHANE + Pressonex: Do not use Pressonex during Halothane anesthesia. (1)

HARMONYL: see Rauwolfia Alkaloids and Reserpine

HARMONYL

+ Alcohol: The action of Harmonyl may be potentiated by alcohol. (1)
+ Anesthesia: Harmonyl should be withdrawn or dosage reduced during the two weeks prior to elective surgery. An unexpected degree of hypotension and bradycardia has been reported in patients undergoing anesthesia while under treatment with rauwolfia alkaloids. When emergency surgical procedures are necessary, parenteral administration of vagal blocking agents to prevent or reverse hypotension and/or bradycardia may be considered. (1, 2) Mechanism: Rauwolfia alkaloids deplete tissue including cardiac muscles of catecholamines. This increases the risk of vascular collapse during anesthesia and surgery. Depletion of catecholamines also increases the hazard of cardiac arrest. (5)
+ Anticonvulsants: In epileptic patients, therapy may necessitate dosage adjustment of anticonvulsant medication. Rauwolfia alkaloids lower the convulsive threshold and shorten seizure latency. This factor should also be taken in account in patients undergoing electroshock treatment. (1, 2)

HARMONYL (continued)

+ Barbiturates: The action of Harmonyl may be potentiated by barbiturates. (1)
+ Electroshock Therapy: When patients on Harmonyl receive electroshock therapy, use lower milliamperage and shorten the duration of stimulus initially, since more prolonged and severe convulsions as well as apnea have been reported with previously well tolerated stimulation. Shock therapy within seven days after giving the drug is hazardous. (1)
+ Enduron: see Enduron + Harmonyl
+ Narcotics: The action of Harmonyl may be potentiated by narcotics. (1)

HAY FEVER PREPARATIONS: see listed under the following agents:

Eutonyl
MAO Inhibitors
Marplan

HEAT: see listed under the following agents:

Cortisporin Otic
Dartal
Inversine
Lidosporin Otic Solution
Phenothiazine
Quide
Repoise
Thorazine
Tindal
Trilafon
Unitensin

HEDULIN: see Anticoagulants

HEDULIN: see listed under the following agents:

Analexin
Haldol

HEDULIN: The following agents may potentiate the hypoprothrombinemic effects of Hedulin: (1)

Antibiotics
Butazolidin
Dietary deficiency of protein
Dietary deficiency of Vitamin C
Penicillin
Phenothiazines
Salicylates

HEDULIN: The following agents may reverse the hypoprothrombinemic effects of Hedulin: (1)

ACTH
Antihistamines
Barbiturates
Corticosteroids
Diet of large amounts of green or leafy vegetables, fish, and fish oils

HEMASTIX + Ascorbic Acid: Ascorbic acid when present in large amounts in urine is the only identified substance which may retard or inhibit the chemical reaction of hemoglobin with the peroxide-orthotolidine system, Therefore Hemastix should not be used as the sole procedure when testing for occult blood in urines from patients taking large therapeutic dosages of ascorbic acid or parenteral antibiotics (Terramycin, Achromycin, Panmycin, Tetracyn, etc.) containing large amounts of ascorbic acid in the urine; microscopy is the procedure of choice. (1) The following injectable antibiotics contain ascorbic acid:

Abbocillin 800 M
Achromycin I.M. and I.V.
Panmycin Parenteral
Steclin Parenteral
Sumycin I.M.
Syntetrin I.M. and I.V.
Terramycin I.M. and I.V.
Tetracyn I.M. and I.V.
Tetrex I.M.

HEMATOPOIETIC DEPRESSANTS + Methotrexate: see Methotrexate + Hematopoietic Depressants

HEMOSTATIC AGENTS + Surgical Absorbable Hemostat: see Surgical Absorbable Hemostat + Hemostatic Agents

HEPARIN: see listed under the following agents:

Cardio-Green
Dicumarol
Panwarfin
Thrombolysin

HEPARIN: see various trade name preparations for additional interactions, e.g., Liquaemin Sodium.

HEPARIN

+ Aspirin: Patients given anticoagulant therapy with heparin depend upon hemostasis to prevent bleeding. These patients should not receive aspirin or Persantine concomitantly, since these drugs inhibit platelet adhesiveness, the basis of the hemostatic mechanism. (154)
+ Persantine: see Heparin + Aspirin
+ Polymixin B: These agents are incompatible when mixed together in an intravenous mixture. (55)
+ Prothrombinopenic Drugs: It has been shown that high concentrations of heparin will cause an apparent lowering of the prothrombin level, but that this does not occur unless the coagulation time (Lee-White) is more than 40 minutes. Thus, during the period of combined heparin-prothrombinopenic therapy (oral coumarin anticoagulants), false prothrombin levels can be avoided if the daily test is performed prior to the administration of the prothrombinopenic drug

HEPARIN (continued)

+ Prothrombinopenic Drugs (continued): and at a time when the coagulation time has fallen to less than 40 minutes.

HEPATOTOXIC DRUGS + Aralen: see Aralen + Hepatotoxic Drugs

HEXADROL + Diabetes: Average doses of Hexadrol will not usually increase insulin requirements in controlled diabetics, but such patients should be observed closely for increased hyperglycemia or glycosuria. Periodic determinations of blood sugar during prolonged therapy are advised. (1, 2)

HEXOBARBITAL: see listed under the following agents:

Aventyl
Phenobarbital

HISTAMINE: see listed under the following agents:

Isordil
Pentritol
Sorbitrate

HUMORSOL

+ Carbonic Anhydrase Inhibitor: In some cases, the concomitant administration of a carbonic anhydrase inhibitor may enhance the effectiveness of Humorsol in controlling intraocular tension. (2)
+ Succinylcholine: see Succinylcholine + Humorsol

HYDELTRASOL Injection

+ Anesthetics, Local: When Hydeltrasol Injection is mixed with a local anesthetic, a fine precipitate will form. While this precipitate will not interfere with the ease of injection or the response to therapy, it is advisable to use such a mixture immediately and discard any unused portion, since this precipitate may settle out on standing and clog the needle. (1)
+ Diabetes: Systemic adrenocortical hormone therapy may cause alteration of glucose metabolism with aggravation of diabetes mellitus. (1, 2)
+ Vasopressors: In cases of shock unresponsive to conventional therapy, vasopressors should be used in conjunction with Hydeltrasol Injection but in a separate syringe. (1)

HYDELTRA-TBA + Diabetes: Hydeltra-TBA should be used with great caution in patients with diabetes mellitus, since it causes disturbances in glucose metabolism. (2)

HYDREA
+ Antineoplastic Agents: see Hydrea + Cytotoxic Drugs
+ Cytotoxic Drugs: Because hematopoiesis may be compromised by other antineoplastic agents, it is recommended that Hydrea be administered cautiously to patients who have recently received chemotherapy with other cytotoxic drugs. See also Hydrea + Depressants, Bone Marrow. (1)
+ Depressants, Bone Marrow: Treatment with Hydrea should not be initiated if bone marrow function is depressed. It should be borne in mind that bone marrow depression is more likely in patients who have recently received radiotherapy or cancer chemotherapy. Patients who have received irradiation therapy in the past may have an exacerbation of post-irradiation erythemia. (1)
+ Irradiation: Because hematopoiesis may be compromised by irradiation, it is recommended that Hydrea be administered cautiously to patients who have recently received radiation therapy. See also Hydrea + Depressants, Bone Marrow. (1)

HYDROCHLOROTHIAZIDE + Singoserp: see Singoserp + Hydrochlorothiazide

HYDROCORTISONE + Diphenylhydantoin: see Diphenylhydantoin + Hydrocortisone

HYDROCORTONE
+ Diabetes: Hydrocortone causes gluconeogenesis; therefore hyperglycemia and glycosuria may occur, glucose tolerance may be altered, and diabetes mellitus may be aggravated. These effects are usually reversible on discontinuation of therapy or sometimes with a decrease in dosage. (1)
+ Diuretics: Concomitant therapy may provoke a dangerous loss of potassium. (1)

HYDRODIURIL: see Diuretics, Thiazide

HYDRODIURIL
+ ACTH: see HydroDIURIL + Steroids
+ Alcohol: Orthostatic hypotension may occur with HydroDIURIL and may be potentiated by alcohol, barbiturates, or narcotics. (1)
+ Antihypertensive Agents: HydroDIURIL potentiates the action of other antihypertensive drugs. Therefore the dosage of these agents, especially the ganglionic blockers, must be reduced by at least 50% as soon as the diuretic is added to the regimen. (1, 2)
+ Barbiturates: see HydroDIURIL + Alcohol

HYDRODIURIL (continued)

+ Diabetes: Insulin requirements in diabetic patients may be increased, decreased, or unchanged. In latent diabetics, HydroDIURIL, in common with other benzothiadiazines, may cause hyperglycemia and glycosuria. (1, 2)
+ Digitalis: During periods of marked diuresis with excessive potassium loss, patients are more sensitive to the development of digitalis toxicity. Caution must therefore be exercised to prevent hypokalemia during digitalis administration. Alteration of dosage requirements for both digitalis and HydroDIURIL must be anticipated. (1, 2)
+ Ganglionic Blocking Agents: see HydroDIURIL + Antihypertensive Agents
+ Levophed: see HydroDIURIL + Pressor Amines
+ Mercurials, Parenteral: HydroDIURIL and parenterally administered mercurials may be given concomitantly; the effects are additive. (2)
+ Narcotics: see HydroDIURIL + Alcohol
+ Pressor Amines: Recent reports indicate that the thiazides may decrease arterial responsiveness to pressor amines (e.g., Levophed). HydroDIURIL decreases arterial responsiveness to norepinephrine, as do the other thiazides, necessitating due care in surgical patients. It is recommended that thiazides be discontinued 48 hours before elective surgery. (1, 2)
+ Steroids: Hypokalemia may develop with HydroDIURIL as with any other potent diuretic, especially during concomitant steroid or ACTH administration. (1, 2)
+ Trisoralen: see Trisoralen
+ Tubocurarine: Thiazide drugs may increase the responsiveness to tubocurarine. (1, 2)

HYDROGEN PEROXIDE + Panafil Ointment: see Panafil Ointment + Hydrogen Peroxide

HYDROMOX: Hydromox is not a thiazide, but it may possess certain characteristics of the thiazides. See Diuretics, Thiazide.

HYDROMOX

+ Adrenocortical Steroids: see Hydromox + Digitalis
+ Alcohol: Orthostatic hypotension may occur with Hydromox and may be potentiated by alcohol, barbiturates, or narcotics. (1)
+ Antihypertensive Agents: Hydromox may be combined with reduced doses of other hypotensive agents such as rauwolfia alkaloids or ganglionic blocking agents. To avoid the development of orthostatic hypotension, the dosage of other hypotensive agents such as gangli-

HYDROMOX (continued)

+ Antihypertensive Agents (continued): onic blocking agents, veratrum alkaloids, and Apresoline must be reduced at least one-half when Hydromox is given concomitantly. (1, 2)
+ Apresoline: see Hydromox + Antihypertensive Agents
+ Barbiturates: see Hydromox + Alcohol
+ Corticotrophin: see Hydromox + Digitalis
+ Diabetes: The thiazide diuretics have produced decreased glucose tolerance as evidenced by hyperglycemia and glycosuria, thus aggravating or provoking diabetes mellitus. Similar effects have been reported with Hydromox. (1, 2)
+ Digitalis: Patients in whom depletion of potassium may occur, such as those receiving digitalis, adrenocortical steroids, or corticotrophin, should take some food daily that is high in potassium content (e.g., orange juice) or a potassium salt. (1, 2)
+ Ganglionic Blocking Agents: see Hydromox + Ganglionic Blocking Agents
+ Narcotics: see Hydromox + Alcohol
+ Norepinephrine: Hydromox may decrease arterial responsiveness to norepinephrine and therefore should be withdrawn 48 hours before elective surgery. If emergency surgery is indicated, pre-anesthetic and anesthetic agents should be administered in reduced dosage. (1)
+ Rauwolfia Alkaloids: see Hydromox + Antihypertensive Agents
+ Trisoralen: see Trisoralen
+ Tubocurarine: Hydromox may increase the responsiveness to tubocurarine. (1)
+ Veratrum Alkaloids: see Hydromox + Antihypertensive Agents

HYDROMOX R: see Hydromox and Reserpine

HYDROPRES: see HydroDIURIL and Reserpine

HYGROTON: see listed under the following agents:

Aldactone
Raudixin
Rauwiloid
Unitensin

HYGROTON

+ ACTH: see Hygroton + Adrenal Corticosteroids
+ Adrenal Corticosteroids: Emphasis should be placed on the detection of potassium depletion, especially when the drug is used in patients receiving adrenal corticosteroids, ACTH, or digitalis (where potassium depletion may potentiate its toxic effects). (1)

HYGROTON (continued)

+ Alcohol: Orthostatic hypotension has been reported with Hygroton and may be potentiated when Hygroton is combined with alcohol, barbiturates, or narcotics. (1)
+ Antihypertensive Agents: Antihypertensive therapy with Hygroton should always be initiated cautiously in patients receiving ganglionic blocking agents or other potent antihypertensive agents. Reduction by one-half of the dosage of these agents may be advisable. Careful and continuous supervision of patients on such regimen is necessary. (1, 2)
+ Apresoline: see Hygroton + Ismelin
+ Barbiturates: see Hygroton + Alcohol
+ Diabetes: A decreased glucose tolerance evidenced by hyperglycemia and glycosuria may develop inconsistently. This condition, usually reversible on discontinuation of therapy, responds to control with antidiabetic treatment. Diabetics and those predisposed should be checked regularly. (1, 2)
+ Digitalis: see Hygroton + Adrenal Corticosteroids
+ Ganglionic Blocking Agents: see Hygroton + Antihypertensive Agents
+ Ismelin: Hygroton potentiates the effects of potent hypotensive agents such as Ismelin and Apresoline. The dosage of these preparations should be reduced by one-half initially and gradually elevated to effective levels. (2)
+ Narcotics: see Hygroton + Alcohol
+ Trisoralen: see Trisoralen

HYKINONE: see Coumadin + Vitamin K

HYPERTENSIN

+ Appetite Suppressants: Patients receiving appetite suppressants may respond to smaller doses of Hypertensin. (1)
+ MAO Inhibitors: Patients receiving MAO inhibitors may respond to smaller doses of Hypertensin. (1)
+ Plasma: see Hypertensin + Whole Blood
+ Pressor Agents: Patients receiving other pressor agents may respond to smaller doses of Hypertensin. (1)
+ Ritalin: While oral Ritalin has little or no effect on blood pressure, effects of pressor agents have been potentiated in pharmacological experiments. Use cautiously with Hypertensin. (1)
+ Serum: see Hypertensin + Whole Blood
+ Stimulants: Patients receiving stimulants may respond to smaller doses of Hypertensin. (1)

HYPERTENSIN (continued)

+ Whole Blood: When whole blood, serum, or plasma is indicated to restore blood volume, it should be given through a Y-tube or a separate needle so that naturally occurring enzymes do not inactivate Hypertensin. (1, 2)

HYPNOTICS: see listed under the following agents:

Ambodryl	Navane
Anti-Nausea Suprettes	Niamid
Benadryl	Orinase
Carbrital	Pacatal
Compazine	Permitil
Dalmane	Phenothiazine
Dartal	Placidyl
Diabinese	Proketazine
Disophrol	Prolixin
Dymelor	Pyribenzamine
Eutonyl	Repoise
Forhistal	Seconal Sodium
Hypoglycemic Drugs, Oral	Stelazine
Magnesium Sulfate Inj.	Taractan
Marplan	Tindal
Mellaril	Trilafon
Narcotic Antagonist	Vesprin

HYPOGLYCEMIC DRUGS, ORAL: see Sulfonylurea, Hypoglycemic

HYPOGLYCEMIC DRUGS, ORAL: see listed under the following agents:

Inderal
Tandearil

HYPOGLYCEMIC DRUGS, ORAL

+ Alcohol: The sulfonylureas may decrease tolerance to alcohol, manifested by an unusual flushing of the skin, particularly of the face and neck, similar to that caused by Antabuse-alcohol reaction. (2)
+ Barbiturates: The sulfonylureas may prolong the effects of barbiturates and other sedatives and hypnotics. (2)
+ Benemid: It has been reported that Benemid potentiates the action of the sulfonylureas, resulting in prolonged hypoglycemia. (2)
+ Butazolidin: It has been reported that Butazolidin may potentiate the action of the sulfonylureas, resulting in prolonged hypoglycemia. (2) <u>Mechanism</u>: It is postulated that Butazolidin slows the rate of metabolism of the oral hypoglycemic drug (18) and/or displaces the

HYPOGLYCEMIC DRUGS, ORAL (continued)

+ Butazolidin (continued): oral hypoglycemic agent from its protein-binding site increasing the concentration of unbound drug. (118)
+ Hypnotics: see Hypoglycemic Drugs, Oral + Barbiturates
+ Phenothiazines: The possibility should be borne in mind that phenothiazine-type tranquilizers may cause jaundice or abnormal results in liver function tests when used with the sulfonylurea derivatives. It also should be recognized that the values of alkaline phosphatase, cephalin floculation, and thymol turbidity tests may be significantly higher in diabetic patients. (2)
+ Sedatives: see Hypoglycemic Drugs, Oral + Barbiturates
+ Sulfonamides, Bacteriostatic: It has been reported that the bacteriostatic sulfonamides may potentiate the action of the sulfonylureas, resulting in prolonged hypoglycemia. (2) Mechanism: Displacement of the oral hypoglycemic drug from its protein-binding site. (50)
+ Thiazide Diuretics: Caution should be exercised when thiazide diuretics are given with the sulfonylureas, since the thiazides have been reported to aggravate the diabetic state and to result in increased requirements for the oral hypoglycemic agent, temporary loss of control, and even secondary failure. (2)

HYPOTENSIVE AGENTS: see Antihypertensive Agents

HYPOTENSIVE AGENTS: see listed under the following agents:

Anhydron
Aventyl
Brevital Sodium
Bristuron
Diuretics, Thiazide
Hydromox
Hygroton
MAO Inhibitor
Marplan
Matulane
Mylaxen
Naturetin
Renese
Vontrol

HYPOTHERMIA: see Heat

HYPOTHERMIA: see listed under the following agents:

Quelicin Chloride Inj.
Succinylcholine

ILETIN, LENTE

+ Alcohol: Bathing, rubbing, or medicated alcohol should not be used for syringe sterilization. (1)
+ Sterilizing Solution: The use of heavily chlorinated water or chemical solutions for sterilizing the syringe prior to the injection of Lente Iletin should be avoided. (1)

ILETIN, REGULAR

+ Alcohol: Bathing, rubbing, or medicated alcohol should not be used for syringe sterilization. (1)

ILOPAN

+ Antibiotics: There have been rare instances of allergic reactions of unknown cause during the concomitant administration of Ilopan and other drugs such as antibiotics, narcotics, and barbiturates. (1)
+ Anticholinesterase Agents: see Ilopan + Neostigmine
+ Barbiturates: see Ilopan + Antibiotics
+ Narcotics: see Ilopan + Antibiotics
+ Neostigmine: Wait 12 hours after neostigmine or other enterokinetic drugs before starting Ilopan. (1) Mechanism: Ilopan is converted to pantothenic acid and thus claimed to produce more acetylcholine; therefore there may be an additive effect.
+ Parasympathomimetic Drugs: Theoretical considerations suggest that if parasympathomimetic drugs have been used, Ilopan should not be administered for 12 hours after these agents have been discontinued. (2) See Ilopan + Neostigmine.
+ Succinylcholine: A waiting period of one hour after cessation of succinylcholine administration is recommended before starting drugs having pantothenic acid activity. Respiratory embarrassment was observed in a patient when Ilopan was administered shortly after cessation of succinylcholine therapy. (1, 2) Mechanism: Ilopan is converted to pantothenic acid and thus claimed to produce more acetylcholine. Acetylcholine acts as a depolarizing agent at the motor end plates augmenting the depolarizing action of succinylcholine.

ILOPAN-CHOLINE: see Ilopan

ILOTYCIN ETHYL CARBONATE DROPS + Fruit Juice: Fruit juice or any other acid drink should not be taken immediately after Ilotycin Drops because the combination may give rise to an unpleasant taste. (1)

ILOTYCIN GLUCEPTATE I.V.

+ Diluents: It is important to note that when the initial solution is made, sterile water, and not saline or other diluent, should be used in order to avoid gel formation and to insure prompt and complete solution. Also, because of preservative substances in sterile water packaged in rubber-stoppered multiple-dose vials, only sterile water from glass-sealed ampoules should be employed. (1)
+ Saline Solution: see Ilotycin Gluceptate I.V. + Diluents
+ Sterile Water for Injection: see Ilotycin Gluceptate I.V. + Diluents

IMMU-G + Live Measles Virus Vaccine: Immu-G may be used and is valuable for modifying the reactions associated with the use of attenuated live measles virus vaccine. The dose is 0.01 cc per pound of body weight given intramuscularly after administering the vaccine but in a separate syringe and in the opposite arm. (1)

IMMUNE GLOBULIN, HUMAN: see listed under the following agents:

Lirugen
M-Vac Measles Virus Vaccine, Live, Attenuated

IMMUNE SERUM GLOBULIN, HUMAN: see listed under the following agents:

Attenuvax, Lyovac
Meruvax
Orimune Poliovirus Vaccine, Live, Attenuated, Trivalent
Pfizer-Vax Measles-L

IMMUNE SERUM GLOBULIN, HUMAN

+ Immunization Agents: The recent administration of gamma globulin and other globulin fractions has the potential of interfering with active immunization agents. The inability to specifically define this potential often leads to paradoxical recommendations. For example, the live attenuated measles vaccine may be given at the same time (and at a different site) as measles immune globulin. However, the vaccine should not be given to individuals who have received immune globulin within the past six weeks. After administering live measles vaccine and modifying gamma globulin, one month should elapse before giving oral poliomyelitis vaccine. (67)
+ Live Measles Virus Vaccine: For modification of the symptoms induced by the concomitant administration of attenuated live measles virus vaccine, Immune Serum Globulin, Human is administered in a dose of 0.01 cc per pound of body weight. This should be injected immediately after the vaccine into the deltoid muscle of the opposite arm.

IMMUNE SERUM GLOBULIN, HUMAN (continued)

+ Live Measles Virus Vaccine: Immune Serum Globulin, Human must not be administered intravenously. The administration of the vaccine before Immune Serum Globulin, Human permits the virus to infect the cells of the injection site prior to the distribution of measles antibody from the Immune Serum Globulin. (1, 67)
+ Poliomyelitis Vaccine, Oral: see Immune Serum Globulin, Human + Immunization Agents

IMMUNIZATION AGENTS: see listed under the following agents:
Immune Serum Globulin, Human
Rubeovax, Lyovac

IMMUNIZATION, ELECTIVE: see listed under the following agents:
Attenuvax, Lyovac
Meruvax

IMURAN + Zyloprim: In patients receiving Zyloprim, the concomitant administration of Imuran will require a reduction in dose to approximately one-third to one-fourth the usual dose of Imuran. (1)

INAPSINE

+ Barbiturates: see Inapsine + CNS Depressants
+ CNS Depressants: Other central nervous system depressants (e.g., barbiturates, narcotics, and other major tranquilizers) given so their actions overlap those of Inapsine, must be used in reduced doses (as low as one-half the dose usually recommended) because of additive or possible potentiating effects. (1)
+ Epinephrine: see Inapsine + Pressor Amines
+ Narcotics: see Inapsine + CNS Depressants
+ Pressor Amines: Since Inapsine is capable of blocking the pressor response to pressor amines, the potential exists for the precipitation of severe hypotension in the immediate postoperative period. If fluid therapy does not correct the hypotension, then the administration of pressor agents other then epinephrine should be considered. (1)
+ Tranquilizers, Major: see Inapsine + CNS Depressants

INDERAL

+ Anesthetics: Inderal should not be used to treat arrythmias associated with anesthetics that produce myocardial depression, e.g., chloroform or ether. (1) Beta-adrenergic blocking agents can produce profound hypotension in anesthetized patients. (14)

INDERAL (continued)

\+ Catecholamine Depleting Drugs: Patients receiving catecholamine depleting drugs such as reserpine should be closely observed when Inderal is introduced into the treatment regimen. The added catecholamine blocking action of Inderal may produce an excessive reduction of the resting sympathetic nervous activity. (1)

\+ Chloroform: see Inderal + Anesthetics

\+ Diabetes: Caution should be exercised in the administration of Inderal to patients subject to spontaneous hypoglycemia or to diabetics (especially labile diabetics) receiving insulin or oral hypoglycemic agents. Because of its beta-adrenergic blocking activity, Inderal may prevent the appearance of premonitory signs and symptoms (pulse rate and pulse pressure changes) of acute hypoglycemia. (1, 42, 75) See Insulin + Inderal.

\+ Epinephrine: The infusion of epinephrine following Inderal blockade produced a decrease in heart rate due to baroreceptor mediated reflex activity. (1)

\+ Ether: see Inderal + Anesthetics

\+ Hypoglycemic Agents, Oral: see Inderal + Diabetes

\+ Insulin: see Inderal + Diabetes

\+ Isuprel: Intravenous infusions of Isuprel (isoproterenol) produced an increase in heart rate, cardiac index, and pressure/time gradient in the femoral artery and a fall in mean blood pressure and total peripheral resistance. Following Inderal, the effects of beta-adrenergic stimulation were blocked for various periods of time up to four hours. (1, 50)

\+ Levophed: Levophed (15 μ/min.) produced a fall in cardiac index of 9% while heart rate decreased and stroke volume rose. After beta-blockade with Inderal, the same dose of Levophed resulted in greater fall of cardiac index. Despite a decrease in heart rate, stroke volume failed to rise. (1)

\+ MAO Inhibitors: see Inderal + Psychotropic Agents, Adrenergic-Augmenting

\+ Meals: When given to fasting patients, Inderal was more readily absorbed and had a shorter duration of action than when given after meals. Inderal should be given before meals. (1)

\+ Pertofrane: see Pertofrane + Inderal

\+ Psychotropic Agents, Adrenergic-Augmenting: Inderal is contraindicated in patients on adrenergic-augmenting psychotropic drugs (including MAO inhibitors) and during the two week withdrawal period for such drugs. (1, 70)

\+ Reserpine: see Inderal + Catecholamine Depleting Drugs

INDOCIN: see listed under the following agents:

Anticoagulants
Dicumarol
Panwarfin
Ponstel

INDOCIN
+ Butazolidin: see Indocin + Steroids
+ Diabetes: It has been reported on rare occasions that Indocin may cause hyperglycemia and glycosuria. (1)
+ Salicylates: see Indocin + Steroids
+ Steroids: The possible potentiation of the ulcerogenic effect of steroids, salicylates, or Butazolidin cannot be ruled out at present. (1, 2)

INFUSION FLUID + Lipomul I.V.: see Lipomul I.V. + Infusion Fluid

INNOVAR: see Sublimaze

INNOVAR + Narcotics: During the course of anesthesia, additional (0.5 to 1.0 cc) Innovar may be administered intravenously when changes in vital signs (increased heart rate, elevated blood pressure, increased respiratory rate, or irregular breathing) indicate lightening of the anesthetic plane. Analgesia usually extends well into the postoperative period; hence, most patients do not require narcotic analgesics during the immediate postoperative period. However, because of the long lasting potentiating effects of Innovar, such agents, when required, must be administered in doses reduced to as low as one-fourth or one-half of that usually recommended. (1)

INSECTICIDES, NEUROTOXIC + Quelicin Chloride Inj.: see Quelicin Chloride Inj. + Insecticides, Neurotoxic

INSECTICIDES, ORGANOPHOSPHATE + Succinylcholine: see Succinylcholine + Insecticides, Organophosphate

INSECTICIDES, PHOSPHOROUS: see listed under the following agents:

Anticholinesterase Agents
Dartal
Levoprome
Mellaril
Navane
Permitil
Phenothiazines
Prolixin
Quide
Repoise
Serentil
Sparine
Thorazine
Tindal
Torecan
Trilafon
Vesprin

INSECTICIDES, POLYPHOSPHATE: see listed under the following agents:
- Anectine
- Floropryl
- Succinylcholine

INSULIN: see Diabetes

INSULIN: see listed under the following agents:
- Alcohol
- Anturane
- Butazolidin
- Dymelor
- Eutonyl
- Isuprel
- Mao Inhibitor
- Marplan
- Nardil
- Niamid
- Orinase
- Parnate
- Regitine
- Tandearil
- Tolinase

INSULIN + Inderal: Potentiation of insulin hypoglycemia may occur following antagonism of the metabolic effects of catecholamine by Inderal. (154)

INVERSINE: see Alkaline Agents + Acidic Media

INVERSINE: see Alkaline Agents + Acidifying Agents (Urinary)

INVERSINE: see also Ganglionic Blocking Agents

INVERSINE: see listed under the following agents:
- Diuretics, Thiazide
- HydroDIURIL
- Naqua

INVERSINE
- + Anesthesia: The action of Inversine may be potentiated by surgery and anesthesia; therefore the dosage of Inversine must be altered accordingly. (1)
- + Antibiotics: Patients with chronic pyelonephritis receiving antibiotics and sulfonamides should not be treated with ganglionic blockers. (1)
- + Antihypertensive Agents: When Inversine is given with other hypotensive drugs (e.g., reserpine, oral diuretics), the dosage of these agents, as well as Inversine, should be reduced to avoid excessive hypotension. However, thiazides should be continued in their usual dosage while that of Inversine is decreased by at least 50%. (1, 2)
- + Diuretics, Oral: see Inversine + Antihypertensive Agents
- + Epinephrine: Inversine potentiates the vascular action of epinephrine as do other ganglionic blocking agents. (1)

INVERSINE (continued)

+ Excessive Heat: Excessive heat may potentiate the effects of Inversine. (1)
+ Laxatives: Constipation should not be treated with bulk laxatives. (1)
+ Pressor Amines: Pressor amines may be used to counteract excessive hypotension. Since patients being treated with ganglionic blockers are more than normally reactive to pressor amines, small doses of the latter are recommended to avoid excessive response. (1, 2)
+ Reserpine: see Inversine + Antihypertensive Agents
+ Salt Depletion: Salt depletion resulting from diminished intake or increased excretion due to diarrhea, vomiting, or diuretics may potentiate the action of Inversine. (1) See Inversine + Thiazides.
+ Saluretic Agents: see Inversine + Thiazides
+ Sulfonamides: see Inversine + Antibiotics
+ Thiazides: see Inversine + Antihypertensive Agents. When a saluretic agent is administered with Inversine, liberalization of salt intake may be required to prevent development of a low salt syndrome by this combination therapy. Care should be exercised to prevent the development of hypokalemia when chlorothiazide or hydrochlorothiazide is added to the regimen. (1, 55)
+ Vigorous Exercise: Vigorous exercise may potentiate the effects of Inversine. (1)

IODIDES: see listed under the following agents:

Neo-Vagisol
Vlem-Dome Liquid Concentrate

IODINE + Merthiolate: see Merthiolate + Iodine

ION EXCHANGE RESINS + Modumate: see Modumate + Ion Exchange Resins

IPECHAR POISON CONTROL KIT

+ Corrosive Substance: Ipecac-induced emesis is contraindicated if a corrosive substance has been swallowed. A vomitus containing a corrosive substance may rupture the esophagus. (1)
+ Petroleum Product: Ipecac-induced emesis is contraindicated if a petroleum product has been swallowed. A vomitus containing a petroleum product may pass into the lungs and damage them. (1)

IPHYLLIN + CNS Stimulants: Caution must be exercised in the concomitant use of other central nervous system stimulating drugs. (1)

IRCON + Meals: Although the absorption of iron is best when taken between meals, gastrointestinal disturbances may be controlled by reducing the dose and giving the preparation shortly after meals. (2)

IRON: see listed under the following agents:

Astrafer
Cuprimine
Fuadin
Jectofer
Neomycin
pH
Zyloprim

IRON + BAL: BAL is contraindicated in the treatment of iron overdosage because it forms a toxic complex with iron. (1)

IRON PREPARATIONS: see listed under the following agents

Jectofer
Rencal

IRRADIATION: see also X-Ray Therapy and Radiation Therapy

IRRADIATION: see listed under the following agents:

Attenuvax, Lyovac
Biavax
Cendevax
Hydrea
Lirugen
Measles Virus Vaccine, Live, Attenuated
Measles Virus Vaccine, Live, Attenuated (Edmonston B & Schwarz Strain)
Meruvax
Mumpsvac, Lyovac
M-Vac Measles Virus Vaccine, Live, Attenuated
Pfizer-Vax Measles-L
Rubeovax, Lyovac
Succinylcholine
Sucostrin
Thioguanine
Vercyte

ISMELIN: see listed under the following agents:

Aldomet
Antidepressants, Tricyclic
Aquatag
Aventyl
Bristuron
Diuretics, Thiazide
Esidrix
Eutonyl
Exna
Hygroton
MAO Inhibitor
Naturetin
Pertofrane
Raudixin
Rauwiloid
Serpasil
Sinequan
Singoserp
Tofranil
Unitensin
Vivactil

ISMELIN

\+ Alcohol: Orthostatic hypertension is frequent especially during the initial dosage adjustment of Ismelin. It is most marked in the morning and is accentuated by hot weather, alcohol, or exercise. (1)

ISMELIN (continued)

+ Amphetamines: The concomitant administration of Ismelin and either amphetamine, dextroamphetamine, methamphetamine or appetite suppressants will reverse the hypotensive effect of Ismelin and possibly cause a hypertensive crisis. Mechanism: The amphetamines not only block the uptake of Ismelin into sites of action in nerve endings but actually displace the Ismelin and Norepinephrine at those sites. (1, 2, 34, 46, 55, 72, 130)
+ Anesthesia: Ismelin depletes tissues including cardiac muscle of catecholamines. This increases the risk of vascular collapse during anesthesia and surgery. It is, therefore, advisable to stop Ismelin at least two weeks, preferably three weeks, before surgery is performed, whether under general or local anesthesia. Depletion of catecholamines also increases the hazards of cardiac arrest -- another reason for discontinuing Ismelin well before surgery. If emergency surgery is necessary, give anesthesia cautiously. (1, 2, 5, 55)
+ Antidepressants, Tricyclic: The tricyclic antidepressants (e.g., Aventyl, Elavil, Norpramin, Pertofrane, Tofranil, Vivactil) will reverse the antihypertensive effect of Ismelin. Mechanism: It is suggested that the tricyclic antidepressants will block the uptake of Ismelin at the action site within the sympathetic nerve endings. (34, 44, 46, 72, 130)
+ Antihistamines: see Ismelin + Phenothiazines
+ Appetite Supressants: see Ismelin + Amphetamines
+ Apresoline: Concurrent administration of Ismelin and Apresoline produce further reduction in blood pressure. The Ismelin should be added gradually to patients receiving Apresoline. (1, 2)
+ Aramine: see Ismelin + Wyamine
+ Aventyl: see Ismelin + Antidepressants, Tricyclic
+ Cocaine: The uptake of Ismelin is prevented by cocaine thereby preventing the antihypertensive effect of Ismelin. Mechanism: Ismelin is prevented from reaching its site of action at the sympathetic nerve endings. (104, 118)
+ Dextroamphetamine: see Ismelin + Amphetamines
+ Digitalis: If digitalis is used with Ismelin, it should be remembered that both drugs slow the heart rate. (1, 2)
+ Diuretics, Thiazide: Ismelin may be added gradually to thiazides and/or Apresoline. Thiazide diuretics enhance the effectiveness of Ismelin and may reduce the incidence of edema. When thiazide diuretics are added to the regimen in patients on Ismelin, it is usually necessary to reduce the dosage of Ismelin. (1, 2, 34)

ISMELIN (continued)

+ Elavil: see Ismelin + Antidepressants, Trycyclic
+ Ephedrine: see Ismelin + Stimulants
+ Ganglionic Blockers: In many cases ganglionic blockers will have been stopped before Ismelin is started. It may be advisable, however, to withdraw the blocker gradually to prevent a spiking blood pressure response during the transfer period. (1)
+ Levophed: Direct-acting vasopressors such as Levophed given to patients to counteract severe hypotension in a patient who has been on Ismelin will induce a much greater response than would be expected under normal circumstances. This antihypertensive works to prevent the uptake of norepinephrine into inactivation sites and therefore potentiates the effects of the vasopressors. (1, 55, 108)
+ MAO Inhibitors: Wait one week after discontinuing MAO inhibitors before starting Ismelin. Do not use Ismelin concomitantly with MAO inhibitors. (1) Mechanism: see MAO Inhibitor + Ismelin.
+ Methamphetamine: see Ismelin + Amphetamines
+ Niamid: Niamid reversed the hypotensive action of Ismelin in hypertensive patients. Mechanism: It is believed that Niamid blocks the uptake of Ismelin into sites of action in nerve endings. (50) See also Ismelin + MAO Inhibitors.
+ Norpramin: see Ismelin + Antidepressants, Tricyclic
+ Pertofrane: see Ismelin + Antidepressants, Tricyclic
+ Phenothiazines: Animal studies have revealed that some phenothiazines (e.g., Thorazine) and certain antihistamines (e.g., Pyribenzamine) block the action of Ismelin. The clinical significance of these antagonisms remains to be determined. (46, 55)
+ Pyribenzamine: see Ismelin + Phenothiazines
+ Rauwolfia Derivatives: Ismelin and rauwolfia derivatives should be used together very cautiously because concurrent use may cause excessive postural hypotension, bradycardia, and depression. (1, 2)
+ Ritalin: see Ismelin + Stimulants
+ Stimulants: Mild stimulants (e.g., ephedrine, Ritalin) may decrease the hypotensive effect of Ismelin. Mechanism: It is postulated that ephedrine and Ritalin act at the same site as Ismelin and blocks the uptake of Ismelin at sympathetic nerve endings and perhaps displaces Ismelin from these sites. (1, 2, 46, 72)
+ Sympathomimetics, Direct-Acting: see Ismelin + Levophed
+ Sympathomimetics, Indirect-Acting: see Ismelin + Wyamine, Aramine, Amphetamines, Stimulants
+ Thorazine: see Ismelin + Phenothiazines
+ Tofranil: see Ismelin + Antidepressants, Tricyclic

ISMELIN (continued)

+ Vasopressor Agemts: Because of the possibility of augmented response to vasopressor agents, they should be administered cautiously. (1)
+ Vivactil: see Ismelin + Antidepressants, Tricyclic
+ Wyamine: Catecholamine depleting agents like Ismelin may render patients refractory to pressor agents that depend upon catecholamine release for their activity (e.g., Wyamine and Aramine). (50)

ISONIAZID: see listed under the following agents:

Demerol
Diphenylhydantoin
Haldrone
Modumate

ISONIAZID + Adrenergic Agents: Adrenergic agents may aggravate the side effects of isoniazid. (1)

ISOPROTERENOL

+ Epinephrine: When isoproterenol is used in conjunction with epinephrine, there is evidence of augmented cardiac arrhythmias and even cardiac standstill. (71)
+ Medihaler-Epi: see Medihaler-Epi + Isoproterenol

ISORDIL

+ Acetylcholine: see Isordil + Norepinephrine
+ Alcohol: An occasional individual exhibits marked sensitivity to the hypotensive effects of nitrite, and severe responses (nausea, vomiting, weakness, restlessness, pallor, perspiration, and collapse) can occur even with the usual therapeutic dose. Alcohol may enhance this effect. (1)
+ Antihypertensive Agents: Potent antihypertensive agents should be given cautiously, since their use with nitrates may produce severe hypotension. (2)
+ Norepinephrine: Isordil can act as a physiological antagonist to norepinephrine, acetylcholine, histamine, and many other agents. (1)

ISUFRANOL: see Isuprel

ISUPREL

+ Digitalis: Administration of Isuprel is contraindicated in patients with tachycardia caused by digitalis intoxication. (1)
+ Epinephrine: Do not administer Isuprel with epinephrine as both drugs are cardiac stimulants and their combined effects may induce serious arrhythmias. They may be alternated if desired, provided a four-hour interval elapses. (1)

ISUPREL (continued)

+ Inderal: see Inderal + Isuprel

J

JECTOFER + Iron Preparations: Serious reactions to Jectofer are more likely to occur in patients who are also taking oral iron preparations. Under these conditions transferrin becomes rapidly saturated and the iron entering the blood cannot be bound to protein. It is this unbound iron fraction that appears to be responsible for the acute toxic symptoms. Hence, Jectofer should not be used concomitantly with oral iron therapy. (1, 2)

KANTREX: see listed under the following agents:

Anectine	Magnesium Sulfate Inj.
Coly-Mycin M Inj.	Staphcillin Sol.
Garamycin	Succinylcholine

KANTREX

+ Anesthesia-Muscle Relaxants: Neuromuscular paralysis with respiratory depression may occur when Kantrex is administered intraperitoneally concomitantly with anesthesia and muscle relaxing drugs. Instillation should therefore be postponed until the patient has recovered fully from the effects of anesthesia and muscle relaxants. (1, 2) Mechanism: Kantrex acts as a competitive blocking agent and also causes a reduction in the release of acetylcholine by the motor-nerve impulse. (131)
+ Antibacterial Agents: Kantrex injection should not be physically mixed with other antibacterial agents but each should be administered separately in accordance with its recommended route of administration and dosage schedule. (1)
+ Dihydrostreptomycin: see Kantrex + Ototoxic Drugs
+ Muscle Relaxants: see Kantrex + Anesthesia-Muscle Relaxants
+ Neomycin: see Kantrex + Ototoxic Drugs
+ Ototoxic Drugs: In the administering of Kantrex, consideration should be given to the possibility of the cumulative ototoxic effects of other ototoxic drugs such as streptomycin, dihydrostreptomycin, and neomycin when these drugs are administered concurrently or in series. (1)
+ Streptomycin: see Kantrex + Ototoxic Drugs

KAYEXALATE + Digitalis: Serious potassium deficiency may occur if dosage range is exceeded. When potassium blood levels are lowered, the action of digitalis, particularly its toxic effects, is likely to be exaggerated. (1)

KEFLIN

+ Acidic Solution: The addition of Keflin to solutions having a pH below 4 is not advised. (1)
+ Alkaline Earth Metals: see Keflin + High Molecular Weight Metals
+ Basic Solution: The addition of Keflin to solutions having a pH above 7 is not advised. (1)

KEFLIN (continued)

+ Benedict's Solution: A false positive reaction for glucose in the urine may occur with Benedict's or Fehling's solution or with Clinitest tablets but not with Tes-Tape. (1)
+ Benemid: Benemid blocks the tubular excretion of Keflin. (88)
+ Clinitest Tablet: see Keflin + Benedict's Solution
+ Fehling's Solution: see Keflin + Benedict's Solution
+ High Molecular Weight Compounds: In general, Keflin is not compatible with compounds of high molecular weight or with alkaline earth metals. (1)

KEMADRIN: see listed under the following agents:

Dopar
Haldol
Laradopa
Levodopa

KENACORT + Diabetes: Kenacort, like other glucocorticoids, may aggravate diabetes so that higher insulin dosage may become necessary, or it may precipitate the manifestations of latent diabetes mellitus. (1)

KENACORT PARENTERAL + Diabetes: Kenacort, like other glucocorticoids, may aggravate diabetes so that higher insulin dosage may become necessary, or it may precipitate the manifestations of latent diabetes mellitus. (1)

KESSO-PEN + Meals: Potassium penicillin G should be given orally, preferably one hour before or two to three hours after meals. In an acidic media penicillin effectiveness is decreased. (1)

KESSO-TETRA

+ Food, Calcium-Containing: see Kesso-Tetra + Milk
+ Meals: Oral forms of tetracycline should be given one hour before or two hours after meals. (1)
+ Milk: Pediatric dosage form (oral) should not be given with milk formulas or other calcium-containing foods, and should be given at least one hour prior to feeding. (1)

KETAJECT

+ Barbiturates: Barbiturates and Ketaject, being chemically incompatible because of precipitate formation, should not be injected from the same syringe. Mechanism: Ketaject has a pH of 3.5 to 5.5, and the barbiturates are strongly alkaline. See also Ketaject + Narcotics. (1)

KETAJECT (continued)

+ Narcotics: Prolonged recovery time may occur if barbiturates and/or narcotics are used concurrently with Ketaject. (1)

KETALAR

+ Barbiturates: Barbiturates and Ketalar, being chemically incompatible because of precipitate formation, should not be injected from the same syringe. Prolonged recovery time may occur if barbiturates are used concurrently with Ketalar. (1) Mechanism: Ketalar has a pH of 3.5 to 5.5, and the barbiturates are strongly alkaline.
+ Narcotics: Prolonged recovery time may occur if narcotics are used concurrently with Ketalar. (1)

KIE TABLETS & SYRUP + Digitalis: Use with caution in patients receiving digitalis due to the ephedrine in the KIE. (1)

KONAKION + Anticoagulants, Coumarin: Excessive doses of Konakion may cause temporary refractoriness to anticoagulants of the coumarin type; therefore the minimum effective dose should be used if continuation of anticoagulant therapy is intended. Konakion will reverse the hypoprothrombinemic effect caused by the coumarin-type anticoagulants. (1)

KYNEX + Trisoralen: see Trisoralen

L

LACTATED RINGER'S INJECTION + Papaverine Hydrochloride Inj.: see Papaverine Hydrochloride Inj. + Lactated Ringer's Inj.

LANOXIN

+ Digitalis: see Lanoxin + Digoxin
+ Digitoxin: see Lanoxin + Digoxin
+ Digoxin: If the patient has been given digoxin during the previous week or any less rapidly excreted drug of the digitalis group (digitoxin, digitalis) during the previous two weeks, the dose of Lanoxin injection must be reduced accordingly. (1)
+ Diuretic Agents: Potassium depletion usually caused by the use of diuretic agents or electrolyte manipulations sensitizes the heart to digitalis intoxication and may produce arrhythmias without actual overdosage of the drug. Under these conditions, it may be necessary to reduce the recommended dosage. (1)

LARODOPA: see Levodopa

LARODOPA

+ Aldomet: see Larodopa + Antihypertensive Agents
+ Anesthetics, General: In the event general anesthesia is required, Larodopa may be discontinued 24 hours before surgery. While to date no untoward reactions from general anesthesia have been encountered, it is suggested that local or regional anesthesia be used wherever possible. Where general anesthesia is unavoidable, cardiorespiratory function should be carefully monitored. (1)
+ Antihypertensive Agents: Careful monitoring of patients receiving antihypertensive agents is required, particularly Aldomet, reserpine, or ganglionic blocking agents. (1)
+ Antiparkson Drugs: Larodopa may be used concomitantly with other antiparkinson drugs such as Cogentin, Artane, or Kemadrin in which case the usual dose of each may need to be reduced. (1)
+ Artane: see Larodopa + Antiparkinson Drugs
+ Cogentin: see Larodopa + Antiparkinson Drugs
+ Diabetes: The control of the diabetic patient may be adversely affected by treatment with Larodopa. Therefore its use in diabetics should be attended with frequent monitoring of the patient and proper adjustment of the antidiabetic regimen. (1)

LARODOPA (continued)
+ Ganglionic Blocking Agents: see Larodopa + Antihypertensive Agents
+ Kemadrin: see Larodopa + Antiparkinson Drugs
+ MAO Inhibitor: Monoamine oxidase inhibitors and Larodopa should not be given concomitantly, and these inhibitors must be discontinued two weeks before initiating therapy with Larodopa. (1)
+ Psychoactive Drugs. If the concomitant administration of psychoactive drugs is deemed necessary, such drugs should be administered with caution and patients should be carefully observed for unusual side effect. (1)
+ Pyridoxine Hydrochloride: It has been reported that pyridoxine hydrochloride (Vitamin B_6) in oral doses of 10 to 25 mg rapidly reverses the antiparkinson effects of levodopa. This should be considered before recommending multivitamin preparations containing pyridoxine hydrochloride. (1)
+ Reserpine: see Larodopa + Antihypertensive Agents

LARGON: see Phenothiazines

LARGON
+ Analgesics: see Largon + CNS Depressants
+ Barbiturates: see Largon + CNS Depressants
+ CNS Depressants: Largon enhances the effects of central nervous system depressants; therefore the dose of barbiturates should be eliminated or reduced by at least one-half in the presence of Largon. The dose of meperidine, morphine, and other analgesic depressants should be reduced by one-quarter to one-half. (1, 2)
+ Epinephrine: The pressor response to epinephrine is usually reduced and may even be reversed in the presence of Largon. If it is necessary to administer a vasopressor agent to patients receiving Largon, Levophed appears to be the most suitable. (1, 2)
+ Meperidine: see Largon + CNS Depressants
+ Morphine: see Largon + CNS Depressants

LASIX
+ Aldactone: Aldactone will enhance the action of Lasix; in addition, this agent tends to prevent excessive excretion of potassium. (37, 89)
+ Antihypertensive Agents: In edematous hypertensive patients being treated with antihypertensive agents, care should be taken to reduce the dose of these drugs when Lasix is administered, since Lasix potentiates the hypotensive effect of antihypertensive medications. (1, 89)

LASIX (continued)

+ Benemid: Although Benemid reduces the excretion of Lasix, it does not interfere with the diuresis. However, Benemid by reducing the excretion of Lasix, will prolong the duration of its action. These observations may be of importance as Lasix may reduce the secretion of uric acid by this route; the hyperuricemia so produced may be treated by the administration of Benemid. (37)
+ Curare and Derivatives: see Lasix + Tubocurarine
+ Diabetes: Although no pronounced effect on carbohydrate metabolism has been demonstrated, periodic checks on urine and blood glucose should be made in diabetic patients receiving Lasix. Occasional increase in blood glucose have occurred during therapy with Lasix. (1, 37, 89)
+ Digitalis: Excessive loss of potassium due to the Lasix in patients receiving digitalis glycosides may precipitate digitalis toxicity. (1, 89)
+ Dyrenium: Dyrenium will enhance the action of Lasix; in addition, this agent tends to prevent excessive excretion of potassium. (37, 89)
+ Levophed: see Lasix + Pressor Amines
+ Ototoxic Drugs: Cases of reversible deafness and tinnitus have been reported following the injection of Lasix. These adverse reactions occurred when Lasix was given at doses exceeding several times the usual therapeutic doses of one to two ampuls (20 to 40 mg). Transient deafness is more likely to occur in patients with severe impairment of renal function and in patients who are also receiving drugs known to be ototoxic. (1)
+ Pressor Amines: Since the Lasix molecule is a sulfonamide type, related to the thiazides, it may decrease arterial responsiveness to pressor amines (e.g., Levophed). Consequently, it is suggested that Lasix be discontinued one week prior to surgery. (1, 89)
+ Salicylates: Patients receiving high doses of salicylates, as in rheumatic disease, in conjunction with Lasix may experience salicylate toxicity at lower doses because of competitive excretory sites. (1)
+ Steroids: Care should be exercised in patients receiving potassium depleting steroids and Lasix as it may cause hypokalemia. (1, 89)
+ Tubocurarine: Sulfonamide diuretics have been reported to enhance the effect of tubocurarine. Great caution should be exercised in administering curare or its derivatives to patients undergoing therapy with Lasix, and it is advisable to discontinue Lasix for one week prior to any elective surgery. (1, 89)

LAXATIVES + Inversine: see Inversine + Laxatives

LAXATIVES, BULK + Ansolysen: see Ansolysen + Laxatives, Bulk

LENETRAN
+ Alcohol: Patients should avoid the concomitant use of Lenetran and alcohol, since the effects may be additive. (2)
+ Monoamine Oxidase Inhibitors: see Lenetran + Psychotropic Agents
+ Phenothiazines: see Lenetran + Psychotropic Agents
+ Psychotropic Agents: Other psychotropic agents, particularly phenothiazines or monoamine oxidase inhibitors, that are known to potentiate the action of other drugs should not be given with Lenetran. (2)

LERITINE
+ Anesthetics: see Leritine + Narcotics
+ Narcotics: Special caution should be observed when Leritine is used with other narcotics, sedatives, or anesthetic agents, since these agents may enhance respiratory depression. (1)
+ Sedatives: see Leritine + Narcotics

LETTER
+ Cytomel: see Cytomel + Letter
+ Diabetes: Letter should be used with caution in diabetic patients. (1)

LEUCOVORIN CALCIUM INJECTION
+ Folic Acid Antagonist: Routine simultaneous therapy is not recommended, since the effect of the folic acid antagonist may be completely nullified. (1)
+ Methotrexate: see Methotrexate + Leucovorin Calcium Injection

LEUKERAN
+ Bone Marrow Depressants: Leukeran is contraindicated for four weeks after a course of treatment with another drug that depresses the bone marrow because of the danger of damaging the bone marrow irreversibly. (2)
+ Cytotoxic Drugs: Cytotoxic drugs render the bone marrow more vulnerable to damage and, therefore, this drug should not be used within four weeks of a full course of chemotherapy. (1)
+ Radiation Therapy: Radiation therapy renders the bone marrow more vulnerable to damage and, therefore, this drug should not be used within four weeks of a full course of radiation therapy. (1, 2)

LEVANIL
+ Monoamine Oxidase Inhibitors: see Levanil + Psychotropic Agents
+ Phenothiazines: see Levanil + Psychotropic Agents
+ Psychotropic Agents: Other psychotropic agents, particularly phenothiazines or monoamine oxidase inhibitors, that are known to potentiate the action of other drugs, should not be given with Levanil. (2)

LEVODOPA: see also Dopar and Larodopa

LEVODOPA (continued)

+ Aldomet: see Levodopa + Antihypertensive Agents
+ Antihypertensive Agents: Careful and frequent monitoring for patients receiving antihypertensive drugs, particularly Aldomet, reserpine, or ganglionic blocking agents. (152)
+ Antiparkinsonism Drugs: L-dopa may be used concomitantly with other antiparkinsonism drugs such as Cogentin, Artane, or Kemadrin, in which case the usual dose of each may need to be reduced. (152)
+ Artane: see Levodopa + Antiparkinsonism Drugs
+ Cogentin: see Levodopa + Antiparkinsonism Drugs
+ Ganglionic Blocking Agents: see Levodopa + Antihypertensive Agents
+ Kemadrin: see Levodopa + Antiparkinsonism Drugs
+ MAO Inhibitor: Monoamine oxidase inhibitors and L-dopa should not be given concomitantly, and these inhibitors must be discontinued two weeks prior to initiating therapy with L-dopa. (152)
+ Psychoactive Drugs: If the concomitant administration of psychoactive drugs is deemed necessary, such drugs should be carefully observed for unusual side effects. It has been reported that L-dopa is not effective in the treatment of drug-induced extrapyramidalism. (152)
+ Pyridoxine Hydrochloride: It has been reported that pyridoxine hydrochloride (vitamin B_6) in oral doses of 10 mg to 25 mg rapidly reverses the antiparkinson effects of L-dopa. This should be considered before recommending multivitamin preparations containing pyridoxine hydrochloride. (152)
+ Reserpine: see Levodopa + Antihypertensive Agents

LEVO-DROMORAN: see Alkaline Agents + Acidifying Agents (Urinary)

LEVOID + Cytomel: see Cytomel + Levoid

LEVOPHED: see listed under the following agents:

Anhydron
Aquatag
Aventyl
Bristuron
Chloroform
Cyclopropane
Diuretics, Thiazide
Diuril
Enduron
Esidrix
Ether
Exna
Fluormar
Fluothane
HydroDIURIL
Inderal
Ismelin
Lasix
Naqua
Naturetin
Penthrane
Phenothiazine
Regitine
Renese

LEVOPHED (continued)

Reserpine
Saluron
Tofranil

LEVOPHED

+ Blood, Whole: Whole blood or plasma, if indicated, should be administered separately (e.g., by use of a Y-tube from separate bottles). (1)
+ Cyclopropane: Cyclopropane and Fluothane anesthetics increase cardiac autonomic irritability and therefore seem to sensitize the myocardium to the action of intravenously administered epinephrine and Levophed. Hence the use of Levophed during cyclopropane and Fluothane anesthesia is generally considered contraindicated because of the risk of producing ventricular tachycardia or fibrillation. (1)
+ Dibenzyline: see Levophed + Regitine
+ Fluothane: see Levophed + Cyclopropane
+ Plasma: see Levophed + Blood, Whole
+ Regitine: Alpha-stimulating agents like Levophed are competitively antagonized by Regitine or Dibenzyline. (50)
+ Saline Solution: Administration of Levophed in saline solution is not recommended. (1)

LEVOPROME: see Phenothiazine

LEVOPROME

+ Acetylsalicylic Acid: see Levoprome + Analgesic Drugs
+ Alcohol: Potentiation of alcohol has been reported following the use of the phenothiazine family of drugs usually during their administration as psychotherapeutic agents. (1)
+ Analgesic Drugs: Levoprome has been demonstrated to exert an additive effect when used with a variety of other analgesic drugs (e.g., morphine, acetylsalicylic acid, meperidine). (1, 90) See also Levoprome + CNS Depressants
+ Anesthetics, General: see Levoprome + CNS Depressants
+ Antihistamines: Potentiation of antihistamines has been reported following the use of the phenothiazine family of drugs usually during their administration as psychotherapeutic agents. (1)
+ Antihypertensive Agents: Levoprome should not be used concurrently with antihypertensive drugs, including monoamine oxidase inhibitors. Concomitant use may cause severe hypotension. (1, 90)
+ Atropine: Levoprome should be used with caution when used concomitantly with atropine, scopolamine, and succinylcholine in that tachycardia and fall in blood pressure may occur and undesirable central

LEVOPROME (continued)

+ Atropine (continued): nervous system effects such as stimulation, delerium, and extrapyramidal symptoms may be aggravated. (1, 90)
+ Barbiturates: see Levoprome + CNS Depressants
+ CNS Depressants: Levoprome exerts additive effects with central nervous system depressants, including narcotics, barbiturates, and general anesthetics. Consequently, the dosage of Levoprome and of each central nervous system depressant drug should be reduced and critically adjusted when used concomitantly or when sequence of use results in overlapping of drug effects. (1, 90)
+ Epinephrine: see Levoprome + Vasopressor Agents
+ MAO Inhibitors: see Levoprome + Antihypertensive Agents
+ Meperidine: see Levoprome + Analgesic Drugs
+ Meprobamate: see Levoprome + Sedative Drugs
+ Morphine: see Levoprome + Analgesic Drugs
+ Narcotics: see Levoprome + CNS Depressants, and Analgesics
+ Phosphorous Insecticides: Potentiation of phosphorous insecticides has been reported following the use of the phenothiazine family of drugs during their administration as psychotherapeutic agents. (1)
+ Reserpine: see Levoprome + Sedative Drugs
+ Scopolamine: see Levoprome + Atropine
+ Sedative Drugs: Levoprome has been demonstrated to exert an additive effect when used with a variety of sedative drugs (e.g., reserpine, meprobamate, barbiturates). (1) See Levoprome + CNS Depressants, and Antihypertensive Agents
+ Succinylcholine: see Levoprome + Atropine
+ Trisoralen: see Trisoralen
+ Vasopressor Agents: If a vasopressor agent is needed to treat hypotension caused by Levoprome, Neosynephrine and Vasoxyl are suitable vasopressor agents; however, epinephrine should not be used, since a paradoxical decrease in blood pressure may result. Levophed should be reserved for hypotension not reversed by other vasopressors. (1, 90)

LEVOTHYROXINE + Cytomel: see Cytomel + Levothyroxine

LIBRIUM: see Benzodiazepines

LIBRIUM

+ Alcohol: see Librium + CNS Depressants
+ Anticoagulants: Although clinical studies have not established a cause and effect relationship, physicians should be aware that variable effects on blood coagulation have been reported very rarely in patients receiving oral anticoagulants and Librium. (1)

LIBRIUM (continued)

+ CNS Depressants: As in the case of other central nervous system acting drugs, patients receiving Librium should be cautioned about possible combined effects with alcohol or other central nervous system depressants, since the effects may be additive. (1, 2)
+ MAO Inhibitors: see Librium + Psychotropic Agents
+ Phenothiazines: see Librium + Psychotropic Agents
+ Psychotropic Agents: In general, the concomitant administration of Librium and other psychotropic agents is not recommended. If such combination therapy seems indicated, careful consideration should be given to the pharmacology of the agents to be employed, particularly when the known potentiating compounds such as monoamine oxidase inhibitors and phenothiazines are to be used. (1, 55)

LIDOSPORIN OTIC SOLUTION + Heat: Patients should be cautioned to avoid heating Lidosporin Otic Solution above body temperature to prevent loss of potency of the antibiotic. (1)

LIGHT, COLD QUARTZ: see listed under the following agents:
Loroxide Lotion
Vanoxide Lotion

LIGHT, ULTRAVIOLET: see listed under the following agents:
Loroxide Lotion
Vanoxide Lotion

LINCOCIN + Meals: For optimal absorption it is recommended that nothing be given by mouth except water for a period of one to two hours before and after oral administration of Lincocin. (1)

LIPAN + Alcohol: Alcoholic beverages should be avoided, since they negate the activity of Lipan. (1)

LIPOMUL I. V.

+ Blood: Lipomul I. V. must not be mixed with blood or any other infusion fluid or drugs or given simultaneously through the same tubing. (1)
+ Drugs: see Lipomul I. V. + Blood
+ Infusion Fluid: see Lipomul I. V. + Blood

LIQUAEMIN SODIUM

+ Antihistamines: see Liquaemin Sodium + Digitalis
+ Digitalis: Certain drugs diminish the clinical effectiveness of heparin; these include digitalis, tetracycline, antihistamines, and nicotine. (1)
+ Nicotine: see Liquaemin Sodium + Digitalis
+ Tetracycline: see Liquaemin Sodium + Digitalis

LIQUAMAR
+ Acetylsalicylic Acid: see Liquamar + Salicylates
+ Blood, Whole: Fresh whole blood or phytonadione (vitamin K_1) will counteract the anticoagulant effect of Liquamar. (1)
+ PAS: see Liquamar + Salicylates
+ Phytonadione: see Liquamar + Blood, Whole
+ Salicylates: Dosage should be carefully controlled when administered to patients who have been receiving large doses of salicylic acid or its derivatives (acetylsalicylic acid, PAS, etc.), since these compounds may affect the prothrombin value and increase the tendency to hemorrhage. (1)

LIRUGEN
+ Alkylating Agents: see Lirugen + Steroids
+ Antimetabolites: see Lirugen + Steroids
+ Blood, Whole: see Lirugen + Immune Globulin, Human
+ Immune Globulin, Human: Defer vaccination if the subject has received human immune globulin, plasma, or whole blood within the preceding six weeks. (1)
+ Irradiation: see Lirugen + Steroids
+ Plasma: see Lirugen + Immune Globulin, Human
+ Poliomyelitis Vaccine: see Lirugen + Vaccines, Live Virus
+ Smallpox Vaccine: see Lirugen + Vaccines, Live Virus
+ Steroids: Live measles vaccine should not be given to patients undergoing treatment with steroids, irradiation, alkylating drugs, or antimetabolites. (1)
+ Vaccines, Live Virus: The effects of administering Lirugen concurrently with other live virus vaccines (poliomyelitis or smallpox) have not been definitely established. It would therefore not be advisable to administer Lirugen during such vaccinations. (1)

LISTICA
+ Alcohol: Patients should avoid the concomitant use of Listica and alcohol, since the effects may be additive. (2)
+ MAO Inhibitors: see Listica + Psychotropic Agents
+ Phenothiazines: see Listica + Psychotropic Agents
+ Psychotropic Agents: Other psychotropic agents, particularly phenothiazines or monoamine oxidase inhibitors, that are known to potentiate the action of other drugs should not be used with Listica. (2)

LITHANE: see Lithium Carbonate

LITHANE + Diuretics: Diuretics should not be used concomitantly with lithium carbonate. Decreased tolerance to lithium has been reported

LITHANE + Diuretics (continued): to ensue from protracted sweating or diarrhea. If such symptoms occur, supplemental fluid and salt should be administered. (1)

LITHIUM CARBONATE + Diuretics: Lithium decreases sodium reabsorption by the renal tubules which could lead to sodium depletion. It is essential, therefore, for the patient to maintain a normal diet, including salt and an adequate fluid intake (2500 to 3000 cc), at least during the initial stabilization period. Decreased tolerance has been reported from protracted sweating or diarrhea. If such symptoms occur, supplemental fluid and salt should be administered. Diuretics should not be used with Lithium therapy. (153)

LITHONATE: see Lithium Carbonate

LITHONATE + Diuretics: Lithium decreases sodium reabsorption by the renal tubules which could lead to sodium depletion. Therefore it is essential for the patient to maintain a normal diet, including salt and an adequate fluid intake (2500 to 3000 cc), at least during the initial stabilization period. Decreased tolerance to lithium has been reported to ensue from protracted sweating or diarrhea. If such symptoms occur, supplemental fluid and salt should be administered. Diuretics should not be used concomitantly with lithium carbonate therapy. (1)

LIXAMINOL + Xanthine Preparations: No other xanthine preparation should be administered concurrently with Lixaminol. (1)

LIXAMINOL AT + Xanthine Preparations: No other xanthine preparation should be administered concurrently with Lixaminol AT. (1)

LOMOTIL
+ Barbiturates: It is recommended that patients receiving a combination of barbiturates and Lomotil be observed closely for the appearance of undesirable manifestations such as barbiturate potentiation. (1)
+ Narcotics: Because of the structural similarity of Lomotil and drugs with a definite addiction potential, Lomotil should be administered with considerable caution to patients who are also receiving such addicting drugs. (1)

LORFAN
+ Anesthetics: see Lorfan + Barbiturates
+ Barbiturates: If used without a narcotic, Lorfan may cause respiratory depression. Lorfan is not effective against respiratory depression due to barbiturates, anesthetics, or other non-narcotic agents. (1, 2)
+ Non-Narcotic Agents: see Lorfan + Barbiturates

LORIDINE

+ Antibiotics: Extemporaneous mixtures with other antibiotics are not recommended. (1)
+ Antibiotics, Nephrotoxic: Because of the nephrotoxic potential of Loridine, caution should be observed when used with other antibiotics having nephrotoxic potential. (1)
+ Anticoagulants: There have been some reports of prolonged prothrombin time after use of Loridine. (91)

LOROPHYN SUPPOSITORIES and JELLY + Douche: Do not douche for at least six hours after use of either Lorophyn Suppositories or Jelly. (1)

LOROXIDE LOTION + Cleansers, Abrasive: Concomitant use of Ultraviolet, Cold Quartz and harsh abrasive cleansers is not recommended. (1)

LUFYLLIN

+ CNS Stimulants: Caution must be exercised when used concomitantly with other central nervous system stimulating drugs. (1)
+ Xanthine Preparations: Caution must be exercised when used concomitantly with other xanthine-containing formulations. (1)

LYTREN POWDER

+ Electrolyte Fluids: see Lytren Powder + Milk
+ Fruit Juice: see Lytren Powder + Milk
+ Milk: Do not mix Lytren with milk, fruit juice, or other electrolyte-containing fluid. (1)

M

MACRODANTIN

+ Drugs Impairing Renal Function: Macrodantin should not be given along with drugs which may produce impaired renal function. Peripheral neuropathy has occurred in patients with impaired renal function. (1)
+ Trisoralen: see Trisoralen

MAGNESIUM SULFATE INJECTION

+ Antibiotics, Curare-Like: The magnesium ion depolarizes the myoneural junction membrane, thereby precluding passage of the transmitter substance to activate muscle fiber. The neuromuscular-blocking type of antibiotics (e.g., neomycin, streptomycin, kanamycin, bacitracin and vancomycin) product an additive effect with the magnesium ion, and the combination must be avoided. Neomycin and streptomycin have the greatest blocking effects, particularly if given intravenously or intraperitoneally. (150)
+ Bacitracin: see Magnesium Sulfate Injection + Antibiotics, Curare-Like
+ Kanamycin: see Magnesium Sulfate Injection + Antibiotics, Curare-Like
+ Neomycin: see Magnesium Sulfate Injection + Antibiotics, Curare-Like
+ Neuromuscular Blocking Agents: The magnesium ion potentiates the effects of neuromuscular blocking agents (e.g., d-Tubocurarine, Syncurine, or Succinylcholine). Mechanism: The magnesium ion decreases the amount of acetylcholine liberated from the motor nerve terminals. It appears that calcium ions are needed for the release of acetylcholine and that magnesium ions interferes with this release. (151)
+ Streptomycin: see Magnesium Sulfate Injection + Antibiotics, Curare-Like
+ Succinylcholine: see Magnesium Sulfate Injection + Neuromuscular Blocking Agents
+ Syncurine: see Magnesium Sulfate Injection + Neuromuscular Blocking Agents
+ Tubocurarine: see Magnesium Sulfate Injection + Neuromuscular Blocking Agents
+ Vancomycin: see Magnesium Sulfate Injection + Antibiotics, Curare-Like

MAGNESIUM SULFATE INJECTION (Eli Lilly)
+ Anesthetics: see Magnesium Sulfate Injection (Eli Lilly) + Barbiturates
+ Barbiturates: When barbiturates, narcotics, or other hypnotics (or systemic anesthetics) are to be given in conjunction with magnesium, their dosage should be adjusted with caution because of the additive central depressive effects of magnesium. (1)
+ Hypnotics: see Magnesium Sulfate Injection (Eli Lilly) + Barbiturates
+ Narcotics: see Magnesium Sulfate Injection (Eli Lilly) + Barbiturates

MALTSUPEX + Diabetes: Precaution: In diabetes mellitus, allow for carbohydrate content in Maltsupex. (1)

MANDALAY + Dolonil: see Dolonil + Mandalay

MANDELAMINE
+ Acidity: An acid urine is essential for antibacterial action with maximum efficacy occurring at pH 5.5 or less. In an acid urine, mandelic acid exerts its antibacterial action and also contributes to the acidification of the urine. The methenamine component in an acid urine is hydrolyzed to ammonia and the bactericidal formaldehyde. (1)
+ Alkalizing Drugs: Since an acid urine is essential for antibacterial activity with maximum efficacy occurring at pH 5.5 or below, restriction of alkalizing foods and medication is thus desirable. (1)
+ Alkalizing Foods: see Mandelamine + Alkalizing Drugs

MAO INHIBITOR: see listed also under individual trademarked preparations. (77)

MAO INHIBITOR: see listed under the following agents:

Adrenergic Agents
Aldomet
Amphaplex
Amphetamine
Antidepressants, Tricyclic
Aramine
Atarax
Aventyl
Benzedrine
Benzodiazepines
Demerol
Desbutal
Desoxyn
Dexedrine
Diabinese
Diuretics
Diuretics, Thiazide
Dopar
Dopram
Drinalfa
Edrisal
Elavil
Emivan
Eutonyl
Furoxone
Gerilets Filmtab
Hypertensin
Inderal

MAO INHIBITOR (continued)

Ismelin
Larodopa
Lenetran
Levanil
Levodopa
Levoprome
Librium
Listica
MAO Inhibitor
Marplan
Mellaril
Meperidine
Norpramin
Obedrin LA
Obetrol
Orinase
Paredrine
Parnate
Pertofrane
Phantos
Phenothiazines
Placidyl
Preludin
Pre-Sate
Prinadol
Quiactin
Ritonic
Sinequan
Softran
Solacen
Span-RD
Striatran
Sublimaze
Tegetrol
Tenuate
Tepanil
Theptine
Thorazine
Tofranil
Tolinase
Trancopal
Trepidone
Trisoralen
Tybatran
Ultran
Valium
Vasopressors
Vi-Dexemin
Vistaril
Vivactil
Wyamine

MAO INHIBITOR: The following agents may be potentiated by concomitant use with MAO inhibitors:

Acetanilid (33)
Aminopyrine (33)
Chloral Hydrate (55, 115)
Codeine
Ether (33); see also Marplan + Ether
Morphine
Muscle Relaxants; see also Niamid + Muscle Relaxants
Procaine (33); see also Nardil + Procaine

MAO INHIBITOR

+ Adrenergic Blockers: see MAO Inhibitor + Vasopressor
+ Aldomet: see MAO Inhibitor + Ismelin
+ Alpha-methyl-m-tyrosine: see MAO Inhibitor + Ismelin

MAO INHIBITOR (continued)

+ Amino Acids: Adverse effects may occur in patients receiving MAO inhibitors after ingestion of amino acids (see list below) that can be decarboxylated to active amines. Reactions after the ingestion of amino acids have been associated with hypertensive crisis. (105)
    - 3, 4-Dihydroxyphenylalanine
    - Histidine
    - 5-Hydroxytryptophan
    - Phenylalanine
    - Tryptophan
    - Tyrosine
+ Anesthetic Agents: Hypotension, hypertensive crisis, hyperthermia, convulsions, coma, and potentiation of atropine, corticosteroids, and Arfonad have all been observed as untoward reactions when anesthesia or anesthetic agents are given to patients on MAO inhibitors. (6, 55, 105, 115)
+ Ansolysen: see MAO Inhibitor + Vasopressor and also Parnate + Ansolysen
+ Antidepressants, Tricyclic: Serious reactions and even fatalities have been reported with the combination of a MAO inhibitor and tricyclic antidepressants (e.g., Aventyl, Elavil, Norpramin, Pertofrane, Tofranil, Vivactil) given in therapeutic doses. Symptoms have ranged from dizziness, nausea, vomiting, and excitation to coma, hyperpyrexia, convulsions, and circulatory collapse. Mechanism: Imipramine-like compounds are known to inhibit the uptake of amines in monoadrenergic nerve endings (monoadrenergic neurons release norepinephrine, dopamine, or serotonin at the nerve terminals) and to sensitize adrenergic and tyraminergic receptors to the action of norepinephrine and serotonin. MAO inhibitors increase the amount of these amines available extraneuronally to receptor sites. (2, 26, 55, 59, 105, 115)
+ Antihypertensive Agents: MAO inhibitors may potentiate the hypotensive effect of antihypertensive agents. (33, 55, 105) See Marplan + Antihypertensive Agents and also Nardil + Antihypertensive Agents.
+ Antiparkinsonism Agents: MAO inhibitors may potentiate the effects of antiparkinsonism agents. (55, 105)
+ Arfonad: see MAO Inhibitor + Anesthetic Agents. (6)
+ Atropine: see MAO Inhibitor + Anesthetic Agents. (6)
+ Aventyl: see MAO Inhibitor + Antidepressants, Tricyclic
+ Caffeine: see MAO Inhibitor + Ritalin and also Eutonyl + Caffeine
+ CNS Depressants: The following classes of drugs may be potentiated by MAO inhibitors and may cause hypotension, coma, and shock: (105)

MAO INHIBITOR (continued)

+ CNS Depressants (continued):
  - Alcohol (2, 33, 55, 105)
  - Analgesics (2)
  - Anesthetic Agents (6, 55, 105, 115); see also MAO Inhibitor + Atropine
  - Antihistamines
  - Barbiturates (2, 33, 105, 115). Mechanism: It is postulated that MAO inhibitors inhibit the hepatic microsomal enzymes which metabolize barbiturates, causing prolonged sedation.
  - Hypnotics
  - Narcotics (33, 105)
  - Sedatives
  - Tranquilizers (55)

+ Cocaine: Concomitant administration may cause hyperexcitation. (33) Mechanism: Cocaine may release the accumulated norepinephrine from peripheral and/or central stores. (115)

+ Coffee: Concomitant administration has been reported to cause hyperexcitability. (105) See also MAO Inhibitor + Ritalin

+ Corticosteroids: see MAO Inhibitor + Anesthetic Agents (6)

+ Diabetes: The hypoglycemic action of insulin may be potentiated or prolonged following treatment with MAO inhibitors. (107)

+ Diuretics: see Parnate + Diuretics

+ Diuretics, Thiazide: Concomitent use may cause an augmented hypotensive effect due to an additive effect. (2, 55, 105, 115) See also Nardil + Diuretics, Thiazide and Niamid + Diuretics, Thiazide.

+ Elavil: see MAO Inhibitor + Antidepressants, Tricyclic

+ Food: The concomitant use of MAO inhibitor drugs with foods that contain tyramine or DOPA (see list below) have been reported to cause severe headache, severe hypertension, cardiac arrhythmias, intracranial bleeding, circulatory bleeding, and death has occurred. Tyramine is a pressor amine which releases norepinephrine from the tissues. The response to tyramine is increased manyfold when a patient is taking MAO inhibitors; this is because the tyramine is not being metabolized and is accumulating in the adrenergic nerves. The naturally occurring norepinephrine, at the same time, is not being destroyed by monoamine oxidase. DOPA is a precursor of the catecholamine dopamine, and is also metabolized by monoamine oxidase. DOPA in food accumulates in the presence of a MAO inhibitor and the same response is obtained as with tyramine accumulation. (2, 6, 33, 55, 77, 105, 115)

MAO INHIBITOR (continued)

+ Food (continued):
  - Bananas (serotonin)
  - Beer (DOPA and tyramine)
  - Broad Beans (DOPA)
  - Canned Figs (tyramine)
  - Cheese (except cream cheese and cottage cheese contains tyramine)
  - Chicken Livers
  - Pickled Herring
  - Wine (tyramine)
  - Yeast Products (tyramine)
  - Yogurt (DOPA)

+ Insulin: see MAO Inhibitor + Diabetes (33, 58, 105, 115)

+ Ismelin: Drugs such as Ismelin, Aldomet, and alpha-methyl-m-tyrosine have both reserpine and tyramine-like actions. The effects of norepinephrine released by these drugs may, therefore, be potentiated after inhibition of monoamine oxidase. Concomitant administration may cause a hypertensive reaction, and central excitation is also possible. Mechanism: see MAO Inhibitor + Reserpine and MAO Inhibitor + Food. (55, 115)

+ Levophed: see MAO Inhibitor + Sympathomimetics, Direct-Acting. (6) See also MAO Inhibitor + Vasopressors.

+ MAO Inhibitor: The comcomitant administration of monoamine oxidase inhibitors may cause agitation, tremor, opisthotonus, coma, and hyperpyrexia due to an additive effect. The amphetamine-like MAO inhibitors (e.g., Parnate, Nardil) if used with a MAO inhibitor may cause hypertensive crisis (mainly systolic hypertension). Clinical picture resembles pheochromocytoma or subarachnoid bleeding because of intracerebral bleeding. Mechanism: There is an additive effect of the MAO inhibitor as well as release of norepinephrine from peripheral and/or central stores. (55, 105, 115)

+ Meperidine: MAO inhibitors potentiate the effects of meperidine causing hypertension or hypotension, respiratory depression, fever, excitation, rigidity, coma, or shock. (2, 115) Mechanism: MAO inhibitors inhibit the hepatic microsomal enzymes which metabolize meperidine and release norepinephrine as well. Dependent upon which action is predominant the patient may suffer any of the above symptoms. (26, 55, 105)

+ Norpramin: see MAO Inhibitor + Antidepressants, Tricyclic

+ Pertofrane: see MAO Inhibitor + Antidepressants, Tricyclic

+ Phenothiazines: The effects of phenothiazines may be potentiated by

MAO INHIBITOR (continued)

+ Phenothiazines (continued): MAO inhibitors causing extrapyramidal symptoms or hypotension. (2, 6, 55, 59, 78, 105) Mechanism: It is postulated that MAO inhibitors inhibit the hepatic microsomal enzymes which metabolize phenothiazines. (115)
+ Phenoxene: see MAO Inhibitor + Antiparkinsonism Agents and Nardil + Phenoxene
+ Psychotherapeutic Agents: Many psychotherapeutic agents are potentiated by MAO inhibitors
+ Psychotropic Agents: see Marplan + Psychotropic Agents
+ Rauwolfia Compounds: see Niamid + Rauwolfia Compounds
+ Regitine: see MAO Inhibitor + Vasopressor and also Parnate + Regitine
+ Reserpine: Concomitant administration of a MAO Inhibitor and reserpine causes a reduction and reversal of the reserpine effect. (59) Mechanism: The major portion of norepinephrine released by reserpine is metabolized intraneuronally by monoamine oxidase and enters the circulation as deaminated inactive metabolites. Following inhibition of monoamine oxidase, a large proportion of the norepinephrine released by reserpine can reach receptor sites in an active form. (115)
  Reserpine-released amines, which are normally metabolized by monoamine oxidase will be present in high concentrations in the brain and will cause central stimulation. Therefore a paradoxical exciting effect of reserpine has been observed in patients treated with MAO inhibitors. (55) See also Eutonyl + Reserpine.
+ Ritalin: Concomitant administration of Ritalin or caffeine may cause central additive stimulation. Mechanism: For drugs passing the blood-brain barrier central excitation is possible because MAO inhibitors will increase the catecholamine concentration in the brain and be additive with the Ritalin or caffeine. (115) See Eutonyl + Ritalin.
+ Stimulants: see Niamid + Stimulants
+ Sulfonylurea, Hypoglycemic: MAO inhibitors may potentiate or prolong the effects of oral hypoglycemic sulfonylureas.
+ Sympathomimetics, Direct-Acting: The pressor effects of Levophed and Vasoxyl which are not metabolized by monoamine oxidase (direct-acting catecholamines are largely destroyed by catechol-O-methyl transferase) were augmented to a degree but were not prolonged. This augmentation occurred only after the development of postural hypotension. Even in patients who have developed orthostatic hypotension after MAO inhibitors, hypertensive reactions may still take place. Moreover, the response to amines which are not

MAO INHIBITOR (continued)

+ Sympathomimetics, Direct-Acting (continued): metabolized by monoamine oxidase can now be augmented due to denervation supersensitivity of adrenergic receptor sites to any directly acting amine. (115)
+ Sympathomimetics, Indirect-Acting: Concomitant use of the following indirect-acting sympathomimetics in a patient receiving a MAO inhibitor is likely to result in excessively high blood pressure, severe headache, cardiac arrhythmias, chest pain, intracranial bleeding, circulatory failure, and death has occurred. Mechanism: Monoamine oxidase within the nerve endings is primarily responsible for the inactivation of norepinephrine before it is released from its storage site. Administration of a MAO inhibitor will increase the level of norepinephrine at its storage site because production of norepinephrine is continued without the usual rate of destruction. Indirect-acting sympathomimetics will release this increased amount of norepinephrine from its storage site (neuron terminal) to act at the receptor sites causing a hypertensive crisis. (2, 6, 33, 55, 77, 115, 117)
    - Amphetamine
    - Anorexiants
    - Cold Remedies
    - Dextroamphetamine
    - Ephedrine
    - Hay Fever Preparations
    - Mephentermine (Wyamine)
    - Metaraminol (Aramine, Pressorol, Pressonex)
    - Methamphetamine
    - Phenylepinephrine (Neosynephrine)
    - Phenylpropanolamine (Propadrine)
    - Tyramine
    - Vasoconstrictors, Nasal
+ Tofranil: see MAO Inhibitor + Antidepressants, Tricyclic
+ Vasopressor: Caution must be used in the administration of vasopressors (e.g., norepinephrine) or adrenergic blockers (e.g., Regitine) in the treatment of hypotensive or hypertensive reactions due to MAO inhibitors. (6) The use of adrenergic blockers, such as Ansolysen or Regitine, may produce an excessive hypotensive effect.
+ Vasoxyl: see MAO Inhibitor + Sympathomimetics, Direct-Acting
+ Vivactil: see MAO Inhibitor + Antidepressants, Tricyclic

MARAX: see Atarax

MARPLAN: see MAO Inhibitor

MARPLAN
+ Alcohol: see Marplan + Sedatives. (See also MAO Inhibitor + Alcohol.)
+ Anorexiants: see Marplan + Proprietary Drugs. Mechanism: see MAO Inhibitor + Sympathomimetics, Indirect-Acting.
+ Antidepressants: see Marplan + MAO Inhibitor. Mechanism: see MAO Inhibitor + Antidepressants, Tricyclic.
+ Aventyl: see Marplan + Tofranil
+ Barbiturates: see Marplan + Sedatives. Mechanism: see MAO Inhibitor + CNS Depressants
+ Cocaine: see Marplan + Sedatives. Mechanism: see MAO Inhibitor + Cocaine.
+ Cold Remedies: see Marplan + Proprietary Drugs. Mechanism: see MAO Inhibitor + Sympathomimetics, Indirect-Acting.
+ Elavil: see Marplan + Tofranil
+ Ether: see Marplan + Sedatives
+ Foods, Tyramine-Containing: Rare instances of hypertension have been reported following the combined use of some monoamine oxidase inhibitors and cheese. Patients on such therapy should be cautioned against the ingestion of foods high in tyramine such as cheese and herring. (1) Mechanism: see MAO Inhibitor + Food.
+ Hay Fever Preparations: see Marplan + Proprietary Drugs. Mechanism: see MAO Inhibitor + Sympathomimetics, Indirect-Acting.
+ Hypotensive Agents: see Marplan + Sedatives
+ Insulin: see Marplan + Sedatives. (See also MAO Inhibitor + Diabetes.)
+ MAO Inhibitor: Concomitant administration of Marplan and other antidepressants is contraindicated; in addition, other amine-oxidase inhibitors or antidepressants should not be used for a period of at least seven days before or seven days after Marplan administration. (1) Mechanism: see MAO Inhibitor + MAO Inhibitor.
+ Meperidine: see Marplan + Sedatives. Mechanism: see MAO Inhibitor + Meperidine.
+ Norpramin: see Marplan + Tofranil
+ Opiates: Marplan in large doses can potentiate the effects of opiates.(2) (See also MAO Inhibitor + CNS Depressants.)
+ Parnate: see Parnate + Marplan
+ Pertofrane: see Marplan + Tofranil
+ Phenylephrine: see Marplan + Sedatives. Mechanism: see MAO Inhibitor + Sympathomimetics, Indirect-Acting.
+ Pressor Agents: see Marplan + Proprietary Drugs. Mechanism: see MAO Inhibitor + Sympathomimetics, Indirect-Acting.
+ Procaine: see Marplan + Sedatives

MARPLAN (continued)

+ Proprietary Drugs: Patients should be warned against taking proprietary drugs that contain pressor agents (e.g., certain cold remedies, hay fever preparations, or anorexiants) or to drink alcoholic beverages, since a hypertensive crisis may occur. (2)
+ Psychotropic Agents: Concomitant use of Marplan and other psychotropic agents should be avoided because of the possibility of potentiation and a lowering of the margin of safety. If such combination therapy seems indicated, careful consideration should be given to the pharmacology of all agents to be employed. The effects of Marplan may persist for a substantial period after discontinuation of the drug, and this should be borne in mind when another drug is prescribed following Marplan. To avoid potentiation, the physician wishing to terminate treatment with Marplan and begin therapy with another agent should do so after gradual reduction of the Marplan dosage. (1, 2)
+ Sedative-Hypnotics: see Marplan + Sedatives
+ Sedatives: Special caution should be exercised when Marplan is used in conjunction with the following classes of drugs: sedatives, hypotensive agents, ether, meperidine, barbiturates, cocaine, procaine, insulin, phenylephrine, alcohol, sedative-hypnotics. Marplan may potentiate these drugs. (1, 2)
+ Sympathomimetic Amines: Large doses of Marplan can potentiate the effects of sympathomimetic amines. (2) <u>Mechanism</u>: see MAO Inhibitor + Sympathomimetics, Direct-Acting and MAO Inhibitor + Sympathomimetics, Indirect-Acting.
+ Tofranil: Tofranil, Pertofrane, Norpramin, Elavil, Aventyl, or Vivactil should not be given with or at least two weeks after discontinuing the use of Marplan. (2) <u>Mechanism</u>: see MAO Inhibitor + Antidepressants, Tricyclic.
+ Trisoralen: see Trisoralen
+ Vivactil: see Marplan + Tofranil

MATULANE: Matulane exhibits some amine-oxidase inhibitory activity. See MAO Inhibitor. (1)

MATULANE

+ Alcohol: Ethyl alcohol should not be used concomitantly with Matulane, since there may be an Antabuse-alcohol type reaction. (1)
+ Antidepressant Drugs: see Matulane + Sympathomimetic Drugs. (See also MAO Inhibitor + Antidepressants, Tricyclic.)
+ Antihistamines: see Matulane + CNS Depressants
+ Barbiturates: see Matulane + CNS Depressants
+ Bone Marrow Depressants: see Matulane + Radiation

MATULANE (continued)

+ CNS Depressants: To minimize central nervous system depression and possible synergism, barbiturates, antihistamines, narcotics, hypotensive agents, or phenothiazines should be used with caution. (1)
+ Elavil: see Matulane + Antidepressant Drugs
+ Foods: see Matulane + Sympathomimetic Drugs. (See also MAO Inhibitor + Foods.)
+ Hypotensive Agents: see Matulane + CNS Depressants
+ Narcotics: see Matulane + CNS Depressants
+ Phenothiazines: see Matulane + CNS Depressants
+ Radiation: If radiation or chemotherapeutic agents known to have bone marrow depressant activity has been used, an interval of one month or longer without such therapy is recommended before starting treatment with Matulane. The length of this interval may also be determined by evidence of bone marrow recovery based on successive bone marrow studies. (1)
+ Sympathomimetic Drugs: Because Matulane exhibits some monoamine oxidase inhibitory activity, sympathomimetic drugs, antidepressant drugs (e.g., Elavil, Tofranil), and other drugs and foods with known high tyramine content, such as ripe cheese and bananas, should be avoided. (1) (See MAO Inhibitor + Sympathomimetics, Direct-Acting and MAO Inhibitors + Sympathomimetics, Indirect-Acting.)
+ Tofranil: see Matulane + Antidepressant Drugs
+ Trisoralen: see Trisoralen

MAXIPEN + Food: Absorption of Maxipen is greater if the drug is given when the stomach is empty. (2)

MEALS: see also Food

MEALS: see listed under the following agents:

Dynapen
Inderal
Ircon
Kesso-Pen
Kesso-Tetra
Lincocin
Mysteclin F
Ostensin
Panmycin
Pathocil
Pentids
Principen
Retet
Rondomycin
Signemycin
Sugracillin
Sumycin
Tegopen
Terramycin
Tetrachel
Tetracyn
Tetrex
Toleron
Unipen
Urobiotic
Veracillin

MEASLES IMMUNE GLOBULIN: see listed under the following agents:
Measles Virus Vaccine, Live, Attenuated (Edmonston B)
Measles Virus Vaccine, Live, Attenuated (Schwarz)

MEASLES VACCINE + Sterneedle: see Sterneedle + Measles Vaccine

MEASLES VIRUS VACCINE, INACTIVATED + Adrenocorticosteroids: see Adrenocorticosteroids + Measles Virus Vaccine, Inactivated

MEASLES VIRUS VACCINE, INACTIVATED, ALUMINUM PHOSPHATE ADSORBED

+ ACTH: Adrenocorticotrophin (ACTH) and adrenal corticosteroids may suppress the antibody response to the vaccine. Therefore, if possible, it would seem advisable to avoid administration of the vaccine concomitantly with these hormones. (1)

+ Adrenal Corticosteroids: see Measles Virus Vaccine, Inactivated, Aluminum Phosphate Adsorbed + ACTH

MEASLES VIRUS VACCINE, LIVE, ATTENUATED: see listed under the following agents:
Adrenocorticosteroids
Gamastan
Gammagee
Gamulin
Immu-G
Immune Serum Globulin (Human)
Orimune Poliovirus Vaccine, Live, Oral, Trivalent
Poliovirus Vaccine, Live, Oral, Trivalent (Pfizer)

MEASLES VIRUS VACCINE, LIVE, ATTENUATED

+ Alkylating Agents: see Measles Virus Vaccine, Live, Attenuated + Steroids

+ Antimetabolites: see Measles Virus Vaccine, Live, Attenuated + Steroids

+ Blood, Whole: see Measles Virus Vaccine, Live, Attenuated + Plasma

+ Irradiation: see Measles Virus Vaccine, Live, Attenuated + Steroids

+ Measles Immune Globulin: see Measles Virus Vaccine, Live, Attenuated + Plasma

+ Plasma: Vaccination should be deferred if the child has received plasma or whole blood or if more than 0.02 ml of measles immune globulin per kilogram (0.01 ml/lb.) of body weight has been administered during the preceding six weeks, since these substances may block antibody formation. (92)

+ Steroids: This vaccine is contraindicated in patients receiving agents

MEASLES VIRUS VACCINE, LIVE, ATTENUATED (continued)

+ Steroids (continued): that depress resistance (e.g., steroids, irradiation, alkylating agents, and antimetabolites). (92)

MEASLES VIRUS VACCINE, LIVE, ATTENUATED (Edmonston B)

+ Alkylating Agents: see Measles Virus Vaccine, Live, Attenuated (Edmonston B) + Steroids
+ Antimetabolites: see Measles Virus Vaccine, Live, Attenuated (Edmonston B) + Steroids
+ Irradiation: see Measles Virus Vaccine, Live, Attenuated (Edmonston B) + Steroids
+ Measles Immune Globulin: Previously administered measles immune globulin may block measles antibody formation, therefore the use of live measles vaccine should be deferred if more than 0.02 ml of measles immune globulin per kilogram of body weight has been administered within the preceding six weeks. (2)
+ Steroids: Measles virus vaccine, live, attenuated (Edmonston B) is contraindicated in patients being treated with agents that depress resistance (e.g., steroids, irradiation, alkylating agents, and antimetabolites). (2)

MEASLES VIRUS VACCINE, LIVE, ATTENUATED (Eli Lilly)

+ Alkylating Agents: see Measles Virus Vaccine, Live, Attenuated (Eli Lilly) + Steroids
+ Antimetabolites: see Measles Virus Vaccine, Live, Attenuated (Eli Lilly) + Steroids
+ Antiseptics: see Measles Virus Vaccine, Live, Attenuated (Eli Lilly) + Detergents
+ Blood Transfusion: A recent blood transfusion, because of the probable immune globulin content, may temporarily inhibit the response to vaccination. (1)
+ Detergents: Because detergents, preservatives, and antiseptics will inactivate live measles virus vaccine, only the syringe with sterile diluent supplied with the vaccine should be used for reconstitution and administration. (1)
+ Gamma Globulin: If gamma globulin has been administered within the preceding six weeks, immunization with live measles vaccine should be deferred, since the globulin may inhibit a vaccine take. (1)
+ Irradiation: see Measles Virus Vaccine, Live, Attenuated (Eli Lilly) + Steroids
+ Poliomyelitis Vaccine: see Measles Virus Vaccine, Live, Attenuated (Eli Lilly) + Vaccines, Live

MEASLES VIRUS VACCINE, LIVE, ATTENUATED (Eli Lilly) (continued):

+ Preservatives: see Measles Virus Vaccine, Live, Attenuated (Eli Lilly) + Detergents
+ Smallpox Vaccine: see Measles Virus Vaccine, Live, Attenuated (Eli Lilly) + Vaccines, Live
+ Steroids: This vaccine should not be administered to individuals receiving therapeutic agents which depress resistance such as steroids, alkylating agents, antimetabolites, and irradiation. (1)
+ Vaccines, Live: Reports concerning the concomitant administration with other live vaccines, such as poliomyelitis and smallpox vaccines, have not appeared. Until the safety of dual administration is demonstrated, it is probably preferable not to give live vaccines simultaneously. (1)

MEASLES VIRUS VACCINE, LIVE, ATTENUATED (Philips Roxane)

+ Alkylating Agents: see Measles Virus Vaccine, Live, Attenuated (Philips Roxane) + Steroids
+ Antimetabolites: see Measles Virus Vaccine, Live, Attenuated (Philips Roxane) + Steroids
+ Blood, Whole: Because of the possible presence of measles antibody in whole blood, vaccination within six weeks following transfusion should be avoided. (1)
+ Gamma Globulin: see Measles Virus Vaccine, Live, Attenuated (Philips Roxane) + Steroids
+ Irradiation: see Measles Virus Vaccine, Live, Attenuated (Philips Roxane) + Steroids
+ Steroids: Based on the recommendations of the Ad Hoc Advisory Committee on Measles Control, the vaccine is contraindicated in those patients who receive therapy which suppresses resistance, such as steroids, irradiation, alkylating agents, and antimetabolites; gamma globulin (within the preceding six weeks). (1)

MEASLES VIRUS VACCINE, LIVE, ATTENUATED (Schwarz)

+ Alkylating Agents: see Measles Virus Vaccine, Live, Attenuated (Schwarz) + Steroids
+ Antimetabolites: see Measles Virus Vaccine, Live, Attenuated (Schwarz) + Steroids
+ Blood, Whole: see Measles Virus Vaccine, Live, Attenuated (Schwarz) + Measles Immune Globulin
+ Irradiation: see Measles Virus Vaccine, Live, Attenuated (Schwarz) + Steroids
+ Measles Immune Globulin: Vaccination should be deferred if the child has received more than 0.02 ml of measles immune globulin per

MEASLES VIRUS VACCINE, LIVE, ATTENUATED (Schwarz) (continued)
+ Measles Immune Globulin (continued): kilogram (0.01 ml/lb) of body weight or if whole blood or plasma has been administered during the preceding six weeks, since these substances may block antibody formation. (2)
+ Plasma: see Measles Virus Vaccine, Live, Attenuated (Schwarz) + Measles Immune Globulin
+ Steroids: This vaccine is contraindicated in patients being treated with agents that depress resistance (e.g., steroids, irradiation, alkylating agents, and antimetabolites). (2)

MECOSTRIN: see Tubocurarine

MECOSTRIN
+ Ether: Since ether appreciably potentiates the action of curare, with ether anesthesia the dose of Mecostrin should be about one-third of that normally used with other anesthetic agents. (1, 126)
+ Neostigmine: In marked curare overdosage, Neostigmine may be dangerous, since its vasodepressant action may act synergistically with curare and intensify central depression of vital centers. (1) Neostigmine antagonizes only the skeletal muscular blocking action of the competitive blocking agents and may aggravate such side effects as hypotension or bronchospasm. (131)

MEDIGESIC + Barbiturates: Sedation and respiratory depression may be marked if Medigesic is administered in conjunction with a barbiturate. (1)

MEDIHALER-EPI + Isoproterenol: Isoproterenol and epinephrine may be used interchangeably if the patient becomes unresponsive to one or the other but should not be used concurrently; an interval of four hours should elapse before changing from one to the other compound. (1)

MEDIHALER-ISO + Epinephrine: Isoproterenol and epinephrine may be used interchangeably if the patient becomes unresponsive to one or the other but should not be used concurrently; an interval of four hours should elapse before changing from one to the other compound. (1)

MEDOMIN
+ CNS Depressants: Reduced dosage may be necessary in patients using other CNS depressants or tranquilizers. (1)
+ Tranquilizers: see Medomin + CNS Depressants

MEDROL
+ Diabetes: Glucocorticoid steroids may aggravate diabetes mellitus so that higher insulin dosage may become necessary or manifestations

MEDROL (continued)

+ Diabetes (continued): of latent diabetes mellitus may be precipitated. (1)
+ Diuretics, Carbonic Anhydrase-Inhibiting: see Medrol + Diuretics, Thiazide
+ Diuretics, Thiazide: Medrol has been observed to be effective not only in potentiating the actions of mercurial and carbonic anhydrase inhibiting diuretic agents but also in restoring response to diuretics in patients with resistant cardiac edema. (1)

DEPO-MEDROL + Solutions: Depo-Medrol should not be diluted or mixed with other solutions. (1)

MEGIMIDE + Placidyl: see Placidyl + Megimide

MELLARIL: see Phenothiazines

MELLARIL

+ Alcohol: see Mellaril + CNS Depressants
+ Anesthetics: see Mellaril + CNS Depressants
+ Antihypertensive Agents: Mellaril may potentiate the action of antihypertensive agents, and care should be exercised when used concomitantly. (2)
+ Atropine: Phenothiazines are capable of potentiating atropine. (1)
+ CNS Depressants: Phenothiazines are capable of potentiating central nervous system depressants (e.g., anesthetics, opiates, alcohol, etc.). (1)
+ Epinephrine: The administration of epinephrine should be avoided in the treatment of drug-induced hypotension in view of the fact that phenothiazines may induce a reversed epinephrine effect on occasions. Should a vasoconstrictor be required, the most suitable are Levophed and Neosynephrine. (1)
+ MAO Inhibitor: In common with other phenothiazines, Mellaril is contraindicated in association with or following recent use of MAO inhibitors. (1) Mechanism: see MAO Inhibitor + Phenothiazines.
+ Opiates: see Mellaril + CNS Depressants
+ Phosphorous Insecticides: In common with other phenothiazines, Mellaril is capable of potentiating phosphorous insecticides. (1)
+ Trisoralen: see Trisoralen

MEPERGAN: see Meperidine and Phenergan

MEPERGAN + Barbiturates: Barbiturates are not chemically compatible in solution with Mepergan and should not be mixed in the same syringe. (1)

MEPERIDINE: see Alkaline Agents + Acidic Media

MEPERIDINE: see Alkaline Agents + Acidifying Agents (Urinary)

MEPERIDINE: see also Demerol

MEPERIDINE: see listed under the following agents:

Ammonium Chloride	Marplan
Atarax	Nalline
Eutonyl	Nardil
Largon	Phenergan
Levoprome	Phenothiazines
MAO Inhibitor	Sparine

MEPERIDINE

+ Amphetamines: It has been reported that amphetamines enhance the analgesic effect of meperidine. (106)
+ Antidepressants, Tricyclic: Meperidine-induced respiratory depression may also be enhanced by tricyclic antidepressants of the Tofranil type. (106)
+ Atropine: There may be an additive effect if used concomitantly. (55)
+ MAO Inhibitor: MAO inhibitors potentiate the effects of meperidine causing hypertension or hypotension, respiratory depression, fever, excitation, rigidity, coma, or shock. (2, 55, 106, 115) Mechanism: see MAO Inhibitor + Meperidine.
+ Phenothiazines: Concomitant administration may cause enhanced sedation. (55) If therapeutic doses of meperidine are injected concurrently with phenothiazines, there is marked exaggeration of the respiratory depressant effects; maximal respiratory depression does not occur until 1-1/2 hours and persists close to a maximum of 3-1/2 hours. This is in contrast to maximal depression at one hour with approximately 50% recovery at the end of 3-1/2 hours when meperidine is given alone. (106)

MEPHYTON: see listed under the following agents:

Liquamar
Miradon

MEPHYTON + Anticoagulants: When vitamin K_1 is used to correct excessive anticoagulant-induced hypoprothrombinemia, anticoagulant therapy still being indicated, the patient is again faced with the clotting hazards existing prior to starting anticoagulant therapy. Mephyton is not a clotting agent, but overzealous therapy with vitamin K_1 may restore conditions which originally permitted thromboembolic phenomena. Dosage of both the oral and injectable preparation, therefore, should be kept as

MEPHYTON + Anticoagulants (continued): low as possible, and prothrombin time should be checked regularly as clinical conditions indicate. Temporary resistance to prothrombin-depressing anticoagulants may result, especially when larger doses of Mephyton are used. If relatively large doses have been employed, it may be necessary when reinstituting anticoagulant therapy to use somewhat larger doses of the prothrombin-depressing anticoagulant or to use one which acts on a different principle such as heparin sodium. (1)

MEPROBAMATE: see listed under the following agents:
Alcohol
Anticoagulants
Levoprome
Ostensin
Panwarfin

MEPROBAMATE + Alcohol: Special care should be taken to warn patients taking meprobamate that their tolerance to alcohol may be lowered with resultant slowing of reaction time and impairment of judgment and coordination. (1)

MERCURIALS: see listed under the following agents:
Aquacort Suprettes
Exna
Trinalis Suprettes
Vlem-Dome Liquid Concentrate

MERCURIC OINTMENTS + Selsun: see Selsun + Mercuric Ointments

MERCURIALS, PARENTERAL: see listed under the following agents:
Esidrix
HydroDIURIL

MERTHIOLATE
+ Acids: Merthiolate is incompatible with strong acids, salts of heavy metals, and iodine and should not be used in combination with or immediately following their application. (1)
+ Blood, Whole: Merthiolate, like other antiseptics, is antagonized by whole blood. (1)
+ Heavy Metal Salts: see Merthiolate + Acids
+ Iodine: see Merthiolate + Acids

MERUVAX
+ Alkylating Agents: see Meruvax + Corticosteroids
+ Antimetabolites: see Meruvax + Corticosteroids

MERUVAX (continued)

+ Blood Transfusion: Vaccination with Meruvax should be deferred for at least six weeks following blood transfusions or administration of more than 0.02 cc of human immune serum globulin per pound of body weight. (1)
+ Corticosteroids: Meruvax is contraindicated in patients with malignancies or those receiving therapy with corticosteroids, irradiation, alkylating agents, or antimetabolites because of their alteration of normal defense mechanisms. (1)
+ Immune Serum Globulin, Human: see Meruvax + Blood Transfusions
+ Immunization, Elective: Meruvax should be given at least one month before or after but not at the same time as other elective immunizations. (1)
+ Irradiation: see Meruvax + Corticosteroids

MESTINON: see listed under the following agents:

Atropine
Phospholine Iodide

MESTINON + Atropine: Atropine may be used to abolish or obtund gastrointestinal side effects or other muscarinic reactions of Mestinon, but such use, by masking signs of overdosage, can lead to inadvertent induction of cholinergic crisis (e.g., a state characterized by increasing muscle weakness which, through involvement of the muscles of respiration, may lead to death). (1)

METABOLIC DISEASE + Oracon: see Oracon + Metabolic Disease

METAHYDRIN: see Diuretics, Thiazide

METAHYDRIN

+ Antihypertensive Agents: Metahydrin is known to potentiate most of the nondiuretic hypotensors. Addition of Metahydrin to an established regimen based on a nondiuretic hypotensor should be initiated with caution to avoid precipitous decreases in blood pressure. Dosage of the nondiuretic agent should be reduced as Metahydrin is added, and an optimal dosage level for each agent in the combined therapy should be developed gradually. This same principle holds true when antihypertensive therapy has been initiated with Metahydrin, and it is subsequently deemed advisable to introduce the combined action of a nondiuretic hypotensor. (1, 2)
+ Apresoline: see Metahydrin + Rauwolfia Alkaloids
+ Diabetes: Thiazide drugs, including Metahydrin, may precipitate or aggravate diabetes mellitus. Patients predisposed by heredity to this metabolic disease should be afforded particularly close obser-

METAHYDRIN (continued)

+ Diabetes (continued): vation when being treated with Metahydrin or other thiazide diuretics. (1, 2)
+ Digitalis: Patients taking digitalis are in extra jeopardy in regard to potassium depletion which may be caused by metahydrin. (1, 2)
+ Ganglionic Blocking Agents: see Metahydrin + Rauwolfia Alkaloids
+ Levophed: see Metahydrin + Pressor Amines
+ Pressor Amines: Recent reports indicate that the thiazides may decrease arterial responsiveness to pressor amines (e.g., Levophed). (1, 2)
+ Rauwolfia Alkaloids: Metahydrin therapy potentiates the action of rauwolfia alkaloids, ganglionic blocking agents, Apresoline, and veratrum alkaloids given simultaneously. (1, 2) See also Metahydrin + Antihypertensive Agents.
+ Steroids: Patients on steroid therapy are in extra jeopardy in regard to potassium depletion which may be caused by concomitant usage. (1, 2)
+ Trisoralen: see Trisoralen
+ Tubocurarine: Thiazide drugs, possibly causing a relative intracellular depletion of potassium, tend to potentiate the action of tubocurarine. This should be kept in mind if curare is to be given to a patient who has received drugs containing thiazides. (1, 2)
+ Veratrum Alkaloids: see Metahydrin + Rauwolfia Alkaloids

METALLIC COMPOUNDS + Selsun: see Selsun + Metallic Compounds

METALS: see listed under the following agents:
Mucomyst
Xylocaine

METAL SALTS, HEAVY: see listed under the following agents:
Merthiolate
Thrombin, Topical

METATENSIN: see Metahydrin and Reserpine

METERMATIC VAPO-N-ISO + Epinephrine: Isoproterenol and epinephrine may be used interchangeably if the patient becomes unresponsive to one or the other but should not be used together. (1)

METHADON

+ Barbiturates: see Methadon + Tranquilizers
+ Mercurial Diuretics: Since potent analgesics may decrease urinary output, it may interfere with the action of mercurial diuretics. (1)
+ Tranquilizers: Caution must be observed and doses decreased when administering Methadon to patients receiving tranquilizers, barbiturates, or other agents that depress respiration. (1)

METHAMPHETAMINE: see trademarked name preparations

METHAMPHETAMINE: see listed under the following agents:
Aldomet
Aventyl
MAO Inhibitor
Reserpine

METHEDRINE: see Methamphetamine

METHEDRINE
+ Cyclopropane: Methedrine appears to have a preferential stimulating effect on supraventricular regions of the heart. In the cyclopropane-sensitized heart it gives no evidence of ventricular stimulation but it does increase supraventricular activity and thus may cause sino-auricular tachycardia. (1) See also Cyclopropane + Methedrine.
+ Ergot Alkaloids, Parenteral: Caution should be observed when used closely following the parenteral administration of ergot alkaloids and/or pressor fraction of posterior pituitary extract to prevent excessive rise in blood pressure. (1)
+ Posterior Pituitary Extract, Pressor Fraction: see Methedrine + Ergot Alkaloids, Parenteral

METHENAMINE: see listed under the following agents:
Sulfathiazole
Sulfonamides
Thiosulfil

METHIUM + Fluothane: see Fluothane + Methium

METHOTREXATE
+ p-Aminobenzoic Acid: see Methotrexate + Sulfonamides
+ Hematopoietic Depressants: Avoid concomitant therapy with other agents known to have a depressing effect on the hematopoietic system. (1)
+ Leucovorin: Intramuscularly-administered Leucovorin has been used with intra-arterial perfusion of Methotrexate, in certain instances, in an attempt to diminish undesirable side effects. This technique presents a special problem in that Leucovorin may, in some cases, interfere with the desired therapeutic effects of Methotrexate. (1)
+ Radiation Therapy: Combined Methotrexate and radiation may exert a depressing effect upon the bone marrow. (1)
+ Salicylates: see Methotrexate + Sulfonamides
+ Salicylic Acid: see Methotrexate + Sulfonamides
+ Sulfonamides: Methotrexate is bound in part to plasma albumin, and

METHOTREXATE (continued)

+ Sulfonamides (continued): studies suggest that it is displaced by a number of drugs, including sulfonamides and salicylic acid. Since Methotrexate is often used in doses that are nearly toxic, sulfonamides, whether antibacterial, hypoglycemic, or diuretic, p-aminobenzoic acid, and salicylates should be used with caution; when given concurrently with Methotrexate, pancytopenia has occurred. (1, 55, 79, 118)
+ Sulfonamides, Antibacterial: see Methotrexate + Sulfonamides
+ Sulfonamides, Diuretic: see Methotrexate + Sulfonamides
+ Sulfonamides, Hypoglycemic: see Methotrexate + Sulfonamides

METHOXYFLURANE + Flaxedil: see Flaxedil + Methoxyflurane

METHYLPARABEN: see listed under the following agents:
Aristocort Forte Suspension
Furadantin Sodium Sterile

METHYL-PHENYL HYDANTOIN (Nuvarone): see listed under the following agents:
Paradione
Tridione

METHYLTHIOURACIL: see listed under the following agents:
Anticoagulants
Dicumarol
Panwarfin

METICORTELONE + Diabetes: The physician must weigh possible undesirable effects against anticipated clinical improvement in patients with diabetes mellitus. Although hyperglycemia, glycosuria, and increased insulin requirements usually do not occur in the controlled diabetic patient, close observation should be maintained during therapy. (1)

METICORTEN + Diabetes: Although hyperglycemia, glycosuria, and increased insulin requirements usually do not occur with Meticorten in the controlled diabetic patient, close observation should be maintained during therapy. (1)

METOPIRONE

+ Corticosteroids: All corticosteroid therapy must be discontinued prior to and during Metopirone testing. (1)
+ Meprobamate: see Metopirone + Rela
+ Niamid: see Metopirone + Rela
+ Rela: Several drugs (e.g., Rela, Soma, Thorazine, Meprobamate, Niamid) cause subnormal response to the metyrapone test, possibly

METOPIRONE (continued)

+ Rela (continued): by decreasing the secretion of corticotrophin by the pituitary gland. Moreover, certain drugs (e.g., Thorazine) may interfere with the chemical determination of urinary steroids, and the test procedure should be carried out without any medications whenever possible. (2)
+ Soma: see Metopirone + Rela
+ Thorazine: see Metopirone + Rela

METRAZOL: see listed under the following agents:

Compazine	Stelazine
Phenothiazine	Thorazine
Quide	Tindal

METUBINE IODIDE: see also Tubocurarine

METUBINE IODIDE

+ Barbiturate Solution: Because of the high pH of barbiturate solutions, a precipitate will form when Metubine Iodide is combined with such agents as Brevital Sodium and Pentothal Sodium. Although it is possible to mix solutions of curare and barbiturates for concomitant administration from the same syringe, it is recommended that each compound be given from a separate syringe to assure more uniform and predictable results with each of the drugs. (1)
+ Brevital Sodium: see Metubine Iodide + Brevital Sodium. (See also Brevital Sodium + Metubine Iodide.)
+ Neostigmine: In the event of overdosage of curare, neostigmine may be of some benefit. However, the use of neostigmine may be dangerous if an excessive amount of curare has been given, since excessive curare shock and neostigmine both tend to lower blood pressure and to produce symptoms of shock. The two together may exert a dangerous effect. (1)
+ Pentothal Sodium: see Metubine Iodide + Pentothal Sodium
+ Tensilon: see Tensilon + Metubine Iodide

METYCAINE HYDROCHLORIDE INJ.

+ Epinephrine: Addition of epinephrine to the Metycaine Hydrochloride solution in concentrations of 1:300,000 to 1:500,000 may be expected to potentiate anesthesia somewhat. (1)
+ Sulfates: Since sulfates are incompatible with the calcium in Ringer's Injection, vasoconstrictor drugs should be added to this ampoule only in the form of hydrochlorides. (1)
+ Vasoconstrictors: see Metycaine Hydrochloride Inj. + Sulfates

MIDICEL: see Sulfonamides, Long-Lasting

MIDICEL: see Trisoralen

MILK: see Listed under the following agents:

Achromycin
Aureomycin
Declomycin
Dulcolax
Feosol Elixir
Kesso-Tetra
Lytren Powder
Mysteclin F
Panmycin
Retet
Rondomycin
Signemycin
Steclin
Sumycin
Terramycin
Tetrachel
Tetracyn
Tetramax
Tetrex

MILPREM: see Miltown

MILTOWN: see also Meprobamate

MILTOWN + Alcohol: Investigators differ concerning effects of combining meprobamate with alcohol. It is possible that effects of excessive alcohol consumption may be increased by Miltown. Therefore appropriate caution should be exercised with patients prone to excessive drinking. (1)

MILTRATE: see Miltown

MINERALOCORTICOSTEROIDS + Tubocurarine: see Tubocurarine + Mineralocorticosteroids

MINERAL OIL: see listed under the following agents:

Coumadin
Sintrom
Sulfasuxidine
Sulfathalidine
Tromexan
Vacuetts Suppositories

MINTEZOL + Diabetes: Mintezol has caused hyperglycemia on occasion. (1)

MIO-PRESSIN: see Rauwolfia Alkaloids, Veratrum Preparations, and Dibenzyline

MIO-PRESSIN

+ Anesthesia: If possible, Mio-Pressin should be discontinued for two weeks prior to elective surgery, since certain anesthetic agents potentiate the hypotensive action of rauwolfia. (1)

+ Diuretics: When Mio-Pressin and diuretics with hypotensive properties are used concomitantly, the dosage of Mio-Pressin may be reduced

MIO-PRESSIN (continued)

+ Diuretics (continued): according to the patient's response. (1)

MIOTICS: see listed under the following agents:

Diamox
Ethamide

MIRADON

+ Vitamin K_1: see Miradon + Whole Blood
+ Whole Blood: Fresh whole blood or Vitamin K_1 (phytonadione) will counteract the activity of this anticoagulant. (2)

MITHRACIN

+ Anticoagulants: Mithracin is contraindicated in patients with thrombocytopenia, thrombocytopathy, coagulation disorder, or an increased susceptibility to bleeding due to other causes. Mithracin has been reported to cause a depression of prothrombin content and elevation of clotting time and bleeding time. (1)
+ Bone Marrow Depressants: Mithracin should not be administered to any patient with impairment of bone marrow function. (1)

MODERIL: see Rauwolfia Alkaloids

MODERIL

+ Anesthetics: To avoid hypotension in patients who are given anesthetics during treatment with a rauwolfia alkaloid, the dosage should be reduced or the drug discontinued two weeks prior to surgery. (2)
+ Anticonvulsants: Moderil should be used with caution in patients with epilepsy, and the dosage of anticonvulsant drugs may have to be adjusted. (2)

MODUMATE

+ Barbiturates: see Modumate + Narcotics
+ Diuretics: see Modumate + Narcotics
+ Ion Exchange Resins: see Modumate + Narcotics
+ Isoniazid: see Modumate + Narcotics
+ Narcotics: Other drugs should not be given concomitantly with Modumate if their use can be avoided because a wide variety of compounds, including narcotics, barbiturates, isoniazid, ion exchange resins, and diuretics (chlorothiazide, Diamox, Aldactone), may produce ammonia or interfere with its catabolism or excretion. (2)

MONOAMINE OXIDASE INHIBITOR: see MAO Inhibitor

MORPHINE: see listed under the following agents:

- Dilantin
- Eskay's Neuro Phosphate
- Largon
- Levoprome
- MAO Inhibitors
- Organophosphates (Poisoning)
- Parnate
- Phenergan
- Phenothiazines
- Protopam
- Quelicin Chloride Inj.
- Succinylcholine
- Unitensin

MOTILYN

+ Parasympathomimetic Drugs: Theoretical considerations suggest that if parasympathomimetic drugs have been used Motilyn should not be administered for 12 hours after these agents have been discontinued. (2) Mechanism: Dexpanthenol is converted to pantothenic acid and thus claimed to produce more acetylcholine thus there may be an additive effect.
+ Succinylcholine: Respiratory embarrassment was observed in one patient when Motilyn was administered shortly after cessation of succinylcholine therapy; thus it may be advisable to wait one hour after discontinuing succinylcholine before using Motilyn. (2) Mechanism: Motilyn is converted to pantothenic acid and thus claimed to produce more acetylcholine. Acetylcholine acts as a depolarizing agent at the motor end plates augmenting the depolarizing action of succinylcholine.

MUCOMYST

+ Antibiotics: Mucomyst should not be mixed with all antibiotics. For example, the antibiotics, tetracycline hydrochloride, erythromycin lactobionate, oleandomycin phosphate were found to be incompatible when mixed in the same solution. These agents may be nebulized from separate solutions if administration of these agents is desirable. (1)
+ Erythromycin Lactobionate: see Mucomyst + Antibiotics
+ Metals: Certain materials used in nebulization equipment react with Mucomyst. The most reactive of these are certain metals (notably iron and copper) and rubber. These are included in the component parts of much of the nebulization equipment presently available. (1)
+ Oleandomycin Phosphate: see Mucomyst + Antibiotics
+ Rubber: see Mucomyst + Metals
+ Tetracycline Hydrochloride: see Mucomyst + Antibiotics

MUMPS VACCINE: see also Mumpsvax Lyovac

MUMPS VACCINE + Adrenocorticosteroids: see Adrenocorticosteroids + Mumps Vaccine

MUMPSVAX LYOVAC
+ Alkylating Agents: see Mumpsvax Lyovac + Corticosteroids
+ Antimetabolites: see Mumpsvax Lyovac + Corticosteroids
+ Antiseptics: Caution: excessive antiseptic remaining on the skin may inactivate the vaccine. (1)
+ Corticosteroids: Mumpsvax is contraindicated in patients receiving corticosteroids, irradiation, alkylating agents, or antimetabolites because the normal defense mechanisms are suppressed. (1)
+ Irradiation: see Mumpsvac Lyovac + Corticosteroids
+ Vaccines: Do not administer Mumpsvax with other vaccines but allow one month to elapse between elective immunizations. (1)

MUREL INJECTABLE + Phenobarbital Sodium Sol.: Murel Injectable and phenobarbital sodium in solution are chemically incompatible when mixed in the same syringe. (1)

MUSCLE RELAXANTS: see listed under the following agents:

Alvodine
Dopram
Kantrex Inj.
MAO Inhibitor
Niamid
Penthrane
Quelicin Chloride Inj.
Tubocurarine

MUSCLE RELAXANTS, CURARIFORM + Coly-Mycin M Inj.: see Coly-Mycin M Inj. + Muscle Relaxants, Curariform

MUSCLE RELAXANTS, NON-DEPOLARIZING: see listed under the following agents:

Antibiotics
Succinylcholine

MUSTARGEN
+ Chemotherapeutic Agents: see Mustargen + X-Ray Therapy
+ X-Ray Therapy: Precaution must be observed with nitrogen mustard and x-ray therapy or other chemotherapy in alternating courses. Hematopoietic function is characteristically depressed by either form of therapy, and neither nitrogen mustard nor x-ray therapy subsequent to the drug should be given until bone marrow function has recovered. In particular, irradiation of such areas as sternum, ribs, and vertebrae shortly after a course of nitrogen mustard may lead to hematological complication. (1)

M-VAC MEASLES VIRUS VACCINE, LIVE, ATTENUATED

+ Alkylating Agents: see M-Vac Measles Virus Vaccine, Live, Attenuated + Steroids
+ Antimetabolites: see M-Vac Measles Virus Vaccine, Live, Attenuated + Steroids
+ Antiseptics: see M-Vac Measles Virus Vaccine, Live, Attenuated + Diluents
+ Detergents: see M-Vac Measles Virus Vaccine, Live, Attenuated + Diluents
+ Diluents: Do not mix the vaccine with any substance other than the diluent supplied. Mixture with immune serum globulin or traces of residual antiseptics, detergents, preservatives, etc. have been found to be responsible for inactivation of the vaccine. Use only the syringe and diluent supplied with the package. (1)
+ Gamma Globulin: This vaccine is contraindicated in the presence of recent gamma globulin administration when more than 0.01 ml per pound of body weight was used, or whole blood transfusion, either of which was administered within the preceding six weeks. (1)
+ Immune Serum Globulin: see M-Vac Measles Virus Vaccine, Live, Attenuated + Diluents
+ Irradiation: see M-Vac Measles Virus Vaccine, Live, Attenuated + Steroids
+ Poliomyelitis Vaccine: see M-Vac Measles Virus Vaccine, Live, Attenuated + Vaccines, Live
+ Preservatives: see M-Vac Measles Virus Vaccine, Live, Attenuated + Diluents
+ Smallpox Vaccine: see M-Vac Measles Virus Vaccine, Live, Attenuated + Vaccines, Live
+ Steroids: This vaccine is contraindicated in the presence of therapy which depresses resistance to disease such as steroids, irradiation, alkylating agents, and antimetabolites. (1)
+ Vaccines, Live: Concomitant administration of this vaccine and other live vaccines, such as poliomyelitis and smallpox vaccine, is not advised. (1)
+ Whole Blood: see M-Vac Measles Virus Vaccine, Live, Attenuated + Gamma Globulin

MYCHEL: see also Chloromycetin

MYCHEL + Bone Marrow Depressants: Concurrent therapy with other drugs that may cause bone marrow depression should be avoided. (1)

MYCIFRADIN: see Neomycin

MYCIFRADIN (continued)

+ Antibiotics, Ototoxic: The toxic effects of neomycin on the eighth nerve are principally auditory; neomycin should not be given concurrently with or in series with other ototoxic antibiotics, such as streptomycin or dihydrostreptomycin, because the eighth nerve toxicity of such antibiotics is likely to be additive. (1)
+ Dihydrostreptomycin: see Mycifradin + Antibiotics, Ototoxic
+ Streptomycin: see Mycifradin + Antibiotics, Ototoxic

MYCOSTATIN VAGINAL TABLETS + Douche: Adjunctive measures such as therapeutic douches are unnecessary and sometimes inadvisable. (1)

MYDRIATICS + Diamox: see Diamox + Mydriatics

MYLAXEN: see listed under the following agents:
Succinylcholine
Tubocurarine

MYLAXEN

+ Antibiotics: see Mylaxen + Quinidine
+ Cardiac Glycosides: see Mylaxen + Quinidine
+ Cholinergic Agents: see Mylaxen + Quinidine
+ Ether: The Mylaxen-induced neuromuscular block was potentiated by ether. (143)
+ Flaxedil: The interaction of Mylaxen and Flaxedil were not consistent. (143)
+ Hypotensive Agents: see Mylaxen + Quinidine
+ Quinidine: There is little knowledge concerning the interaction of Mylaxen with other drugs that surgical patients may receive (e.g., quinidine, cardiac glycosides, antibiotics, hypotensive agents, cholinergic agents). It is possible prolongation of the action of such drugs must be kept in mind. (2)
+ Succinylcholine: Although Mylaxen markedly prolongs the action of succinylcholine, the depth of respiration at the termination of surgery remains good. Side effects such as cardiac irregularities, hypotension, and bronchospasm reported with the use of some muscle relaxants, in particular when administered concomitantly with fluorinated anesthetics, have only occasionally been observed with the combined use of Mylaxen-succinylcholine by the procedure recommended by the manufacturer. (1, 143)
+ Syncurine: The neuromuscular effects of Mylaxen and Syncurine were mutually antagonistic. (143)
+ Tubocurarine: Mylaxen and d-tubocurarine always increased the neuromuscular effects of each other. (143)

MYODIGIN: see also Digitalis

MYODIGIN
+ Calcium: The administration of calcium intravenously may induce digitalis intoxication. (1)
+ Diuretic Agents: Potassium depletion, usually caused by the use of diuretic agents or electrolyte manipulation, sensitizes the heart to digitalis and may produce arrhythmias even with recommended doses. In this event, reduction of dosage may be necessary and administration of potassium may be indicated. (1)

MYSTECLIN F
+ Food, Calcium-containing: see Mysteclin F + Milk
+ Meals: Oral dosage forms of tetracycline should be given at least one hour before or two hours after meals. (1)
+ Milk: Pediatric dosage forms of tetracycline should not be given with milk formulas or calcium-containing foods and should also be given at least one hour prior to feeding or meals. (1) Mechanism: Tetracyclines are chelating agents that form insoluble chelates with multivalent cations (Ca^{++}, Mg^{++}, Al^{+++}) and gastrointestinal absorption is inhibited.
+ Trisoralen: see Trisoralen

MYTELASE CHLORIDE: see listed under the following agents:
Phospholine Iodide
Succinylcholine

MYTELASE CHLORIDE
+ Atropine: see Mytelase Chloride + Belladonna Derivatives
+ Belladonna Derivatives: Since belladonna derivatives (atropine, etc.) may suppress the parasympathomimetic (muscarinic) symptoms of excessive gastrointestinal stimulation, leaving only the more serious symptoms of fasciculation and paralysis of voluntary muscle as a sign of overdosage, routine administration of atropine with Mytelase is contraindicated. (1)
+ Cholinergic Agents: Because Mytelase has a more prolonged action than any other antimyasthenic drug, simultaneous administration with other cholinergics is contraindicated except under strict medical supervision. The overlap in duration of action of several drugs complicates dosage schedules. Thus when whensferring to Mytelase Chloride all other cholinergics should be suspended until the patient has been stabilized. (1)

N

NALLINE

+ Anesthetics: Nalline does not counteract mild respiratory depression and may increase it. Respiratory depression due to such agents such as cyclopropane, ethyl ether, nitrous oxide, anesthetics, or barbiturates, or to traumatic or other causes, is not counteracted by Nalline but may increase it. (1)
+ Barbiturates: see Nalline + Anesthetics
+ Cyclopropane: see Nalline + Anesthetics
+ Ethyl Ether: see Nalline + Anesthetics
+ Meperidine: Nalline should not be mixed with solutions of meperidine; precipitation caused by the buffer in the Nalline will result. (1)
+ Nitrous Oxide: see Nalline + Anesthetics
+ Non-Narcotic Drugs: In patients who have received both narcotic and non-narcotic drugs and developed respiratory depression, Nalline should be administered with great caution, since it may increase respiratory depression caused primarily by the non-narcotic drug. If the first dose of Nalline does not have an affect on the depression, it is probable that the non-narcotic drug is the offender and further use of Nalline is contraindicated unless it is ascertained that the narcotic is the causative agent.
  If the respiratory depression caused by Darvon is present, supportive artificial respiration, Nalline, or Lorfan should be administered. (1)

NAQUA: see Diuretics, Thiazide

NAQUA

+ ACTH: see Naqua + Digitalis
+ Adrenocortical Steroids: see Naqua + Digitalis
+ Antihypertensive Agents: Naqua has definite antihypertensive action when given alone and augments the antihypertensive action of other agents. The addition of other antihypertensive agents, usually after one to two weeks (if response is inadequate or requires augmentation), should be made cautiously to prevent the occurrence of acute hypotensive episodes and their usual dosage reduced by half, since Naqua potentiates the antihypertensive action of ganglionic blocking agents such as Inversine and rauwolfia alkaloids. This drug is expected to have similar action with veratrum alkaloids and Apresoline.

NAQUA (continued)

+ Antihypertensive Agents (continued): Naqua should be added to these agents when they are being administered as the primary drug in a similarly cautious manner to avoid precipitation of acute blood pressure falls. (1, 2)
+ Apresoline: see Naqua + Antihypertensive Agents
+ Diabetes: Thiazides may effect glucose metabolism and aggravate diabetes mellitus or precipitate it in a latent case. (1, 2)
+ Digitalis: In patients who are prone to develop disturbances due to low levels of cellular potassium such as those on digitalis therapy or under treatment with certain adrenocortical steroids or ACTH, note should be taken of serum potassium levels despite the fact that there is less likelihood of hypokalemia with Naqua than with chlorothiazide or hydrochlorothiazide. (1, 2)
+ Ganglionic Blocking Agents: see Naqua + Antihypertensive Agents
+ Inversine: see Naqua + Antihypertensive Agents
+ Levophed: see Naqua + Pressor Amines
+ Pressor Amines: Recent reports indicate that the thiazides may decrease arterial responsiveness to pressor amines (e.g., Levophed). (1, 2)
+ Rauwolfia Alkaloids: see Naqua + Antihypertensive Agents
+ Trisoralen: see Trisoralen
+ Tubocurarine: Recent reports indicate that the thiazides may augment the paralyzing action of tubocurarine in surgical patients. (1, 2)
+ Veratrum Alkaloids: see Naqua + Antihypertensive Agents

NAQUIVAL: see Naqua and Reserpine

NAQUIVAL + Anesthesia: Consideration should be given to stopping Naquival therapy at least two weeks before elective surgery or electroconvulsive therapy, since hypotension or shock may result from general anesthesia. Patients undergoing emergency surgery should be watched for precipitous falls in blood pressure. (1)

NARCOTIC ANTAGONISTS: see also Lorfan and Nalline

NARCOTIC ANTAGONISTS

+ Anesthetics: see Narcotic Antagonists + Barbiturates
+ Barbiturates: Narcotic antagonists should not be given to patients with respiratory depression induced by barbiturates or anesthetic agents. Moreover, the narcotic antagonists should be given with caution to patients who develop respiratory depression after having received morphine-like analgesics with sedatives, hypnotics, or anesthetics, since they will not reverse respiratory depression due to the latter agents. (2)

NARCOTIC ANTAGONISTS (continued)
+ Hypnotics: see Narcotic Antagonists + Barbiturates
+ Sedatives: see Narcotic Antagonists + Barbiturates

NARCOTICS: see listed under the following agents:

Anhydron
Anticoagulants
Apodol
Atarax
Bristuron
Coumadin
Dartal
Diuretics, Thiazide
Diuril
Esidrix
Eutonyl
Furoxone
Harmonyl
HydroDIURIL
Hydromox
Hygroton
Ilopan
Inapsine
Innovar
Ketaject
Ketalar
Leritine
Levoprome
Lomotil
Magnesium Sulfate Inj.
Marplan
Matulane
Modumate
Nardil
Parnate
Penthrane
Pentothal Sodium
Phenothiazine
Probanthine with Dartal
Quide
Regitine
Repoise
Sintrom
Stelazine
Sublimaze
Talwin
Temaril
Thorazine
Tindal
Trilafon
Tromexan
Tubocurarine Chloride Inj.
Valium
Valium Inj.
Vesprin

NARDIL: see MAO Inhibitor

NARDIL: see listed under the following agents:

Demerol
Parnate

NARDIL
+ Alcohol: see Nardil + CNS Depressants
+ Aldomet: Aldomet, dopamine, and tryptophan used concomitantly with a MAO inhibitor may result in an unusual elevation of blood pressure --Hypertensive Crisis--and precludes their concomitant use with Nardil. (1) Mechanism: see MAO Inhibitor + Aldomet.
+ Amphetamines: see Nardil + Sympathomimetic Agents

NARDIL (continued)

+ Ansolysen: see Nardil + Regitine
+ Anticonvulsants: Because its affect on the convulsive threshold may be variable, adequate precautions should be taken when treating epileptic patients. (1)
+ Antidepressant, Tricyclic: Nardil should not be administered together with or in rapid succession with tricyclic antidepressants (e.g., Elavil, Norpramin, Pertofrane, Tofranil, Aventyl). At least ten days should elapse between the discontinuation of Nardil therapy and the institution of another antidepressant regimen. (1, 2) Mechanism: see MAO Inhibitor + Antidepressant, Tricyclic.
+ Antihypertensive Agents: see Nardil + CNS Depressants
+ Antiparkinsonism Agents: Great care is required if Nardil is administered in conjunction with antiparkinsonism agents. (1) (See also MAO Inhibitor + Antiparkinsonism Agents.)
+ Aventyl: see Nardil + Antidepressant, Tricyclic
+ Barbiturates: see Nardil + Psychotropic Agents. Mechanism: see MAO Inhibitor + CNS Depressants.
+ Chlorothiazide: Nardil has been shown to potentiate chlorothiazide. (1)
+ CNS Depressants: Nardil should not be used in combination with some CNS depressants such as narcotics (e.g., meperidine: death has been reported in patients who have received meperidine and Nardil concomitantly) and alcohol, or with antihypertensive agents. (1, 2) See also MAO Inhibitor + CNS Depressants.
+ Cocaine: see Nardil + Psychotropic Agents. Mechanism: see MAO Inhibitor + Cocaine.
+ Dopamine: see Nardil + Aldomet
+ Drugs, Over-the-Counter, Cold: see Nardil + Sympathomimetic Agents
+ Drugs, Over-the-Counter, Hay Fever: see Nardil + Sympathomimetic Agents
+ Drugs, Over-the-Counter, Reducing: see Nardil + Sympathomimetic Agents
+ Drugs, Proprietary: see Nardil + Sympathomimetic Agents
+ Elavil: see Nardil + Antidepressant, Tricyclic
+ Ether: see Nardil + Psychotropic Agents
+ Foods: Foods with a high tyramine content (aged cheese, beers, chianti wine, pickled herring, the pods of broad beans, chicken livers, yeast extracts) used concomitantly with a MAO inhibitor may result in an unusual elevation of blood pressure --Hypertensive Crisis-- and preclude their concomitant use with Nardil. (1, 2) Mechanism: see MAO Inhibitor + Food

NARDIL (continued)

+ Insulin: see Nardil + Psychotropic Agents. (See also MAO Inhibitor + Diabetes.)
+ MAO Inhibitors: Nardil should not be administered together with or in rapid succession to other MAO inhibitors (e.g., Niamid, Parnate, Eutonyl, Marplan). At least ten days should elapse between the discontinuation of another MAO inhibitor and the institution of Nardil, because hypertensive crises and convulsive seizures may occur. (1) Mechanism: see MAO Inhibitor + MAO Inhibitor
+ Meperidine: see Nardil + CNS Depressants. Mechanism: see MAO inhibitor + Meperidine.
+ Narcotics: see Nardil + CNS Depressants
+ Norpramin: see Nardil + Antidepressant, Tricyclic
+ Opiates: see Nardil + Psychotropic Agents and CNS Depressants
+ Pertofrane: see Nardil + Antidepressant, Tricyclic
+ Phenoxene: Nardil has been shown to potentiate Phenoxene. (1)
+ Phenylephrine: see Nardil + Sympathomimetic Agents
+ Procaine: see Nardil + Psychotropic Agents
+ Psychotropic Agents: Large doses of Nardil can potentiate the effects of certain drugs, and care should be taken when the following agents are administered concomitantly: psychotropic agents, ether, barbiturates, cocaine, procaine, opiates, insulin. (1, 2)
+ Regitine: If a hypertensive crisis occurs, Nardil should be discontinued immediately and therapy to lower blood pressure should be instituted immediately. On the basis of present evidence, Regitine or Ansolysen is recommended. (The dosage reported for Regitine is 5 mg i.v. and for Ansolysen 3 mg s.c.) Care should be taken to administer these drugs slowly in order to avoid producing excessive hypotensive effect. Fever should be managed by means of external cooling. (1)
+ Sympathomimetic Agents: The unusual potentiation of MAO inhibitors by sympathomimetic drugs (amphetamine--including over-the-counter cold, hay fever, and reducing preparations containing vasoconstrictors and phenylephrine) may result in an unusual elevation of blood pressure--Hypertensive Crisis--and preclude their concomitant use with Nardil. (1, 2, 53) Mechanism: see MAO Inhibitor + Sympathomimetics, Indirect-Acting.
+ Tofranil: see Nardil + Antidepressant, Tricyclic
+ Trisoralen: see Trisoralen
+ Tryptophan: see Nardil + Aldomet
+ Vasoconstrictors: see Nardil + Sympathomimetic Agents

NARONE

+ Anticoagulants: Narone will cause an aggravation of prothrombin deficiency. (1)
+ Thorazine: Do not use Narone with Thorazine because of severe resulting hypothermia. (1)

NASAL DECONGESTANTS: see listed under the following agents:
Eutonyl
Furoxone

NATURETIN: see Diuretics, Thiazide

NATURETIN

+ ACTH: see Naturetin + Steroids
+ Aldomet: see Naturetin + Antihypertensive Agents
+ Anesthetic Agents: see Naturetin + Norepinephrine
+ Antihypertensive Agents: If Naturetin is used with other antihypertensive agents, lower maintenance doses for each drug are usually sufficient. When used concomitantly with ganglionic blocking agents, veratrum, Apresoline, Ismelin, Aldomet, or rauwolfia alkaloids, the dosage of these latter preparations must be decreased to avoid a precipitous drop in blood pressure, particularly of the orthostatic type. The dosage of ganglionic blocking agents must be immediately reduced by about 50%. Adjustment in dosage may subsequently be made as warranted by the patient's response. A similar reduction is necessary when one or more of these agents is added to an established Naturetin regimen. (1, 2)
+ Apresoline: see Naturetin + Antihypertensive Agents
+ Diabetes: Therapy with benzothiadiazines may result in increased glycemia and glycosuria in diabetic patients and may occasionally precipitate acidosis. It may also unmask a diabetic predisposition in apparently normal individuals. (1, 2)
+ Digitalis: Although significant hypokalemia occurs infrequently, at times it may become significant. This effect may become additive with the use of thiazides in conjunction with other agents, as digitalis. Patients who have been on digitalis therapy may exhibit bizarre cardiac arrhythmias of digitalis intoxication when subjected to potassium loss. (1, 2)
+ Diuretics, Mercurial: Electrolyte imbalance is more likely to develop during concomitant therapy with mercurial diuretics. The naturetic effect of thiazides with mercurials is additive. (1)
+ Ganglionic Blocking Agents: see Naturetin + Antihypertensive Agents
+ Ismelin: see Naturetin + Antihypertensive Agents
+ Levophed: see Naturetin + Pressor Amines

NATURETIN (continued)

+ Norepinephrine: Since thiazides may decrease arterial responsiveness to norepinephrine, thiazides should be withdrawn 48 hours prior to surgery. If emergency surgery is indicated, preanesthetic and anesthetic agents should be administered in reduced dosage. (1) See Naturetin + Pressor Amines.
+ Preanesthetic Agents: see Naturetin + Norepinephrine
+ Pressor Amines: Recent reports indicate that thiazides may decrease arterial responsiveness to pressor amines (e.g., Levophed). (2) See Naturetin + Norepinephrine.
+ Rauwolfia Alkaloids: see Naturetin + Antihypertensive Agents
+ Steroids: Particular care should be exercised in administering a potent diuretic to patients taking steroids or ACTH, since hypokalemia may occur. (1, 2)
+ Trisoralen: see Trisoralen
+ Tubocurarine: Recent reports indicate that the thiazides may augment the paralyzing action of tubocurarine in surgical patients. (2)
+ Veratrum: see Naturetin + Antihypertensive Agents

NAVANE: In view of the structural and pharmacological similarities to the phenothiazines, the possibility exists that Navane may cause any of the reactions of that group of drugs. (1, 93) See also Phenothiazines.

NAVANE

+ Alcohol: see Navane + Antihypertensive Agents
+ Analgesics: see Navane + CNS Depressants
+ Anesthetics, General: see Navane + Antihypertensive Agents
+ Anticholinergic Agents: see Navane + Antihypertensive Agents
+ Anticoagulants: It has been observed that Navane may decrease prothrombin time. (93)
+ Anticonvulsants: Navane may lower the seizure threshold and should be used cautiously in patients with a history of convulsive disorders. (1, 93)
+ Antihistamines: see Navane + CNS Depressants
+ Antihypertensive Agents: Because Navane may potentiate the actions of antihypertensive agents, anticholinergic agents, and central nervous system depressants (e.g., general anesthetics, hypnotics, and alcohol), care should be exercised if Navane is added to the regimen of patients receiving such agents. (93)
+ Atropine: Though not reported in the present clinical experience with Navane, potentiation of atropine has been observed with phenothiazines. (1)
+ Barbiturates: see Navane + CNS Depressants

NAVANE (continued)

+ CNS Depressants: See Navane + Antihypertensive Agents. Though not reported in the present clinical experience with Navane, potentiation of CNS depressants (e.g., opiates, analgesics, antihistamines, barbiturates) have been observed with the phenothiazine drugs. (1)
+ Epinephrine: Navane may reverse the action of epinephrine and cause profound hypotension. (1, 93)
+ Hypnotics: see Navane + Antihypertensive Agents
+ Opiates: see Navane + CNS Depressants
+ Phosphorous Insecticides: Though not reported in the present clinical experience with Navane, potentiation of phosphorous insecticides have been observed with the phenothiazine drugs. (1)
+ Trisoralen: see Trisoralen

NEBAIR

+ Epinephrine: see Nebair + Sympathomimetic Agents
+ Sympathomimetic Agents: Use Nebair with care in patients concomitantly treated with epinephrine or other sympathomimetic agents. (1)

NEGGRAM: see Acidic Agents + Alkaline Media

NEGGRAM: see Acidic Agents + Alkanizing Agents (Urinary)

NEGGRAM

+ Benedict's Solution: see NegGram + Clinitest Reagent Tablets
+ Clinitest Reagent Tablets: When Benedict's or Fehling's solutions or Clinitest Reagent Tablets are used to test the urine of patients taking NegGram, a false-positive reaction for glucose may be obtained due to the liberation of glucuronic acid from the metabolites excreted. However, a colorimetric test for glucose based on an enzyme reaction (using, for example, Clinistix Reagent Strips or Tes-Tape) does not give a false-positive reaction to NegGram glucuronide. (1)
+ Fehling's Solution: see NegGram + Clinitest Reagent Tablets
+ Trisoralen: see Trisoralen

NEBUTAL SODIUM INJ.

+ Cocaine: Nembutal intravenously is effective for symptomatic control of convulsions in poisoning with such drugs as cocaine, strychnine, or picrotoxin. In such cases the dosage should be kept to a minimum and timed to avoid adding the depressant action of a barbiturate to the depression that characteristically follows convulsions. (1)
+ Picrotoxin: see Nembutal Sodium Inj. + Cocaine
+ Respiratory Depressants: Since Nembutal is a depressant drug, caution is required when administered to patients who may have a decreased

NEMBUTAL SODIUM INJ. (continued)
+ Respiratory Depressants: tolerance, such as the recent administration of other respiratory depressants. (1)
+ Strychnine: see Nembutal Sodium Inj. + Cocaine

NEOBIOTIC: see Neomycin

NEOCYLATE with COLCHICINE
+ Gold: Neocylate with Colchicine should not be used during gold therapy since PABA may aggravate dermatitis and fever. (1)
+ Sulfonamides: Neocylate with Colchicine therapy should not be administered concurrently with sulfonamides since PABA is absorbed by bacterial pathogens preferentially over sulfonamides and so inhibits their action. (1)

NEOCYLONE: Neocylone contains PABA; therefore see Neocylate with Colchicine.

NEOCYLONE + Diabetes: The prednisolone content of this product may cause alteration of glucose metabolism with aggravation of diabetes mellitus including hyperglycemia and glycosuria. (1)

NEOCYTEN: Neocyten contains PABA; therefore see Neocylate with Colchicine.

NEO-IOPAX + Benemid: see Benemid + Neo-Iopax

NEOMYCIN: see listed under the following agents:

Anectine
Coly-Mycin M. Inj.
Flaxedil
Kantrex
Magnesium Sulfate Inj.
Quelicin Chloride Inj.
Succinylcholine
Sucostrin
Tubocurarine
Tubocurarine Chloride Inj.

NEOMYCIN
+ Anesthesia: Since a curare-like neuromuscular block has been reported with parenteral neomycin, the possibility of an interaction between this neomycin effect and other agents used by the anesthesiologist should be considered if a patient with poor renal function has received large doses of neomycin parenterally. (1)
+ Antibiotics: see Neomycin + Penicillin V Potassium
+ Carotene: see Neomycin + Iron
+ Drugs, Curare-Type: Intraperitoneal instillation of neomycin may intensify the paralyzing effects of curare-type drugs. (1)
+ Glucose: see Neomycin + Iron
+ Iron: Individuals treated daily with 4 to 6 gm of neomycin orally can

NEOMYCIN (continued)

+ Iron (continued): cause an interference with absorption of carotene, glucose, and iron. (1)
+ Nephrotoxic Drugs: see Neomycin + Ototoxic Agents
+ Nutrients: see Neomycin + Penicillin V Potassium
+ Ototoxic Agents: Since parenteral neomycin may cause serious ototoxicity and nephrotoxicity, this possibility should be considered if prolonged, high dosage, oral therapy is planned. (1)
+ Penicillin V Potassium: Oral Neomycin has been reported to produce a transient and reversible gastrointestinal malabsorption of certain nutrients and of potassium penicillin V as well. The underlying mechanism which causes the malabsorption and the effect of neomycin on intestinal absorption of other antibacterial antibiotics is not yet known. It is suggested that when concomitant systemic penicillin therapy is indicated in a patient receiving the tablets, the penicillin should be given parenterally rather than orally to insure its full effectiveness. (1)

NEOPLASTIC CHEMOTHERAPY + Thio-Tepa: see Thio-Tepa + Neoplastic Chemotherapy

NEOSTIGMINE: see listed under the following agents:

Anectine
Cozyme
Flaxedil
Ilopan
Mecostrin
Metubine Iodide
Phospholine Iodide
Quelicin Chloride In.
Succinylcholine
Sucostrin
Surbex-T Sol.
Tubocurarine
Tubocurarine Chloride Inj.

NEOSYNEPHRINE: see Phenylephrine

NEOSYNEPHRINE Parenteral

+ Halothane: During Halothane anesthesia vasopressors may cause serious cardiac arrhythmias and therefore should only be used with great caution or not at all. (1)
+ Oxytocic Drugs: In obstetrics oxytocics and vasopressor drugs should not be used simultaneously because severe persistent hypertension may occur. (1)

NEOTHALIDINE: see Neomycin and Sulfathaladine

NEO-VAGISOL

+ Arsenic: see Neo-Vagisol + Iodides
+ Iodides: Vaginal preparations containing iodides or arsenic must not

NEO-VAGISOL (continued)

+ Iodides (continued): be used concurrently or within three days of insertion of Neo-Vagisol. A severe burn can be caused by the reaction of the mercury in the Neo-Vagisol with iodides. (1)

NEPHENALIN + Epinephrine: Do not administer Nephenalin with epinephrine though they may be alternated. (1)

NEUROMUSCULAR BLOCKING AGENTS: see also Anectine; Antibiotics, Curare-Like; Curare; Curare and Derivatives; Curariform Drugs; Curare-Like Drugs; Curare-Type Muscle Relaxants; Flaxedil; Mecostrin; Metubine Iodide; Muscle Relaxants, Curariform; Muscle Relaxants, Non-Depolarizing; Mylaxen; Mytelase; Quelicin Chloride Injection; Succinylcholine; Sucostrin; Tubocurarine; and Tubocurarine Chloride Injection

NEUROMUSCULAR BLOCKING AGENTS + Magnesium Sulfate Injection: see Magnesium Sulfate Injection + Neuromuscular Blocking Agents

NEUTRAPEN: see listed under the following agents:

Dimocillin-RT
Pathocil
Resistopen
Staphcillin
Unipen

NIAMID: see MAO Inhibitor

NIAMID: see listed under the following agents:

Demerol
Parnate

NIAMID

+ Alcohol: see Niamid + Pressor Agents. (See also MAO Inhibitor + CNS Depressants.)
+ Analgesics: see Niamid + Narcotics
+ Antidepressants, Tricyclic: see Niamid + Tofranil
+ Antihypertensive Agents: see Niamid + Psychotropic Agents
+ Aventyl: see Niamid + Tofranil
+ Barbiturates: see Niamid + Psychotropic Agents. Mechanism: see MAO Inhibitor + CNS Depressants.
+ Cocaine: see Niamid + Psychotropic Agents. Mechanism: see MAO Inhibitor + Cocaine
+ Diuril: Hypotensive reactions are rare when Niamid is administered alone but are not uncommon when Diuril or closely related drugs are given concomitantly. (2)

NIAMID (continued)

+ Diuretics, Thiazide: see Niamid + Diuril
+ Ether: see Niamid + Psychotropic Agents
+ Food: Patients receiving a monoamine oxidase inhibitor have experienced hypertensive crises after the ingestion of cheese, particularly strong or aged varieties. Thus, patients should be warned not to eat cheese. (2) Mechanism: see MAO Inhibitor + Food.
+ Hypnotics: see Niamid + Narcotics
+ Insulin: see Niamid + Psychotropic Agents. (See also MAO Inhibitor + Insulin.)
+ Isoniazid: As with Marsalid (iproniazid), sensitive strains of M. tuberculosis can be made resistant to isoniazid in vitro by repeated exposure to sublethal concentrations. Organisms with such artificially induced resistance apparently are not virulent in the experimental animal. Existing data do not indicate whether or not resistance of M. tuberculosis to isoniazid may be induced in the human host with Niamid therapy; nevertheless, it is recommended that Niamid be withheld in the depressed patient with coexisting tuberculosis who may need isoniazid. (1)
+ MAO Inhibitor: Do not use Niamid concomitantly with other MAO inhibitors. (1) Mechanism: see MAO Inhibitor + MAO Inhibitor.
+ Meperidine: see Niamid + Psychotropic Agents. Mechanism: see MAO Inhibitor + Meperidine.
+ Muscle Relaxants: see Niamid + Narcotics
+ Narcotics: Since Niamid may necessitate a reduction in the dosage of narcotics, analgesics, muscle relaxants, sedatives, hypnotics, and stimulants, the dosage of each of such agents used in combination with Niamid should be titrated to individual patient's need. (1)
+ Norpramin: see Niamid + Tofranil
+ Opiates: see Niamid + Psychotropic Agents
+ Pertofrane: see Niamid + Tofranil
+ Phenylephrine: see Niamid + Psychotropic Agents. Mechanism: see MAO Inhibitor + Sympathomimetics, Indirect-Acting
+ Pressor Agents: Patients should be warned against taking proprietary drugs that contain pressor agents (e.g., certain cold remedies, hay fever preparations, or anorexiants) or to drink alcoholic beverages as a hypertensive crisis may occur. (2) Mechanism: see MAO Inhibitor + Sympathomimetics, Indirect-Acting.
+ Procaine: see Niamid + Psychotropic Agents
+ Proprietary Drugs: see Niamid + Pressor Agents
+ Psychotropic Agents: Since large doses of Niamid can potentiate the effects of certain drugs, care should be taken when the following

NIAMID (continued)

+ Psychotropic Agents (continued): agents are administered concomitantly: meperidine, cocaine, procaine, phenylephrine and other sympathomimetic amines, opiates, antihypertensive agents, insulin, psychotropic agents, and ether. (2)
+ Rauwolfia Alkaloids: Since rauwolfia compounds are believed to act by releasing "bound" serotonin and the MAO inhibitors increase serotonin levels by the inhibition of its destruction, on theoretical grounds, caution should be exercised when these compounds are administered simultaneously. (1) See also MAO Inhibitor + Reserpine.
+ Sedatives: see Niamid + Narcotics
+ Stimulants: see Niamid + Narcotics
+ Sympathomimetic Amines: see Niamid + Psychotropic Agents. Mechanism: see MAO Inhibitor + Sympathomimetics, Indirect-Acting and also Sympathomimetics, Direct-Acting.
+ Tofranil: Tofranil, Aventyl, Pertofrane, Norpramin, or Vivactil should not be given with or for at least two weeks after discontinuing the use of Niamid. Rare instances have been reported of reactions (including atropine-like effects and muscular rigidity) occuring when Tofranil and other similar tricyclic antidepressants were administered during or shortly after treatment with certain other drugs that inhibit monoamine oxidase. (1, 2) Mechanism: see MAO Inhibitor + Antidepressant, Tricyclic.
+ Vivactil: see Niamid + Tofranil

NICALEX

+ Antihypertensive Agents: Patients receiving antihypertensive agents of the adrenal-blocking type should be watched for signs of postural hypotension. (1)
+ Diabetes: An occasional side effect is decreased glucose tolerance (resembling diabetic-type curve) sometimes with glycosuria. Some diabetic patients may require adjustment of diet and insulin dosage in the event of decreased tolerance. (1, 12)

NICOTINE + Liquaemin Sod.: see Liquaemin Sod. + Nicotine

NICOTRON + Vasodilators, Coronary: Caution is advised when there is concomitant administration of a coronary vasodilator. (1)

NILEVAR: see listed under the following agents:

Anticoagulants
Dicumarol
Panwarfin

NISENTIL
+ Anesthetics, General: see Nisentil + Barbiturates
+ Barbiturates: The depressant effects of Nisentil are potentiated in the presence of drugs such as barbiturates, general anesthetic agents, and some phenothiazines; accordingly, the dosage of Nisentil should be reduced when used concomitantly with such agents. (1)
+ MAO Inhibitor: Narcotic analgesics should be used with extreme caution in patients receiving MAO inhibitors because of the possibility of hypotension. (1)
+ Phenothiazines: see Nisentil + Barbiturates

NITOMAN + Reserpine: see Reserpine + Nitoman

NITROGEN MUSTARD: see listed under the following agents:
Fungizone I.V.
Triethylene Melamine (TEM)

NITROGLYCERIN + Alcohol + Nitroglycerin

NITROUS OXIDE: see listed under the following agents:
Nalline
Succinylcholine

NOCTEC + Anticoagulants, Coumarin: Chloral hydrate may increase the rate of metabolism of concomitantly administered coumarin anticoagulants, thus reducing the effectiveness of this class of drugs. Upon withdrawal of chloral hydrate, the rate of metabolism of the anticoagulant drug will decrease and its plasma level will rise with the possibility of a sudden increase of anticoagulant effects (i.e., development of bleeding tendency and hemorrhage). The prothrombin time of patients on anticoagulant therapy must be followed continually; however, when chloral hydrate is added to or subtracted from the therapeutic regimen, or when changes in dosage of chloral hydrate are contemplated, the effect of the sedative on prothrombin time deserves special attention. (1) See Anticoagulants + Chloral Hydrate.

NOLUDAR
+ Alcohol: see Noludar + CNS Depressants
+ CNS Depressants: As in the case of other central nervous system acting drugs, patients receiving Noludar should be cautioned about possible combined effects with alcohol and other CNS depressants. (1)

NOREPINEPHRINE: see also Levophed

NOREPINEPHRINE: see listed under the following agents:

Aquatag	Naturetin
Diuril	Oretic
Enduron	Pentritol
Esidrix	Priscoline
Esimil	Renese
Exna	Ritalin
HydroDIURIL	Sinequan
Hydromox	Sorbitrate
Isordil	Vivactil

NORFLEX + Darvon: Since mental confusion, anxiety, and tremors have been reported in patients receiving orphenadrine and Darvon concurrently, it is recommended that Norflex not be given in combination with Darvon. (1)

NORGESIC: This preparation contains orphenadrine; therefore see Norflex.

NORINYL: see Contraceptives, Oral

NORINYL + Diabetes: A decreased glucose tolerance has been observed in a small percentage of patients on oral contraceptives. The mechanism of this decrease is obscure. For this reason, diabetic patients should be carefully observed while receiving Norinyl therapy. (1)

NORISODRINE

+ Ephedrine: see Norisodrine + Sympathomimetic Agents
+ Epinephrine: If epinephrine is to be injected, isoproterenol should be withheld until the effects of the injection have subsided. (1) See also Norisodrine + Sympathomimetic Agents.
+ pH: In neutral or alkaline solution, isoproterenol may become red on exposure to air. Accordingly, the saliva or sputum may appear pink or red in color after oral inhalation of this drug. This is not harmful to the patient. (1)
+ Sympathomimetic Agents: Although ephedrine and other sympathomimetic agents are pharmacologically compatible with isoproterenol, the dosage of such simultaneously employed agents should be reduced to avoid excessive response and the occurrence of cardiovascular symptoms. (1)

NORISODRINE AEROTROL: see Norisodrine

NORISODRINE SULFATE POWDER for INHALATION: see Norisodrine

NORISODRINE SULFATE SYRUP with CALCIUM IODIDE: see Norisodrine

NORISODRINE SUPPOSITORIES: see Norisodrine

NORLESTRIN: see Contraceptives, Oral

NORLESTRIN + Diabetes: In patients with metabolic disorders (including diabetes), careful consideration should be given before administration of the drug and caution exercised if it is administered. (1)

NORPRAMIN: see Antidepressants, Tricyclic

NORPRAMIN: see listed under the following agents:

Eutonyl
Ismelin
Marplan
Nardil
Niamid
Parnate
Trisoralen

NORPRAMIN

+ Amphetamine: Norpramin impairs the hydroxylation of amphetamine (in animals) in the liver thereby increasing the level of circulating amphetamine. (114) (See Norpramin + Sympathomimetic Drugs.)
+ Anticholinergic Agents: Patients receiving Norpramin and anticholinergic drugs simultaneously are known to experience enhanced atropine-like side effects. (Norpramin has anticholinergic potentiating effects.) (1, 2)
+ Diuretics, Thiazide: Special care should be taken when Norpramin is used with other agents that lower blood pressure (e.g., the thiazide diuretics, phenothiazine compounds, and vasodilators). (2)
+ MAO Inhibitor: Norpramin should not be given within two weeks of treatment with a monoamine oxidase inhibitor, since this combination may produce severe atropine-like reactions, tremors, hyperpyrexia, generalized clonic convulsions, delerium, and even death. (1, 2) Mechanism: see MAO Inhibitor + Antidepressant, Tricyclic.
+ Phenothiazines: see Norpramin + Diuretics, Thiazide
+ Sympathomimetic Drugs. Use Norpramin with caution in patients receiving sympathomimetic drugs as potentiation of these drugs may occur resulting in behavioral and/or cardiovascular toxicity. (Norpramin has an epinephrine potentiating effect.) (1)
+ Thyroid Hormone: Use Norpramin with caution in patients receiving thyroid hormone as potentiation of this agent may occur resulting in behavioral and/or cardiovascular toxicity. (1)
+ Vasodilators: see Norpramin + Diuretics, Thiazide

NORQUEN: see Contraceptives, Oral

NORQUEN (continued)

+ Diabetes: A decrease in glucose tolerance has been observed in a small percentage of patients on oral contraceptives. The mechanism of this decrease is obscure. For this reason, diabetic patients should be carefully observed while receiving Norquen therapy. (1)

NOVOBIOCIN + Penicillin: see Penicillin + Novobiocin

NOVOCAIN

+ Barbiturates: Some investigators have advised the slow intravenous infusion of soluble ultra-short acting barbiturates for the treatment of convulsions due to Novocain overdosage. Barbiturates, however, should not be administered unless convulsions actually occur. (1)
+ Epinephrine: In the treatment of circulatory collapse due to Novocain, epinephrine should not be given in the presence of anoxia because of the risk of causing ventricular fibrillation. (1)

NUMORPHAN

+ Analgesics: see Numorphan + Anesthetic Agents
+ Anesthetic Agents: As with other narcotics, respiratory depression may occur, especially when other analgesics and/or anesthetic drugs with depressant action have been given just previously. (1)
+ Barbiturates: Since barbiturates increase the possibility of respiratory depression, the usual barbiturate dosage should be reduced for concurrent administration. (1)

NUTRIENTS + Neomycin: see Neomycin + Nutrients

NUVARONE: see Methyl-Phenyl Hydantoin

O

OBEDRIN-LA + MAO Inhibitors: Obedrin-LA should not be given concurrently with monoamine oxidase inhibitors. (1) Mechanism: see MAO Inhibitor + Sympathomimetics, Indirect-Acting.

OBETROL + MAO Inhibitors: Concomitant administration with a monoamine oxidase inhibitor is contraindicated. Certain MAO inhibitors may potentiate the action of Obetrol. (1) Mechanism: see MAO Inhibitor + Sympathomimetics, Indirect-Acting.

OLEANDOMYCIN PHOSPHATE + Mucomyst: see Mucomyst + Oleandomycin Phosphate

OPIATES: see listed under the following agents:

Amytal Sodium Sterile
Atarax
Compazine
Marplan
Mellaril
Nardil
Navane
Niamid
Permitil
Phenothiazine
Prolixin
Serentil
Sparine
Taractan
Temaril
Thorazine
Torecan
Vesprin
Vistaril

OPTIPHYLLIN + Theophylline: Optiphyllin should not be administered concomitantly with other theophylline-containing drugs. (1)

ORACON: see Contraceptives, Oral

ORACON + Diabetes: A decrease in glucose tolerance has been observed in a small percentage of patients on oral contraceptives. The mechanism of this decrease is obscure. For this reason, diabetic patients should be carefully observed while receiving Oracon therapy. (1, 2)

ORAGRAFIN

+ Cholecystographic Media I. V.: There is some evidence that the administration of Oragrafin can interfere with subsequent intravenous cholangiography. Also, adverse reactions from intravenous cholangiography may be more frequent and more severe when the injection is immediately preceded by Oragrafin. Similar effects on intravenous cholangiography are produced by other cholecystographic media. (2)

ORAGRAPHIN (continued)
+ Cholografin: Oragrafin sometimes temporarily blocks the hepatic excretion of Cholografin. (2)

ORENZYME + Anticoagulants: Orenzyme should be used with caution in patients with abnormality of the blood clotting mechanism. (1)

ORETIC: see Diuretics, Thiazide

ORETIC
+ Antihypertensive Agents: When other antihypertensive agents are to be added to the regimen, this should be accomplished gradually. Ganglionic blocking agents should be given at only one-half the usual doses, since their effect is potentiated by pretreatment with Oretic. (1)
+ Diabetes: Elevated blood sugar has been associated with the use of thiazide drugs. (1)
+ Digitalis: Hypokalemia may occur during therapy with hydrochlorothiazide. Potassium depletion can be hazardous in patients taking digitalis. Myocardial sensitivity to digitalis is increased in the presence of reduced serum potassium and signs of digitalis intoxication may be produced by formerly tolerated doses of digitalis. (1)
+ Ganglionic Blocking Agents: see Oretic + Antihypertensive Agents
+ Norepinephrine: see Oretic + Pressor Agents
+ Pressor Agents: It has been observed that thiazide drugs may reduce arterial responsiveness to norepinephrine. Accordingly, the dose of vasopressor agents may need to be modified in surgical patients who have been receiving thiazide therapy. (1)
+ Trisoralen: see Trisoralen
+ Tubocurarine: Thiazide drugs may increase the responsiveness to tubocurarine. (1)

ORGANIC PHOSPHATES + Succinylcholine: see Succinylcholine + Organic Phosphates

ORGANOPHOSPHATE PESTICIDES: The following is a partial list of organic phosphate insecticides:

Bis(dimethylamino)phosphorous anhydride
Diazinon EPN
Di-Syston (sulfur analog of demeton)
Hexaethyl Tetraphosphate (HETP)
Malathion
Metacide

ORGANOPHOSPHATE PESTICIDES (continued)

Methyl Parathion
Octamethyl Pyrophosphoramide (OMPA; Pestox III; Schradan)
Para-Oxon
Parathion
Phosdrin
Sulfotepp
Systox (demeton chlorothion)
TEPP (tetraethyl pyrophosphate)
Tetraethyl Dithionopyrophosphate
Thimet (phorate)

ORGANOPHOSPHATE PESTICIDES

+ Aminophylline: see Organophosphate Pesticides + Morphine
+ Barbiturates: Barbiturates should be used with extreme caution in the treatment of convulsions due to poisoning by these anticholinesterase pesticides because barbiturates are potentiated by anticholinesterases. (2,132)
+ Fluids, Intravenous: see Organophosphate Pesticides + Morphine
+ Morphine: The antidote for poisoning by these anticholinesterase chemicals is large doses of atropine and Protopam. Morphine, theophylline, aminophylline, succinylcholine, reserpine, phenothiazine tranquilizers and large amounts of fluids intravenously are contraindicated. (2,132)
+ Phospholine Iodide: see Phospholine Iodide + Organophosphate Pesticides
+ Reserpine: see Organophosphate Pesticides + Morphine
+ Succinylcholine: see Organophosphate Pesticides + Morphine
+ Theophylline: see Organophosphate Pesticides + Morphine

ORIMUNE POLIOVIRUS VACCINE, LIVE, ORAL, TRIVALENT

+ Cancer Chemotherapeutic Agents: see Orimune Poliovirus Vaccine, Live, Oral,Trivalent + Immune Serum Globulin
+ Diluents: The vaccine may be mixed with distilled water, tap water free of chlorine, Simple Syrup U.S.P., milk, or it may be absorbed on any of a number of foods such as bread, cake, or cube sugar. (1)
+ Immune Serum Globulin: Administration of the vaccine should be postponed in those conditions having a suppressive effect on the immune-response mechanism such as treatment with immune serum globulin or with cancer chemotherapy agents. (1)
+ Measles Vaccine, Live: Concomitant administration of oral poliovirus vaccine and other live vaccines such as live measles vaccine is not

ORIMUNE POLIOVIRUS VACCINE, LIVE, ORAL, TRIVALENT (continued)

+ Measles Vaccine, Live (continued): advised. Administration of such live virus vaccines should be separated by a period of one month or more wherever possible. (1)
+ Vaccines, Live: see Orimune Poliovirus Vaccine, Live, Oral, Trivalent + Measles Vaccine, Live

ORINASE: see listed under the following agents:

Alcohol
Dicumarol
Panwarfin
Sulfabid
Trisoralen

ORINASE

+ Alcohol: Disulfiram (Antabuse)-like reactions after ingestion of alcohol have been reported in patients taking Orinase. (1, 2, 76) See also Alcohol + Orinase
+ Analexin: see Orinase + Insulin. Mechanism: Analexin will enhance the action of Orinase and thereby increase the risk of hypoglycemia by interfering with the metabolism of Orinase. (57)
+ Barbiturates: Orinase may prolong the effects of barbiturates and other sedatives and hypnotics. (2)
+ Benemid: see Orinase + Insulin
+ Beta-Adrenergic Blocking Agents: Response to Orinase is diminished in patients receiving therapy with beta blocking agents. (1) See also Inderal + Diabetes.
+ Butazolidin: see Orinase + Insulin. Mechanism: Displacement of Orinase from its protein binding site, thereby increasing the concentration of free unbound Orinase and inducing a further hypoglycemic effect. (18, 118)
+ DBI: see Orinase + Insulin
+ Dicumarol: see Orinase + Insulin. Mechanism: Dicumarol and other oral anticoagulants potentiate the effect of Orinase by interfering with the metabolism of Orinase. (15, 56)
+ Diuretics, Thiazide: Caution should be observed in administering the thiazide-type diuretics to diabetic patients on Orinase therapy because the thiazides have been reported to aggravate the diabetic state and to result in increased Orinase requirement, temporary loss of control, or even secondary failure. (1, 2)
+ Gantrisin: Concomitant administration of Gantrisin and Orinase may cause severe hypoglycemia. (39) Mechanism: see Orinase + Sulfonamides.

ORINASE (continued)

+ Hypnotics: see Orinase + Barbiturates
+ Insulin: Certain drugs may prolong or enhance the action of Orinase and thereby increase the risk of hypoglycemia. These include insulin, DBI, sulfonamides, Tandearil, salicylates, Benemid, MAO inhibitors, Butazolidin, Dicumarol, and Analexin. (1, 2)
+ MAO Inhibitors: see Orinase + Insulin
+ Phenothiazines: There is the possibility that phenothiazine-type tranquilizers may cause jaundice or abnormal results in liver-function tests when these agents are used with Orinase. (2)
+ Salicylates: see Orinase + Insulin. Mechanism: Displacement of the Orinase from its protein-binding site, thereby increasing the concentration of free unbound Orinase and inducing a further hypoglycemic effect. (118)
+ Sedatives: see Orinase + Barbiturates
+ Sulfabid: Sulfabid can displace Orinase from its protein-binding sites, thereby increasing the concentration of free unbound Orinase and inducing a further hypoglycemic effect. (56, 118)
+ Sulfonamides: see Orinase + Insulin. Mechanism: The half life of Orinase is extended by concomitant administration of sulfonamides, producing hypoglycemia. This is caused by the displacement of the Orinase from its protein-binding site. (18, 39, 50, 56, 58)
+ Tandearil: see Orinase + Insulin. Mechanism: Probably the same as with Butazolidin.

ORIODIDE-131 + Antithyroid Drugs: Sodium Iodide I-131 is contraindicated in patients under treatment with antithyroid drugs (except in the presence of carcinoma of the thyroid). (1)

ORNADE + Pilocarpine: While pilocarpine or similar drugs are sometimes recommended for the relief of dry mouth, many authorities feel that these drugs are not indicated, since they relieve the minor peripheral effects but do not influence the more serious central effects, and thus may merely mask signs of drug activity. (1)

ORTHO-NOVUM: see Contraceptives, Oral

ORTHO-NOVUM + Diabetes: A decrese in glucose tolerance has been observed in a small percentage of patients on oral contraceptives. The reason for this mechanism is obscure. For this reason, diabetic patients should be carefully observed while receiving Ortho-Novum therapy. (1)

OSTENSIN

+ Anesthesia, General: Because of the potentiation by other drugs, the dose of Ostensin must be adjusted prior to surgery involving the use of general anesthesia. (2)
+ Chlorothiazide: The adjunctive use of chlorothiazide may afford additional symptomatic relief and allow for a reduction in the daily dose of Ostensin. (2) See also Ostensin + Diuretic Agents.
+ Diuretic Agents: Reduction of sodium chloride intake by dietary restrictions, or increase in sodium chloride excretion by the use of diuretic agents, will enhance the antihypertensive activity of Ostensin and allow for reduction in the total daily dose. (1)
+ Food: For maximum antihypertensive effects, Ostensin must be ingested in a fasting or near-fasting state. Absorption of orally administered Ostensin, although not complete, is regular and predictable when the drug is taken on an empty stomach. (1)
+ Meprobamate: see Ostensin + Reserpine
+ Reserpine: The action of Ostensin may be potentiated by the use of adjuncts. The adjunctive use of reserpine or meprobamate affords additional symptomatic relief and reduces the dosage requirements of Ostensin. (1)
+ Sodium Chloride: see Ostensin + Diuretic Agents

OVRAL: see Contraceptives, Oral

OVRAL + Diabetes: A decrease in glucose tolerance has been observed in a small percentage of patients on oral contraceptives. The mechanism of this decrease is obscure. For this reason, diabetic patients should be carefully observed while receiving Ovral therapy. (1)

OVULEN: see Contraceptives, Oral

OVULEN + Diabetes: Oral contraceptives should be administered cautiously to diabetic patients, since a decrease in glucose tolerance has been observed in some patients taking these drugs. (1)

OXYCEL

+ Escharotic Agents: Oxycel is not to be used following silver nitrate or other escharotic chemical agents. (1)
+ Silver Nitrate: see Oxycel + Escharotic Agents

OXYTOCICS: See listed under the following agents:

Actospar
Carbocaine
Ergotrate Maleate Inj.
Neosynephrine Parenteral
Pressonex
Spartocin
Vasopressor Agents

OXYTOCIN: see listed under the following agents:

Spartocin

Toscamine

P

PABALATE

+ Anticoagulants: Hypoprothrombinemia has been observed during prolonged and intensive therapy with salicylates. (1)
+ Sulfonamides: When administered concurrently, aminobenzoic acid (PABA) inhibits bacteriostatic action of sulfonamides. (1) Mechanism: PABA is absorbed preferentially over sulfonamides by bacterial pathogens and so inhibits the action of sulfonamides.

PABALATE-HC: see Pabalate, Pabalate-SF, and also Hydrocortisone

PABALATE-SF: see Pabalate

PABALATE-SF + Potassium: When prescribing Pabalate-SF or Pabalate-HC for patients who are receiving concurrent potassium supplementation (e.g., to replace potassium excreted during thiazide therapy), it should be kept in mind that Pabalate-SF and Pabalate-HC also contain potassium. A decrease in supplemental potassium dosage should be considered in order to help avoid hyperkalemia. (1)

PABIRIN BUFFERED TABLETS + Sulfonamides: When administered concurrently, aminobenzoic acid inhibits bacteriostatic action of sulfonamides. (1) Mechanism: PABA is absorbed preferentially over sulfonamides by bacterial pathogens and so inhibits the action of sulfonamides.

PACATAL: see Phenothiazines

PACATAL

+ Alcohol: see Pacatal + CNS Depressants
+ Anesthetics, General: see Pacatal + CNS Depressants
+ Antihypertensive Agents: Pacatal may potentiate the action of antihypertensive agents, and care should be exercised when used concomitantly. (2)
+ CNS Depressants: Pacatal may potentiate the effects of CNS Depressants, and care should be exercised when used concomitantly. (2)
+ Epinephrine: Epinephrine should not be used concomitantly, for Pacatal may reverse its action and cause profound hypotension. Vasopressors such as Levophed may be used. (2)
+ Hypnotics: see Pacatal + CNS Depressants

PAMAQUINE + Atabrine: When pamaquine is given to patients previously treated with Atabrine, the pamaquine is displaced from its protein binding sites in organ tissues and its plasma concentration is increased

PAMAQUINE + Atabrine (continued): five to tenfold. Pamaquine toxicity (gastrointestinal, hematopoietic) will be seen. (118, 119)

PANAFIL OINTMENT + Hydrogen Peroxide: Hydrogen peroxide solution may inactivate papain. (1)

PANALBA: see Panmycin and also Albamycin

PANMYCIN
+ Food, Calcium-containing: see Panmycin + Milk
+ Meals: Oral forms of Panmycin should be given one hour before or two hours after meals. (1) This aids in the absorption of tetracycline.
+ Milk: Oral pediatric dosage forms of Panmycin should not be given with milk formulas or other calcium-containing foods and should be given at least one hour before feeding. (1) Mechanism: Tetracyclines are chelating agents that form insoluble chelates with multivalent cations (Ca^{++}, Mg^{++}, Al^{+++}), and gastrointestinal absorption is inhibited. (55, 81)
+ Trisoralen: see Trisoralen

PANMYCIN PARENTERAL + Hemastix: see Hemastix + Panmycin Parenteral

PANTOTHENIC ACID + Benemid: see Benemid + Pantothenic Acid

PANWARFIN: The following agents may increase the hypoprothrombinemic effect of Panwarfin. For mechanism see Anticoagulant plus each of the following agents:

ACTH
Analexin
Antibiotics (broad spectrum)
Atromid-S (85)
Butazolidin
Chloral Hydrate
Choloxin
Indocin
Methylthiouracil
Nilevar
Quinine (23)
Quinidine (23)
Radio-compounds (some)
Salicylates
Tandearil
X-Ray

PANWARFIN: The following agents may reverse the hypoprothrombinemic effect of Panwarfin. For mechanisms see Anticoagulant plus each of the following agents:

Doriden (48, 84, 85)
Griseofulvin (84)
Meprobamate (84, 85)
Phenobarbital and other Barbiturates (48, 83, 85)

PANWARFIN

+ Anesthesia, Lumbar Block: see Panwarfin + Anesthesia, Regional
+ Anesthesia, Regional: Panwarfin is contraindicated in patients undergoing regional and lumbar anesthesia. (1)
+ Diphenylhydantoin: Panwarfin may be potentiated by, and may elevate the serum hydantoin concentration of, diphenylhydantoin. (1, 84) Mechanism: The anticoagulant is displaced from its protein binding sites in blood plasma thus making increased quantities of free anticoagulant available. The anticoagulant in turn will decrease the metabolism of the diphenylhydantoin thereby increasing plasma concentrations of diphenylhydantoin. See also Diphenylhydantoin + Dicumarol.
+ Heparin: It is important to bear in mind that heparin sodium prolongs the one-stage prothrombin time determinations. Accordingly, when heparin is given with Panwarfin, a period of three to four hours should elapse after the last dose of heparin to obtain a valid prothrombin time. (1)
+ Orinase: Panwarfin may potentiate the hypoglycemic effect of Orinase. (1) Mechanism: The anticoagulant interferes with the metabolism of Orinase. (15, 56)

PAPASE + Anticoagulant: Concomitant administration with anticoagulants is contraindicated. (1, 2)

PAPAVERINE HYDROCHLORIDE INJ. + Lactated Ringer's Inj.: Papaverine Hydrochloride Injection should not be added to Lactated Ringer's Injection because precipitation will result. (1)

PARA-AMINO BENZOIC ACID (PABA)

+ Gold Therapy: PABA should not be used during gold therapy, since PABA may aggravate dermatitis and fever. (1)
+ Methotrexate: see Methotrexate + Para-amino Benzoic Acid
+ Sulfonamides: PABA should not be used concurrently with sulfonamides, since PABA is absorbed by bacterial pathogens preferentially over sulfonamides and so inhibits the action of the sulfonamides. (1)

PARA-AMINOHIPPURIC ACID (PAH) + Benemid: see Benemid + Para-aminohippuric Acid (PAH)

PARA-AMINOSALICYLIC ACID: see listed under the following agents:

Aspirin
Benemid
Diphenylhydantoin
Liquamar

PARIDIONE
+ Drugs: Concurrent administration of other drugs known to cause toxic effects should be avoided or used only with extreme caution. (1)
+ Methyl-Phenyl Hydantoin (Nuvarone): Concomitant usage is not advised, since development of aplastic anemia has been reported. (1)

PARAL + Plastic: Use glass syringe as paraldehyde is not compatible with most plastics. (1)

PARALDEHYDE + Antabuse: see Antabuse + Paraldehyde

PARASYMPATHOMIMETIC DRUGS: see listed under the following agents:
Cozyme
Ilopan
Motilyn
Surbex-T Solution

PARATHION + Sucostrin: see Sucostrin + Parathion

PAREDRINE + MAO Inhibitors: Do not use Paredrine in patients taking MAO inhibitors. (1) Mechanism: see MAO Inhibitor + Sympathomimetics, Indirect-Acting

PARENTERAL FLUIDS + Valium Inj.: see Valium Inj. + Parenteral Fluids

PAREST
+ Alcohol: see Parest + Sedatives
+ Analgesics: see Parest + Sedatives
+ CNS Depressants: see Parest + Psychotropic Drugs
+ Psychotherapeutic Drugs: see Parest + Sedatives
+ Psychotropic Drugs: It is not recommended that Parest be used with psychotropic drugs or other central nervous system depressants. (1)
+ Sedatives: Care should be used during administration of Parest with other sedatives, analgesic or psychotherapeutic drugs, or with alcohol because of the possible potentiation of the effects. (1)

PARNATE: see MAO Inhibitor

PARNATE: see listed under the following agents:
Demerol
Eutonyl
Trisoralen

PARNATE
+ Alcohol: see Parnate + CNS Depressants
+ Aldomet: see Parnate + Sympathomimetics. Mechanism: see MAO Inhibitor + Aldomet

PARNATE (continued)

+ Amphetamines: Parnate should not be administered in combination with amphetamines or reducing preparations containing pressor agents. (1, 2) Mechanism: see MAO Inhibitor + Sympathomimetics, Indirect-Acting.
+ Anesthetics: Although excretion of Parnate is rapid, inhibition of monoamine oxidase may persist for a few days. It is therefore suggested that the drug be discontinued seven days before elective surgery to allow time for recovery of enzymatic activity before anesthetic agents are given. (1) (See also MAO Inhibitor + Anesthetic Agents.)
+ Ansolysen: see Parnate + Regitine
+ Antidepressants, Tricyclic: Parnate should not be administered together or in rapid succession with dibenzazepine derivatives (Elavil, Norpramin, Pertofrane, Tofranil, Vivactil, Aventyl). Hypertensive crisis or severe convulsive seizures may occur in patients receiving such combinations. In patients being transferred to Parnate from a dibenzazepine derivative, allow a medication-free interval of at least a week, then initiate Parnate using half the normal starting dosage for at least the first week of therapy. Similarly, at least a week should elapse between the discontinuance of Parnate and the administration of a dibenzazepine derivative. (1, 2) Mechanism: see MAO Inhibitor + Antidepressant, Tricyclic.
+ Antihistamines: see Parnate + CNS Depressants
+ Antihypertensive Agents: Parnate should not be administered in combination with antihypertensive agents or diuretics because of the possibility of additive hypotensive effect. (1, 59)
+ Antiparkinsonism Drugs: Antiparkinsonism drugs should be used with caution in patients receiving Parnate, since severe reactions have been reported. (1, 2)
+ Aventyl: see Parnate + Antidepressant, Tricyclic
+ Barbiturates: Barbiturates have been reported to relieve myoclonic reactions due to overdosage, but frequency of administration should be carefully controlled because Parnate may prolong barbiturate activity. (1) Mechanism: see MAO Inhibitor + CNS Depressants.
+ CNS Depressants: Parnate should not be used in combination with some central nervous system depressants such as narcotics, alcohol, sedatives, and antihistamines. A marked potentiating effect on these drugs has been reported. (1)
+ Diuretics: see Parnate + Antihypertensive Agents
+ Dopamine: see Parnate + Sympathomimetics. Mechanism: see MAO Inhibitor + Food

PARNATE (continued)

+ Drugs, Over-the-Counter, Cold: Patients should be warned against self-medication with proprietary (over-the-counter) drugs such as cold or hay fever preparations that contain pressor agents or vasoconstrictors. (1) (See also MAO Inhibitor + Sympathomimetics, Indirect-Acting.)
+ Drugs, Over-the-Counter, Hay Fever: see Parnate + Drugs, Over-the-Counter, Cold
+ Drugs, Over-the-Counter, Reducing: see Parnate + Amphetamines
+ Elavil: see Parnate + Antidepressant, Tricyclic
+ Eutonyl: see Parnate + MAO Inhibitor
+ Foods, Tyramine-Containing: Hypertensive crises have sometimes occurred during Parnate therapy after ingestion of foods such as cheese (particularly strong or aged varieties), chianti wine, pickled herring, chicken livers, canned figs, or the pods of broad beans. (1, 2, 20, 53) (See also MAO Inhibitor + Foods.)
+ Insulin: Parnate should be used with caution in patients taking Insulin. (1) (See also MAO Inhibitor + Diabetes and Sulfonylurea, Hypoglycemic.)
+ MAO Inhibitor: Parnate should not be administered together or in rapid succession with other MAO inhibitors (Marplan, Niamid, Eutonyl, Nardil). Hypertensive crises or severe convulsive seizures may occur in patients receiving such combinations. In patients being transferred to Parnate from another MAO inhibitor, allow a medication-free interval of at lease a week, then initiate Parnate using half the normal starting dosage for at least the first week of therapy. Similarly, at least a week should elapse between the discontinuance of Parnate and the administration of another MAO inhibitor. (1, 2) Mechanism: see MAO inhibitor + MAO Inhibitor.
+ Marplan: see Parnate + MAO Inhibitor
+ Meperidine: Parnate potentiates the effects of meperidine. Serious and even fatal reactions (with coma, hypotension, and peripheral vascular collapse) have been reported. (1) Mechanism: see MAO Inhibitor + Meperidine.
+ Narcotics: see Parnate + CNS Depressants
+ Nardil: see Parnate + MAO Inhibitor
+ Niamid: see Parnate + MAO Inhibitor
+ Norpramin: see Parnate + Antidepressant, Tricyclic
+ Pertofrane: see Parnate + Antidepressant, Tricyclic
+ Phenothiazines: When Parnate is combined with phenothiazine derivatives or other compounds known to cause hypotension, the possibility of additive hypotensive effects should be considered. (1, 59) Mechanism: see MAO Inhibitor + Phenothiazines.

PARNATE (continued)

+ Pressor Agents: see Parnate + Amphetamines. When hypotension due to overdosage requires treatment, the standard measures for managing circulatory shock should be initiated. If pressor agents are used, the rate of infusion should be regulated by careful observation of the patient because an exaggerated pressor response sometimes occurs in the presence of MAO inhibitors. (See also MAO Inhibitor + Sympathomimetics, Indirect-Acting and Direct-Acting as well as Parnate + Regitine.) (1)
+ Regitine: If a hypertensive crisis occurs, Patnate should be discontinued and therapy to lower blood pressure should be instituted immediately. On the basis of present evidence, Regitine (5 mg i.v.) or Ansolysen (3 mg s.c.) is recommended. Care should be taken to administer these drugs slowly in order to avoid producing an excessive hypotensive effect. (1, 2)
+ Reserpine, Parenteral: Parenteral reserpine should not be used in the treatment of hypertensive crises. (1) (See also MAO Inhibitor + Reserpine.)
+ Sedatives: see Parnate + CNS Depressants
+ Sympathomimetics: Parnate should not be administered in combination with sympathomimetics and compounds such as Aldomet, dopamine, and tryptophane as this may precipitate hypertension, headache, and related symptoms. (1, 2, 53) Mechanism: see MAO Inhibitor + Sympathomimetics, Indirect-Acting.
+ Tofranil: see Parnate + Antidepressant, Tricyclic
+ Tryptophane: see Parnate + Sympathomimetics. (See also MAO Inhibitor + Amino Acids.)
+ Vasoconstrictors: see Parnate + Amphetamines
+ Vivactil: see Parnate + Antidepressant, Tricyclic

PATHOCIL

+ Meals: Pathocil is best absorbed when taken on an empty stomach, one or two hours before meals. (1)
+ Neutrapen: Penicillinase would probably be ineffective for the treatment of allergic reactions. (1)

PAVERIL PHOSPHATE

+ Acidic Compounds: see Paveril Phosphate + Basic Compounds
+ Basic Compounds: Paveril Phosphate in solution is incompatible with basic and weakly acid compounds. (1)

P-B-SAL-C + Sulfonamides: When p-aminobenzoic acid is administered concurrently with sulfonamides, it inhibits bacteriostatic action of sulfonamides. (1) Mechanism: PABA is absorbed by bacterial pathogens

P-B-SAL-C + Sulfonamides (continued): preferentially over sulfonamides and so inhibits the action of the sulfonamides.

P-B-SAL-C with COLCHICINE: see P-B-Sal-C

P-B-SAL-C with EPINEPHRINE: see P-B-Sal-C

P-B-SAL-C with ESOPRINE: see P-B-Sal-C and Physostigmine

P-B-SAL-C with PREDNISOLONE: see P-B-Sal-C and also Prednisolone

PEGANONE

+ Drugs, Hematopoietic Depressants: Its use in combination with other drugs known to adversely affect the hematopoietic system should be avoided if possible. (1)

+ Phenurone: Peganone is compatible with all commonly employed anticonvulsant medications with the exception of Phenurone. Considerable caution should be exercised if Peganone is administered concurrently with Phenurone since paranoid symptoms have been reported during therapy with this combination. (1)

PENBRITIN + Benemid: Concomitant administration of Benemid in urinary tract infections is contraindicated, since it may decrease the urinary concentration of Penbritin. (2)

PENICILLIN: see listed under the following agents:

Benemid
Butazolidin
Coumadin
Hedulin
pH
Surgical Absorbable Hemostat

PENICILLIN

+ Antibiotics, Bacteriostatic: Penicillin is antagonized by bacteriostatic drugs (e.g., tetracycline, Chloromycetin). The penicillin inhibits the formation of the bacterial cell wall so that when growth takes place the cells die by lysis, but when the cells are not growing they are not killed. If penicillin is combined with tetracycline, the latter prevents multiplication of the cells and therefore interferes with the killing effect of the penicillin. (122)

+ Chloromycetin: see Penicillin + Antibiotics, Bacteriostatic

+ Erythromycin: Erythromycin and Novobiocin give variable results depending on the concentration. In low concentrations they are bacteriostatic and may antagonize the penicillin. In high concentrations they are often bacteriocidal, and when mixed with benzylpenicillin (Penicillin G) in such concentrations they are indifferent or even synergistic. (122)

PENICILLIN (continued)

+ Meals: Buffered penicillin tablets should be given not later than one hour before meals and not sooner than two, or preferably three hours after meals. This method of administration will give maximum absorption of the penicillin. (1)
+ Novobiocin: see Penicillin + Erythromycin
+ Tetracycline: see Penicillin + Antibiotics, Bacteriostatic

PENICILLIN V POTASSIUM + Neomycin: see Neomycin + Penicillin V Potassium

PENTHRANE

+ Antibiotics: see Antibiotics + Penthrane
+ Arfonad: see Penthrane + Ganglionic Blocking Agents
+ Barbiturates: Barbiturates and narcotics should be administered in conservative doses to avoid ventilatory depression. Penthrane causes a sufficient reduction in respiratory minute volume during deep surgical anesthesia to produce a significant respiratory acidosis if ventilation is not adequately assisted. When an intravenous barbiturate is administered to facilitate induction of Penthrane, a greater than usual decrease in blood pressure may occur because both agents depress cardiac output. (1)
+ Epinephrine: In conservative amounts epinephrine has been used both topically and subcutaneously in man without complications during Penthrane anesthesia; however, the possibility of complications should be kept in mind, and epinephrine should always be used with caution during anesthesia. Large intravenous doses of epinephrine have produced ventricular fibrillation in dogs. (1)
+ Flaxedil: see Penthrane + Non-Depolarizing Muscle Relaxants
+ Ganglionic Blocking Agents: Penthrane augments the effects of ganglionic blocking agents (e.g., Arfonad) causing an increased hypotensive effect so that less than the usual dosage of these agents may be needed. (1)
+ Muscle Relaxants, Non-Depolarizing: Penthrane augments the effect of non-depolarizing muscle relaxants (e.g., Flaxedil, tubocurarine) so that less than the usual dosage of these agents may be needed. (1)
+ Narcotics: see Penthrane + Barbiturates
+ Plastics: Polyvinylchloride plastics are extracted by Penthrane. (1)
+ Rubber: Penthrane liquid or vapor will extract rubber (including the rubber components of anesthesia circuits) sufficiently to accelerate aging. Rubber absorbs and releases Penthrane vapor to an extent which may affect induction and prolong recovery time. (1)

PENTHRANE (continued)

+ Tubocurarine: see Penthrane + Non-Depolarizing Muscle Relaxants. See also Tubocurarine Chloride Inj. + Penthrane

PENTIDS + Meals: For maximum absorption of penicillin, dosage should be given on an empty stomach. Thus doses of 200,000 units should be given one-half hour before or two hours after meals. The blood concentrations with doses of 400,000 units is sufficiently high to inhibit sensitive bacteria when the tablets are given without regard to meals but, as can be expected, the resultant concentration will be higher when they are given before meals. (1)

PENTOTHAL SODIUM: see listed under the following agents:

Anectine
Dopram
Metubine Iodide
Succinylcholine
Sucostrin

PENTOTHAL SODIUM

+ Phenothiazines: see Pentothal Sodium + Sedatives
+ Narcotics: see Pentothal Sodium + Sedatives
+ Sedatives: Patients who have developed tolerance to the soporific effects of any barbiturate or who have been receiving large doses of narcotic analgesics, phenothiazines, or other sedatives will show resistance to thiopental and similar drugs. Unfortunately, neither acute nor acquired tolerance to these drugs applies to their cardiovascular effects, and if this is not appreciated the large doses given could lead to prolonged hypotension. (110)
+ Succinylcholine: If muscular relaxation is required, succinylcholine chloride may be combined with Pentothal, or preferably given after the desired effect of Pentothal has been achieved. However, combined administration of a muscle relaxant with Pentothal should be used only by those familiar with the actions of both components. Because of the alkalinity of Pentothal, there is a rapid destruction of succinylcholine when the two agents are mixed. Thus, if they are to be administered simultaneously they should be mixed immediately before use. (1)
+ Tubocurarine: If muscle relaxation is required, tubocurarine chloride may be combined with Pentothal, or preferably given after the desired effect of Pentothal has been achieved. However, combined administration of a muscle relaxant should be used only by those familiar with the actions of both components. (1)

PENTRITOL
+ Acetylcholine: see Pentritol + Norepinephrine
+ Alcohol: An occasional individual exhibits marked sensitivity to the hypotensive effects of nitrite and severe responses (nausea, vomiting, weakness, pallor, perspiration, and collapse) can occur, even with the usual therapeutic dose. Alcohol may enhance this effect. (1)
+ Histamine: see Pentritol + Norepinephrine
+ Norepinephrine: Pentritol can act as a physiological antagonist to norepinephrine, acetylcholine, histamine, and many other agents. (1)

PERCORTEN + Sodium: Patients treated with Percorten will show a decreased excretion of sodium and an increased excretion of potassium. Sodium retention and loss of potassium are accelerated by a high intake of sodium. Hence it may be necessary to restrict intake of sodium and increase that of potassium. (1)

PERIACTIN
+ Alcohol: Patients using this drug should be cautioned against the ingestion of alcohol or other central nervous system depressants. (1)
+ CNS Depressants: see Periactin + Alcohol

PERITHIAZIDE SA: see Thiazides

PERITHIAZIDE SA
+ Digitalis: Perithiazide SA should be administered with caution to those receiving digitalis or ganglionic blocking agents. (1)
+ Ganglionic Blocking Agents: see Perithiazide SA + Digitalis

PERMITIL: see also Phenothiazines

PERMITIL
+ Alcohol: see Permitil + CNS Depressants
+ Analgesics: see Permitil + CNS Depressants
+ Anesthesia: As with other phenothiazines, some patients receiving Permitil who are undergoing surgery should be watched closely for the possible occurrence of hypotension. In addition, lower dosage of CNS depressants and anesthetic agents may be required in these patients. (1)
+ Antihistamines: see Permitil + CNS Depressants
+ Antihypertensive Agents: Permitil may potentiate the action of antihypertensive agents, and care should be exercised when used concomitantly. (1)
+ Atropine: Although potentiation of the effects of atropine, high environmental temperature, and phosphorous insecticides has not been

PERMITIL (continued)

+ Atropine (continued): reported with Permitil, the possibility of the occurrence of such potentiation should be kept in mind. (1)
+ Barbiturates: see Permitil + CNS Depressants
+ Bone Marrow Depressants: Permitil is contraindicated in patients exhibiting leukopenia or other signs of bone marrow depression. (1)
+ CNS Depressants: Since some phenothiazine derivatives are known to potentiate the effects of central nervous system depressants (opiates, analgesics, antihistamines, barbiturates, alcohol), Permitil is contraindicated in patients receiving high dosages of such drugs. (1, 2)
+ Drugs, Toxic: It should be kept in mind that the high emetic potency of Permitil may tend to mask effects of toxic agents. (1)
+ Epinephrine: Epinephrine is contraindicated in hypotension due to Permitil, since other phenothiazine derivatives have been found to reverse the usual action of epinephrine resulting in further lowering of blood pressure. Levophed appears to be the most suitable vasoconstrictor. (1, 2)
+ Opiates: see Permitil + CNS Depressants
+ Phosphorous Insecticides: see Permitil + Atropine

PERSANTINE

+ Digitalis: Some investigators claim that Persantine enhances the effect of digitalis, but others have shown that the drug produces no positive inotropic effect. (2)
+ Heparin: see Heparin + Persantine

PERSISTIN + Salicylates: Other salicylates should not be taken concurrently. (1)

PERTOFRANE: see also Antidepressants, Tricyclic

PERTOFRANE: see listed under the following agents:

Eutonyl	Nardil
Ismelin	Niamid
MAO Inhibitors	Parnate
Marplan	Trisoralen

PERTOFRANE

+ Adrenergic Neuron-Blocking Agents: see Pertofrane + Ismelin
+ Alcohol: see Pertofrane + Central Nervous System Drugs
+ Amphetamine: Pertofrane impairs the hydroxylation of amphetamine (in animals) in the liver thereby increasing the level of circulating amphetamine. (114)
+ Anticholinergic Agents: Although the anticholinergic activity of Pertofrane is weak, in susceptible patients and in those receiving anti-

PERTOFRANE (continued)

\+ Anticholinergic Agents (continued): cholinergic drugs (including antiparkinsonism agents), atropine-like effects may be more pronounced (e.g., paralytic ileus). (1, 2)

\+ Antiparkinsonism Agents: see Pertofrane + Anticholinergic Agents

\+ Central Nervous System Drugs: The concurrent use of other central nervous system drugs or alcohol may potentiate the adverse effects of Pertofrane. Since many such drugs may be used during surgery, it is recommended that Pertofrane be discontinued for as long as the clinical situation will allow prior to elective surgery. Hypertensive episodes have been observed during surgery in patients on Pertofrane therapy. (1)

\+ Ismelin: In some instances, Pertofrane and the parent compound, Tofranil, have been shown to block the pharmacologic action of the antihypertensive, Ismelin, and related adrenergic neuron-blocking agents. (1) Mechanism: see Antidepressants, Tricyclic + Ismelin.

\+ MAO Inhibitors: The use of Pertofrane concomitantly or within two weeks of the administration of MAO inhibitors is contraindicated. Hyperpyretic crises or severe convulsive seizures may occur in patients receiving such combinations. The potentiation of adverse reactions can be serious or even fatal. When it is desired to substitute Pertofrane in patients receiving a monoamine oxidase inhibitor, as long an interval should elapse as the clinical situation allows, with a minimum of 14 days. Initial dosage should be low and increases should be gradual and cautiously prescribed. (1) Mechanism: see MAO Inhibitor + Depressants, Tricyclic.

\+ Phenothiazines: see Pertofrane + Thiazide Diuretics

\+ Surgery: see Pertofrane + Central Nervous System Drugs

\+ Sympathomimetic Compounds: Sympathomimetic compounds may potentiate the effects of Pertofrane. (2)

\+ Thiazide Diuretics: Special care should be taken when Pertofrane is used with other agents that lower blood pressure (e.g., thiazide diuretics and phenothiazine compounds). (2)

\+ Thyroid: Caution should be observed in prescribing the drug in hyperthyroid patients and in those receiving thyroid medication. Transient cardiac arrhythmias have occurred in rare instances. (1)

\+ Vasodilators: Orthostatic hypotension and tachycardia have been observed but seldom require discontinuation of treatment. Patients who require concomitant vasodilating therapy should be carefully supervised, particularly during the initial phase. (1)

PERTOFRANE (Overdosage)

+ Alkalinization: Intravenous potassium is the treatment of choice for controlling arrhythmias and conduction disorders that have been associated with Pertofrane overdosage. Sodium lactate and sodium bicarbonate each act by producing a shift in intracellular potassium and calcium. Therefore intravenous administration of these agents may be of value. However, it should be borne in mind that alkalinization may decrease the renal excretion of Pertofrane. (1)
+ Barbiturates: Parenteral barbiturates have been useful in controlling convulsions due to Pertofrane overdosage. However, since barbiturates may induce respiratory depression, particularly in children, it is advisable to have equipment ready for artificial ventilation and resuscitation whenever these agents are employed. (1)
+ Digitalis: Cardiac Failure: Digitalis may increase the toxic effects of Pertofrane (overdosage) on the myocardium. Therefore it should be administered cautiously and only when cardiac failure is imminent, as evidenced by a rising central venous pressure and the development of hypotension. (1)
+ Inderal: Inderal is contraindicated in the treatment of cardiac arrhythmias by conduction defects (commonly seen after Tofranil overdosage). However, although clinical evidence is not available, Inderal treatment may be indicated in rare instances of cardiac arrhythmias without conduction defects. (1)

PETROLEUM JELLY + Vacuetts Suppos.: see Vacuetts Suppos. + Petroleum Jelly

PETROLEUM PRODUCTS + Ipechar Poison Control Kit: see Ipechar Poison Control Kit + Petroleum Products

PFIZER-VAX MEASLES-L

+ Alkylating Agents: see Pfizer-Vax Measles-L + Steroids
+ Antimetabolites: see Pfizer-Vax Measles-L + Steroids
+ Antiseptics: Because preservatives, antiseptics, and detergents will inactivate measles live virus vaccine, it is important that only the sterile diluent for reconstitution supplied be used in reconstituting this vaccine. (1)
+ Blood Transfusion: This vaccine is contraindicated if a blood transfusion had been administered within the preceding six weeks. The probable immune globulin content may inhibit the response to vaccination. (1)
+ Detergents: see Pfizer-Vax Measles-L + Antiseptics
+ Diluent: see Pfizer-Vax Measles-L + Antiseptics

PFIZER-VAX MEASLES-L (continued)

+ Gamma Globulin: If gamma globulin (immune serum globulin, human), more than 0.01 ml per pound of body weight, has been administered within the preceding six weeks, inoculation should be deferred, since the administered globulin may block vaccine response.

  Systemic manifestations, such as fever (in approximately 12% of subjects) and rash (in 4.5% of subjects) may follow the administration of Pfizer-Vax Measles-L with titred gamma globulin. If present these usually occur 5 to 14 days postvaccination and persist from 1 to 2 days. The fever, on occasion, may reach 104°; most reports of fever were indicated to be lowgrade. (1)
+ Irradiation: see Pfizer-Vax Measles-L + Steroids
+ Poliovirus Vaccine, Live, Attenuated (Sabin): see Pfizer-Vax Measles-L + Vaccines, Live Virus
+ Preservatives: see Pfizer-Vax Measles-L + Antiseptics
+ Smallpox Vaccine: see Pfizer-Vax Measles-L + Vaccines, Live Virus
+ Steroids: Therapy which depresses resistance such as steroids, irradiation, alkylating agents, and antimetabolites are contraindications for use of live measles vaccine. (1)
+ Vaccines, Live Virus: Until further data are available, simultaneous administration of measles virus vaccine, live, attenuated, and poliovirus vaccine, live, attenuated (Sabin), or smallpox vaccination should be avoided. A minimum interval of one month is recommended. (1)

pH: see listed under the following agents:
  Fungizone Intravenous
  Garamycin
  Norisodrine

pH
+ Iron: Iron is best absorbed when the gastric content is highly acid. (50)
+ Penicillin: Penicillin G is destroyed at low pH. (50)

PHANTOS + MAO Inhibitors: Co not use Phantos in patients taking MAO Inhibitors, since Phantos contains a sympathomimetic amine. (1)
  Mechanism: see MAO Inhibitor + Sympathomimetics, Indirect-Acting.

PHENERGAN
+ Analgesics: see Phenergan + CNS Depressants
+ Barbiturates: see Phenergan + CNS Depressants
+ CNS Depressants: Phenergan potentiates the effects of central nervous system depressants; therefore the dose of barbiturates should be

PHENERGAN (continued)

+ CNS Depressants (continued): eliminated or reduced by at least one-half in the presence of Phenergan. The dose of meperidine, morphine, and other analgesic depressants should be reduced by one-quarter to one-half. (1, 2)
+ Meperidine: see Phenergan + CNS Depressants
+ Morphine: see Phenergan + CNS Depressants
+ Trisoralen: see Trisoralen

PHENETHYLAMINE: see listed under the following agents:
Aventyl
Reserpine

PHENINDIONE + Analexin: see Analexin + Phenindione

PHENOBARBITAL: see Acidic Agents + Alkanizing Agents (Urinary)

PHENOBARBITAL: see listed under the following agents:
Anticoagulants
Dicumarol
Diphenylhydantoin
Griseofulvin
Panwarfin
Regitine

PHENOBARBITAL

+ Alcohol: The barbiturates are potentiated by alcohol, a combination that is lethal. (55) Mechanism: The inhibition of drug-metabolizing enzymes by alcohol may contribute to the increased sensitivity of inebriated persons to barbiturates. (47)
+ Androgens: Phenobarbital stimulates the metabolism, by enhancing hydroxylation, of androgens, estrogens, progestational steroids, and glucocorticoids, thus possibly decreasing the action of these agents. (44, 55)
+ Anticoagulants: see Phenobarbital + Warfarin
+ Dicumarol: see Phenobarbital + Warfarin (31, 64, 120)
+ Diphenylhydantoin: Phenobarbital increases the metabolism of diphenylhydantoin thereby decreasing plasma levels and anticonvulsant effect. The fact that phenobarbital has anticonvulsant properties which adds to the pharmacologic properties of the primary antiepileptic agent at least a portion of what it subtracts negates the problem of patient management. (31, 55, 120)
+ Estrogens: see Phenobarbital + Androgens
+ Fulvicin: see Phenobarbital + Griseofulvin
+ Glucocorticoids: see Phenobarbital + Androgens
+ Grifulvin: see Phenobarbital + Griseofulvin
+ Grisactin: see Phenobarbital + Griseofulvin

PHENOBARBITAL (continued)

+ Griseofulvin: Phenobarbital stimulates the metabolism of griseofulvin (e.g., Fulvicin, Grifulvin, Grisactin), decreasing blood levels of griseofulvin and presumably decreasing the antifungal action of the compound. (31, 55, 120)
+ Hexobarbital: The ability of phenobarbital to induce enzymatic inactivation of other drugs is most widespread, thus phenobarbital may reduce the hypnotic effect of hexobarbital. (31)
+ Progestational Steroids: see Phenobarbital + Androgens
+ Warfarin: Phenobarbital increases the metabolism of warfarin and Dicumarol thereby decreasing the effect of these anticoagulants. This effect also occurs with other oral anticoagulants. (55)

PHENOBARBITAL SODIUM SOL. + Murel Inj.: see Murel Inj. + Phenobarbital Sodium Solution

PHENOL: see listed under the following agents:

Aristocort Forte Suspension
Furadantin Sodium Sterile
Surital

PHENOLSULFONPHTHALEIN + Benemid: see Benemid + Phenolsulfonphthalein

PHENOTHIAZINES: Not all of the following interactions have been observed with all of the phenothiazines, but they should be borne in mind when drugs of this class are prescribed. See also under individual trade-named preparations.

PHENOTHIAZINES: see listed under the following agents:

Alcohol
Atarax
Benzodiazepines
Cogentin
Diabinese
Dymelor
Eutonyl
Flurothyl
Hedulin
Hypoglycemic Drugs, Oral
Lenetran
Levanil
Levophed
Librium
Listica
MAO Inhibitors
Matulane
Norpramin
Organophosphate Pesticides
Orinase
Parnate
Pentothal
Pertofrane
Prinadol
Protopam
Quiactin
Softran
Solacen

PHENOTHIAZINES (continued)

Tacaryl
Tigan
Trancopal
Trepidone
Trisoralen
Tybatran
Ultran
Valium
Vistaril
Vivactil

PHENOTHIAZINES

+ Alcohol: see Phenothiazines + CNS Depressants. Mechanism: Potentiation of the effects of alcohol are due to decreased metabolism (depression of alcohol dehydrogenase) of the alcohol by the phenothiazine and also the sensory attenuation. (59)

+ Analgesics: see Phenothiazines + CNS Depressants, Morphine, Narcotics, and Meperidine

+ Anesthetics: see Phenothiazines + CNS Depressants. (133)

+ Anticholinergics: Phenothiazines have atropine-like effects which are additive to anticholinergic drugs possibly causing dry mouth, blurred vision, urinary retention, and even precipitating glaucoma. (55, 134)

+ Antidepressants, Tricyclic: Phenothiazines lower the seizure threshold and when possible should be avoided in epileptic patients. This is particularly pertinent when the patients are taking other agents like the tricyclic antidepressants that also lower the seizure threshold. (50, 55)

+ Antihistamines: see Phenothiazines + CNS Depressants. Concomitant use may cause additive sedative and atropine-like effects.

+ Antihypertensive Agents: If a patient receiving a phenothiazine is also given an antihypertensive agent, watch for signs of potentiation of the antihypertensive agent. Profound hypotension may occur. (2, 55)

+ Aramine: Patients in shock or with severe hypotension due to phenothiazines should not be treated with metaraminol (Aramine, Pressonex, Pressorol) because phenothiazines block part of the effect of metaraminol resulting in increased hypotension. (55)

+ Atropine: Phenothiazines potentiate the effects of atropine. See also Phenothiazines + Anticholinergics.

+ Barbiturates: see Phenothiazines + CNS Depressants

+ CNS Depressants: Phenothiazine preparations potentiate the action of other central nervous system depressants. They should be used cautiously in patients undergoing treatment concomitantly with alcohol, anesthetics and general anesthetics, antihistamines, analgesics, barbiturates, hypnotics, morphine, narcotics, opiates, sedatives, and tranquilizers. (55)

PHENOTHIAZINES (continued)

+ Diuretics, Thiazide: Thiazide diuretics given to a patient on phenothiazines have been known to cause severe hypotension and shock. (55)
+ Drugs, Cardiac-Depressing: see Phenothiazines + Quinidine
+ Epinephrine: see Phenothiazines + Vasopressor Agents
+ Heat: Phenothiazines potentiate the effects of heat.
+ Hypnotics: see Phenothiazines + CNS Depressants
+ Insecticides, Phosphorous: Phenothiazines potentiate the effects of phosphorous insecticides.
+ MAO Inhibitors: The effects of phenothiazines may be potentiated by MAO inhibitors causing extrapyramidal symptoms (Parkinsonism, dystonias) or hypotension. (2, 6, 55, 59, 78, 105) Mechanism: It is postulated that monoamine oxidase inhibitors inhibit the hepatic microsomal enzymes which metabolize phenothiazines. (115)
+ Meperidine: Patients receiving phenothiazines and meperidine concomitantly may exhibit marked respiratory depression. (55) Maximal respiratory depression does not occur until 1-1/2 hours and persists for about 3-1/2 hours after concurrent injection of a phenothiazine and meperidine. (106) See also Phenothiazines + Narcotics.
+ Metrazol: see Phenothiazines + Picrotoxin
+ Morphine: see Phenothiazines + CNS Depressants. The depressant actions of morphine are exaggerated and prolonged by phenothiazines; the mechanism of this supra-additive effect is not fully understood. (106) See also Phenothiazines + Narcotics.
+ Narcotics: see Phenothiazines + CNS Depressants. The depressant actions of narcotics are exaggerated and prolonged by phenothiazines; the mechanism of this supra-additive effect is not completely understood. Certain phenothiazines reduce the amount of narcotic required to produce a given level of analgesia. However, the respiratory depressant effects seem also to be enhanced, the degree of sedation is increased, and the hypotensive effects of phenothiazines become an additional complicating side action. Some phenothiazines enhance the sedative effects, but at the same time seem to be antianalgesic and increase the amount of narcotic required to produce satisfactory relief from pain. (106)
+ Opiates: see Phenothiazines + CNS Depressants. See also Phenothiazines + Narcotics.
+ Picrotoxin: Picrotoxin and Metrazol should be avoided in cases of phenothiazine overdosage as well as any stimulant that may cause convulsions.

PHENOTHIAZINES (continued)

+ Pressonex: see Phenothiazines + Aramine
+ Pressorol: see Phenothiazines + Aramine
+ Quinidine: Phenothiazines have a quinidine-like action on myocardial conduction and pacemaking tissues. Ventricular tachycardia has been reported during phenothiazine therapy. The use of cardiac depressing drugs like quinidine should be avoided in the treatment of such arrhythmias. (50)
+ Reserpine: Concomitant use of phenothiazines and reserpine will cause potentiation of the reserpine. (55)
+ Sedatives: see Phenothiazines + CNS Depressants
+ Vasopressor Agents: Phenothiazines may cause a reversal of pressor effect of vasopressor agents (e.g., epinephrine). This does not apply to Neosynephrine or Levophed.

PHENOTHIAZINE TRANQUILIZERS: see listed under the following agents:
Anesthetics, General
Aventyl

PHENOXENE: see listed under the following agents:
MAO Inhibitors
Nardil

PHENOXENE + Barbiturates: Studies with animals indicate that Phenoxene potentiates the sedative effect of barbiturates. Although such potentiation has not been reported for humans, it should be used with caution if the patient is also taking a barbiturate. (1)

PHENURONE + Peganone: see Peganone + Phenurone

PHENYLEPHRINE: see listed under the following agents:
Furoxone
Marplan
Nardil

PHOSPHOLINE IODIDE

+ Anectine: see Anectine + Phospholine Iodide
+ Anesthetics: Plasma cholinesterase level is reduced in patients on prolonged Phospholine Iodide therapy. Special caution, therefore, must be taken with patients receiving this eyedrop when ester-type, hydrolysable local anesthetics are administered in order to prevent prolonged apnea and toxic reactions. (66)
+ Anticholinesterase Pesticides: Phospholine Iodide, after absorption into the general circulation, may be additive (or possibly synergistic) with anticholinesterase pesticides of either the organophosphate

PHOSPHOLINE IODIDE (continued)

+ Anticholinesterase Pesticides (continued): (parathion, TEPP, malathion, etc.) or carbamate (Sevin, etc.) class. Consequently, patients using Phospholine Iodide eyedrops, who are regularly exposed to such poisons (pesticide manufacturing and formulating plant workers and spray operators, especially those spraying from airplanes), should be warned to observe with special care all recommended protective measures: masks, cleanliness, clothing changes, etc. Symptoms of anticholinesterase poisoning are: headache, fatigue, giddiness, nausea, salivation, sweating, blurred vision, tightness in chest, abdominal cramps, vomiting, and diarrhea. In severe poisoning, difficult breathing, tremors, convulsions, collapse, coma, pulmonary edema, and respiratory failure follow. Glycosuria occurs in 30% of the cases and pupils are constricted in about 80% of the cases, but in the remainder may be dilated. (2, 132)
+ Atropine-Like Agents: The mydriasis and ciliary paralysis produced by atropine-like agents are partially counteracted by Phospholine Iodide. (2)
+ Carbonic Anhydrase Inhibitor: In some cases, the concomitant administration of a carbonic anhydrase inhibitor may enhance the effectiveness of Phospholine Iodide in controlling intraocular tension. (2)
+ Mestinon: see Phospholine Iodide + Neostigmine
+ Mytelase: see Phospholine Iodide + Neostigmine
+ Neostigmine: In patients under treatment for myasthenia gravis with drugs such as neostigmine, Mytelase, or Mestinon, Phospholine Iodide eyedrops should only be used with full awareness of the likelihood of pharmacologic interaction. (1)
+ Succinylcholine: The muscle relaxant, succinylcholine, should not be used prior to general anesthesia in patients who are under treatment with Phospholine Iodide since cholinesterase inhibitors, in general, potentiate the effects of this drug. (2, 21, 66) Mechanism: Cholinesterase inhibitors by preventing destruction of acetylcholine, which acts as a depolarizing agent at the motor end plates, augments the depolarizing action of succinylcholine. (131)
+ Tensilon: In patients who have been using Phospholine Iodide the Tensilon test should only be attempted by a qualified specialist where artificial respiration is available. (1) Mechanism: Additive effect.

PHOTOSENSITIZING AGENTS: see listed under Trisoralen

PHYSOSTIGMINE: see listed under the following agents:
Anectine
Succinylcholine

PHYTONADIONE: see listed under trade-named products.

PHYTONADIONE + Miradon: see Miradon + Phytonadione

PICROTOXIN: see listed under the following agents:

Compazine
Copavin
Darvon
Nembutal Sodium Sol
Phenothiazine
Quide
Stelazine
Thorazine
Tindal

PIL-DIGIS: see also Digitalis

PIL-DIGIS

+ Calcium I.V.: The administration of calcium intravenously may induce digitalis intoxication. (1)
+ Diuretic Agents: Potassium depletion, usually caused by the use of diuretic agents or electrolyte manipulation sensitizes the heart to digitalis and may produce arrhythmias even with recommended doses. In this event, reduction of dosage may be necessary and administration of potassium may be indicated. (1)

PILOCARPINE: see listed under the following agents:

Darbid
Floropryl
Ornade
Pyrdonnal

PITOCIN: see also Oxytocics

PITOCIN + Pitocin: Pitocin should not be used simultaneously by more than one route of administration because of inherent complications of controlling dosage. (1)

PITRESSIN + Methedrine: see Methedrine + Posteriod Pituitary Extract, Pressor Fraction

PLACIDYL

+ Alcohol: see Placidyl + CNS Depressants
+ Antidepressants, Tricyclic: Caution is advised if patients are receiving antidepressants such as Elavil. Transient delirium has been reported with the combination of Elavil and Placidyl. The dosage of Placidyl should be reduced if prescribed with antidepressants. (1)
+ Barbiturates: see Placidyl + CNS Depressants

PLACIDYL (continued)

+ CNS Depressants: Patients who are taking Placidyl should be cautioned about the possible combined exaggerated effects with alcohol, barbiturates, sedative-hypnotics, tranquilizers, or other central nervous system depressants. Such exaggerated effects might result in blurring of vision, paralysis of accommodation, and profound hypnosis. (1)
+ Elavil: see Placidyl + Antidepressants, Tricyclic
+ Eutonyl: see Placidyl + MAO Inhibitor
+ MAO Inhibitor: Caution is advised in prescribing Placidyl for patients being treated with monoamine oxidase inhibitors including Eutonyl. The dosage of Placidyl should be reduced if prescribed with MAO inhibitors. (1)
+ Megimide: The use of Megimide is not recommended in Placidyl overdosage. (1)
+ Sedative-Hypnotics: see Placidyl + CNS Depressants
+ Tranquilizers: see Placidyl + CNS Depressants

PLAQUENIL: The listings under Aralen also apply in general to other preparations containing 4-aminoquinoline compounds such as Plaquenil. See also Aralen.

PLAQUENIL

+ Alcohol: see Plaquenil + Drugs, Hepatotoxic
+ Ammonium Chloride: If serious toxic symptoms occur from overdosage or sensitivity, it has been suggested that ammonium chloride (2 gm four times daily for adults) be administered orally three or four days a week for several months after therapy has been stopped, as acidification of the urine may increase renal excretion of 4-aminoquinoline compounds by 20 to 90%. However, caution must be exercised in patients with impaired renal function and/or metabolic acidosis. (1)
+ Butazolidin: see Plaquenil + Drugs
+ Drugs: The concomitant use of medicaments such as Butazolidin, gold, and other drugs known to cause sensitization and dermatis should be avoided. (1)
+ Drugs, Hepatotoxic: Use Plaquenil with caution in patients in conjunction with known hepatotoxic drugs and in alcoholics. (1)
+ Gold: see Plaquenil + Drugs

PLASMA: see listed under the following agents:

Hypertensin
Levophed
Lirugen
Measles Virus Vaccine, Live, Attenuated (Schwarz)

PLASMA, HUMAN + Attenuvax, Lyovac: see Attenuvax, Lyovac + Plasma, Human

PLASMANATE
+ Alcohol: see Plasmanate + Protein Hydrolysate Solutions
+ Intravenous Fluids: Solutions of plasma protein fraction should not be mixed with other intravenous fluids, but the concomitant administration of other fluids through another vein is not interdicted. (1)
+ Protein Hydrolysate Solution: Protein hydrolysate solutions and those containing ethyl alcohol should not be mixed or infused through the same set used for Plasmanate. (1)

PLASTICS: see listed under the following agents:
Cresatin
Diethylstibesterol in Ethyl Oleate
Paral
Penthrane
Triburon Ointment

PMB-200: see Premarin and Meprobamate

PMB-200 + Alcohol: Patients should be warned that meprobamate may potentiate the effects of alcohol and hence may cause a slowdown of reaction time and impairment of judgment and coordination. (1)

PMB-400: see PMB-200

POLIOMYELITIS VACCINE: see listed under the following agents:
Lirugen
Measles Virus Vaccine, Live, Attenuated

POLIOMYELITIS VACCINE (Eli Lilly)
+ ACTH: Adrenocorticotrophin and adrenal corticosteroids may suppress the antibody response to the vaccine. Therefore, if possible, it would seem advisable to avoid administration of the vaccine concomitantly with these hormones. (1)
+ Adrenal Corticosteroids: see Poliomyelitis Vaccine (Eli Lilly) + ACTH

POLIOMYELITIS VACCINE, LIVE + M-Vac Measles Virus Vaccine, Live, Attenuated: see M-Vac Measles Virus Vaccine, Live, Attenuated + Poliomyelitis Vaccine

POLIOVIRUS VACCINE, LIVE, ATTENUATED (Sabin) + Pfizer-Vax Measles-L: see Pfizer-Vax Measles-L + Poliovirus Vaccine, Live, Attenuated (Sabin)

POLIOVIRUS VACCINE, LIVE, ORAL, MONOVALENT + Measles Virus Vaccine, Live: The American Academy of Pediatrics recommends that poliovirus vaccine, live, oral, monovalent: Type 1, 2, and 3 (Sabin) and live measles virus vaccine be given separately at intervals of at least one month. (1)

POLIOVIRUS VACCINE, LIVE, ORAL, TRIVALENT (Pfizer) + Measles Virus Vaccine, Live: The American Academy of Pediatrics recommends that oral poliovirus vaccine, live, trivalent: Type 1, 2, and 3 (Sabin) and live measles virus vaccine be given separately at intervals of at least one month. (1)

POLYCILLIN + Benemid: Concomitant administration of Benemid in urinary tract infections is contraindicated, since it may decrease the urinary concentration of Polycillin. (2)

POLYMIXIN: see listed under the following agents:

Coly-Mycin M Inj.
Flaxedil
Garamycin
Heparin
Tubocurarine

POLYMIXIN-B SULFATE: see listed under the following agents:

Anectine
Tergemist

PONSTEL

+ Anticoagulants: Ponstel may cause additional lowering of prothrombin concentration in patients in whom the concentration has been initially lowered by anticoagulant therapy. Caution must be observed if Ponstel is administered to patients on anticoagulant therapy and Ponstel should not be given when the prothrombin concentration is in the range of 10 to 20% of normal. (113)
+ Butazolidin: see Ponstel + Salicylates
+ Corticosteroids: see Ponstel + Salicylates
+ Indocin: see Ponstel + Salicylates
+ Salicylates: Concomitant administration of Ponstel and salicylates, Butazolidin, Indocin, or corticosteroids may potentiate the ulcerogenic effect of these drugs. (113)

POSTERIOR PITUITARY EXTRACT, OXYTOCIC FRACTION + Fluothane: see Fluothane + Posterior Pituitary Extract, Oxytocic Fraction

POSTERIOR PITUITARY EXTRACT, PRESSOR FRACTION + Methedrine: see Methedrine + Posterior Pituitary Extract, Pressor Fraction

POTABA + Sulfonamides: Potaba should not be administered to patients taking sulfonamides. (1) Mechanism: Aminobenzoate is absorbed by bacterial pathogens preferentially over sulfonamides, and so inhibits the action of the sulfonamides.

POTASSIUM: see listed under the following agents:
Aldactone
Dyazide
Dyrenium
Pabalate-SF

POTASSIUM CITRATE + Benemid: see Benemid + Potassium Citrate

PREANESTHETIC AGENTS: see listed under the following agents:
Naturetin
Unitensin

PRECEPTIN VAGINAL GEL + Douche: If a douche is desired for cleansing purposes, it should be deferred for at least six hours following intercourse. (1)

PREDISAL + Diabetes: Patients with diabetes may need restabilization if Predisal is administered. (1)

PREDNIS + Diabetes: Diabetics treated with either Prednis or prednisone for a concurrent disease may develop exaggerated hyperglycemia, requiring higher dosage of insulin and/or oral hypoglycemic agents. The diabetic state should be followed carefully and regulated accordingly. The increased insulin need is temporary and is reduced upon cessation of corticotherapy. (1)

PREDNISOLONE (McKesson)
+ Diabetes: The use of prednisolone is contraindicated in the long-term treatment of any condition complicated by diabetes mellitus. (1)
+ Diuretics: If the temporary use of diuretics is indicated, dangerous loss of potassium must be considered. (1)

PREDNISONE (McKesson)
+ Diabetes: The use of Prednisone is contraindicated in long-term treatment of any condition complicated by diabetes mellitus. (1)
+ Diuretics: If the temporary use of diuretics is indicated, dangerous loss of potassium must be considered. (1)

PREDSEM + Diabetes: Corticosteroids should be used with caution in patients with diabetes. (1)

PRELUDIN
+ CNS Stimulants: Preludin should not be used with other central nervous system stimulants, including MAO inhibitors. (1)
+ Diabetes: Preludin does not increase insulin requirements in diabetic patients; in fact, requirements may be reduced as weight decreases. (1)
+ MAO Inhibitors: see Preludin + CNS Stimulants: (See also MAO Inhibitors + Sympathomimetics, Indirect-Acting.)

PREMARIN + Diabetes: Glucose tolerance may decrease during estrogen therapy, hence diabetic patients should be followed closely. (1)

PREMARIN I.V.
+ Acidic Sol.: Premarin I.V. is not compatible with any solution with an acid pH. (1)
+ Ascorbic Acid Sol.: Premarin I.V. is not compatible with ascorbic acid solution. (1)
+ Protein Hydrolysate: Premarin I.V. is not compatible with protein hydrolysate. (1)

PRE-SATE + MAO Inhibitors: Pre-Sate is contraindicated in patients who are receiving monoamine oxidase inhibitors. (1) (See also MAO Inhibitors + Sympathomimetics, Indirect-Acting.)

PRESERVATIVES: see listed under the following agents:
Attenuvax, Lyovac
Measles Virus Vaccine, Live, Attenuated
M-Vac Measles Virus Vaccine, Live, Attenuated
Pfizer-Vax Measles-L

PRESSONEX: see also Aramine

PRESSONEX: see listed under the following agents:
Cyclopropane
Halothane
Phenothiazines

PRESSONEX
+ Cyclopropane: The use of Pressonex in patients under cyclopropane anesthesia should be generally avoided. However, if indicated, it should be administered with caution, avoiding hypertension, to reduce the risk of causing ventricular fibrillation. (1) Mechanism: The retention of carbon dioxide during anesthesia causes an increase in sympathetic nervous activity and a liberation of norepinephrine within the myocardium and specialized conducting tissue. In addition, cyclopropane in some manner "sensitizes" the heart to the

PRESSONEX (continued)

+ Cyclopropane (continued): action of norepinephrine and related catecholamines. (123)
+ Halothane: During Halothane anesthesia, vasopressors may cause serious arrhythmias and therefore should only be used with great caution or not at all. (1) Mechanism: see Pressonex + Cyclopropane.
+ MAO Inhibitors: see MAO Inhibitors + Sympathomimetics, Indirect-Acting
+ Oxytocic Drugs: In obstetrics, oxytocics and vasopressor drugs should not be used simultaneously because severe persistent hypertension may occur. (1)

PRESSOR AGENTS: see listed under the following agents:

Compazine	Parnate
Dibenzyline	Ritalin
Hypertensin	Saluron
Nardil	Stelazine
Niamid	Thorazine

PRESSOR AMINES: see listed under the following agents:

Anhydron	Inversine
Aquatag	Lasix
Bristuron	MAO Inhibitor
Deprol	Marplan
Diuril	Metahydrin
Enduron	Naqua
Esidrix	Naturetin
Exna	Oretic
HydroDIURIL	Renese
Inapsine	Saluron

PRESSOROL: see also Aramine and Pressonex

PRESSOROL

+ Cyclopropane: Pressorol injection should not be used concurrently with cyclopropane anesthesia. (1) Mechanism: The retention of carbon dioxide during anesthesia causes an increase in sympathetic nervous activity and a liberation of norepinephrine within the myocardium and specialized conducting tissue. In addition, cyclopropane in some manner "sensitizes" the heart to the action of norepinephrine and related catecholamines. (123)
+ Phenothiazines: see Phenothiazines + Pressorol

PRIMAQUINE

+ Aralen: Primaquine used in combination with the usual antimalarial doses of Aralen, mild and transient headaches, disturbances of visual accommodation, pruritus, and gastrointestinal complaints are frequently observed. (1)
+ Atabrine: Do not give Primaquine to patients receiving Atabrine, as Atabrine appears to potentiate toxicity. (1)

PRINADOL

+ Anesthetics: see Prinadol + CNS Depressants
+ Barbiturates: see Prinadol + CNS Depressants
+ CNS Depressants: The depressant pharmacological effects of Prinadol are potentiated by central nervous system depressants such as barbiturates, phenothiazines, anesthetics, etc. Therefore the lower dose suggested should always be used when these agents are given concomitantly with Prinadol particularly to elderly or debilitated patients. (1)
+ MAO Inhibitors: As with all narcotic-analgesics, Prinadol should be used with extreme caution in patients taking monoamine oxidase inhibitors. (1) (See MAO Inhibitor + CNS Depressants.)
+ Phenothiazines: see Prinadol + CNS Depressants

PRINCIPEN + Meals: For maximal absorption, Principen should be administered at least two hours after or one-half hour before meals. (1)

PRISCOLINE

+ Epinephrine: Large doses of Priscoline (overdosage) may cause "epinephrine reversal." Therefore avoid using epinephrine and norepinephrine (or use very cautiously), since they may cause further reduction of blood pressure followed by an exaggerated rebound. (1)
+ Norepinephrine: see Priscoline + Epinephrine

PROBANTHINE with DARTAL: see Dartal

PROCAINE: see Alkaline Agents + Acidic Media

PROCAINE: see Alkaline Agents + Acidifying Agents (Urinary)

PROCAINE: see listed under the following agents:

Anectine
MAO Inhibitor
Marplan
Nardil
Niamid
Para-amino Salicylic Acid (PAS)
Quelicin Chloride Injection
Succinylcholine
Sucostrin

PROCAINE (continued)

+ Para-Amino Salicylic Acid (PAS): Procaine antagonizes para-amino salicylic acid. (1)
+ Sulfonamides: Procaine and other local anesthetics inhibit the action of sulfonamides. These local anesthetics should not be employed in any condition in which sulfonamide therapy is being employed. Mechanism: Local anesthetics which are hydrolyzed to form para-aminobenzoic acid, such as Procaine, inhibit the action of sulfonamides by being absorbed preferentially by bacterial pathogens over sulfonamides. (135)

PRODECADRON RESPIHALER: see Decadron and Isuprel

PROGESTATIONAL STEROIDS + Phenobarbital: see Phenobarbital + Progestational Steroids

PROKETAZINE: see Phenothiazines

PROKETAZINE: see listed under the following agents:

Demerol
Trisoralen

PROKETAZINE

+ Alcohol: see Proketazine + CNS Depressants
+ Anesthetics, General: see Proketazine + CNS Depressants
+ Antihypertensive Agents: Proketazine may potentiate the action of antihypertensive agents, and care should be exercised when used concomitantly. (1, 2)
+ CNS Depressants: Proketazine may potentiate the effects of central nervous system depressants, and care should be exercised when used concomitantly (e.g., general anesthetics, hypnotics, alcohol, etc.) (1, 2)
+ Epinephrine: Epinephrine should not be administered concomitantly, since Proketazine may reverse its action and cause profound hypotension. When antihypertensives are given, the use of vasopressors such as Levophed may be indicated if the resulting hypotension is prolonged or severe. (1, 2)
+ Hypnotics: see Proketazine + CNS Depressants

PROLIXIN: see Phenothiazines

PROLIXIN

+ Alcohol: see Prolixin + CNS Depressants
+ Analgesics: see Prolixin + CNS Depressants
+ Anesthetics: Psychotic patients on large doses of phenothiazine drugs who are undergoing surgery should be watched carefully for possible

PROLIXIN (continued)

+ Anesthetics (continued): hypotensive phenomena. Moreover, it should be remembered that reduced amounts of anesthetics or central nervous system depressants may be required. (1)
+ Antihistamines: see Prolixin + CNS Depressants
+ Antihypertensive Agents: Prolixin may potentiate the action of antihypertensive agents, and care should be taken when used concomitantly. (2)
+ Atropine: The effects of atropine may be potentiated in some patients receiving Prolixin. (1)
+ CNS Depressants: Although not a general feature of Prolixin, potentiation of central nervous system depressants (opiates, analgesics, antihistamines, barbiturates, alcohol) may occur. (1, 2) See also Prolixin + Anesthetics.
+ Epinephrine: Epinephrine should not be used to treat hypotensive effects that may be caused by Prolixin, since phenothiazine derivatives have been found to reverse its action, resulting in a further lowering of blood pressure. If a pressor drug is indicated, Levophed is the most suitable drug for this purpose. (1, 2)
+ Hypnotics: Phenothiazine compounds, including Prolixin, should not be used in patients receiving large doses of hypnotics. (1)
+ Insecticides, Phosphorous: Phosphorous insecticides may be potentiated in some patients receiving Prolixin. (1)
+ Opiates: see Prolixin + CNS Depressants
+ Trisoralen: see Trisoralen

PROLOID + Cytomel: see Cytomel + Proloid

PRONESTYL + Digitalis: Caution is required in severe digitalis intoxication where the use of Pronestyl may result in additional depression of conduction and ventricular asystole or fibrillation. (1)

PROPADRINE + MAO Inhibitors: see MAO Inhibitor + Propadrine

PROPYLPARABEN: see listed under the following agents:
Aristocort Forte Suspension
Furadantin Sodium Sterile

PROPYLTHIOURACIL + Coumadin: see Coumadin + Propylthiouracil

PROSTAPHLIN + Food: Oral doses of Prostaphlin should be taken one or two hours before meals to minimize retention of the drug and its destruction by gastric acidity. (2)

PROSTIGMIN
+ Atropine: Atropine may be used to abolish or obtund gastrointestinal side effects or other muscarinic reactions, but such use, by masking signs of overdosage, can lead to inadvertent induction of cholinergic crisis. (1)
+ Cyclopropane: see Prostigmin + Halothane
+ Halothane: Prostigmine should never be administered in the presence of high concentrations of halothane or cyclopropane. (1)

PROTALBA
+ Digitalis: Cardiac arrhythmias may occasionally develop, more often in patients receiving Protalba and digitalis concomitantly, and can usually be reversed by administering atropine. Protalba is contraindicated in patients receiving digitalis. (2)
+ Diuretics: Protalba has been used in combination with oral diuretics and rauwolfia compounds, thus reducing the dosage requirement of Protalba and decreasing the incidence of untoward effects. However, there is some doubt that the reduced dosage in such combination allows the veratrum alkaloids to exert any profound effect. (2)
+ Quinidine: Protalba should be used with care in patients receiving quinidine. (2)

PROTAMINE, ZINC and ILETIN + Sterilizing Solution: The use of heavily chlorinated water or chemical solutions for sterilizing the syringe prior to the injection of Protamine, Zinc and Iletin should be avoided. (1)

PROTEF RECTAL SUPPOSITORIES + Corticosteroids: Protef Rectal Suppositories should be used with caution in patients already receiving corticosteroid therapy orally or parenterally, as 20 to 40% of rectally administered hydrocortisone may be absorbed. (1)

PROTEIN HYDROLYSATE SOLUTION: see listed under the following agents:
Plasmanate
Premarin I.V.

PROTERENOL + Epinephrine: Proterenol should not be administered in conjunction with epinephrine; the two may be alternated, however. (1)

PRO-TET + Tetanus Toxoid: Many physicians advocate beginning active immunization against tetanus by administering simultaneously tetanus toxoid and tetanus immune globulin (human). When this procedure is used, the two products should not be administered in the same extremity or area. However, there is not universal agreement as to whether such action is without danger of the two materials neutralizing each other to

PRO-TET + Tetanus Toxoid (continued): some degree and thus reducing the levels of passive and active immunization. (1)

PROTHROMBINOPENIC DRUGS: see listed under the following agents:
Depo-Heparin Sodium
Heparin

PROTOPAM

+ Aminophylline: see Protopam + Morphine
+ Atropine: Treatment of insecticide poisoning: Atropine should be used as soon as cyanosis has been overcome. Atropine should not be used sooner because of the possibility of inducing ventricular fibrillation in a cyanotic patient. The possibility exists of atropine intoxication occurring in a patient who receives Protopam after large doses of atropine due to removal of accumulated acetylcholine by freshly reactivated cholinesterase. It is unlikely that this would prevent a critical problem. (1)
+ Barbiturates: Treatment of convulsions which interfere with respiration in insecticide poisoning: Give sodium thiopental (2.5% solution) intravenously with more than the usual care; poisoning by anticholnesterases sensitizes the medullary centers to depression by barbiturates. (1) See also Organophosphate Pesticides + Barbiturates.
+ Morphine: Morphine, theophylline, aminophylline, and succinylcholine are contraindicated when Protopam is used. (1)
+ Phenothiazines: see Protopam + Tranquilizers
+ Reserpine: see Protopam + Tranquilizers
+ Succinylcholine: see Protopam + Morphine
+ Theophylline: see Protopam + Morphine
+ Tranquilizers: Tranquilizers of the reserpine or phenothiazine type are to be avoided. (1)

PROVELL MALEATE

+ Digitalis: In some patients receiving the full therapeutic dose of digitalis or single cardiac glycosides, the bradycardia produced by Provell Maleate may be more pronounced than in an undigitalized patient. The presence of cardiac glycosides does not alter the hypotensive action of Provell Maleate. The cardiac slowing and irregularities may be abolished by atropine. It has been suggested that initial digitalization should not be undertaken when the patient is receiving large doses of proveratrine. (1)
+ Quinidine: Caution should be exercised in giving the drug to patients receiving Provell Maleate. (1)

PROVERA + Estrogens: Do not use Provera alone or in combination with estrogens in abnormal uterine bleeding until the possibility of genital malignancy has been eliminated. (1)

PROVEST: see Contraceptives, Oral

PROVEST + Diabetes: A decrease in glucose tolerance has been observed in a significant percentage of patients on oral contraceptives. Mechanism of this decrease is obscure. Therefore observe carefully diabetic patients receiving Provest. (1, 2)

PROZINE: see Equinal and also Sparine

PSYCHOACTIVE DRUGS: see listed under the following agents:

Dopar
Laradopa
Levodopa

PSYCHOPHARMACOLOGIC AGENTS + Symmetrel: see Symmetrel + Psychopharmacologic Agents

PSYCHOTHERAPEUTIC AGENTS: see listed under the following agents:

Parest
Quaalude
Somnofac
Sopor

PSYCHOTROPIC AGENTS: see listed under the following agents:

Atarax
Inderal
Lenetran
Levanil
Librium
Listica
MAO Inhibitors
Marplan
Nardil
Niamid
Parest
Quiactin
Serax
Softran
Solacen
Somnofac
Sopor
Striatran
Trancopal
Trepidone
Tybatran
Ultran
Valium
Vistaril

PURGATIVES: see listed under the following agents:

Sulfasuxidine
Sulfathalidine

PURINETHOL + Zyloprim: see Zyloprim + Furinethol

PYOPEN + Benemid: Oral Benemid has been used to achieve higher and prolonged blood levels. (1)

PYRAZINAMIDE + Diabetes: In rare instances, diabetes mellitus may become more difficult to control during use of Pyrazinamide. (1)

PYRDONNAL + Pilocarpine: While pilocarpine or similar drugs are sometimes recommended for the relief of dry mouth, many authorities feel that these drugs are not indicated in that they relieve the minor peripheral effect but do not influence the more serious central effects, and thus merely mask signs of drug activity. (1)

PYRIBENZAMINE

+ Hypnotics: Give hypnotics and sedatives cautiously to patients receiving Pyribenzamine. (1)

+ Sedatives: see Pyribenzamine + Hypnotics. In the treatment of Pyribenzamine overdosage in patients who exhibit the symptoms of excitation, give short-acting intravenous barbiturates; avoid long-acting sedatives since their effects may coincide with the later depression caused by Pyribenzamine. (1)

PYRIDOXINE HYDROCHLORIDE: see listed under the following agents:

Dopar
Larodopa
Levodopa

PYRILGIN

+ Anticoagulants: Pyrilgin can cause aggravation of prothrombin deficiency. (1)

+ Thorazine: Pyrilgin should not be used concomitantly with Thorazine. The potentiating effect of Thorazine modifies the antipyretic action of Pyrilgin and severe hypothermia can result. (1)

Q

QUAALUDE
+ Alcohol: see Quaalude + Sedatives and also Thorazine
+ Analeptics: Some investigators have used analeptics with varying success in treatment of overdosage. Analeptics have increased tonic-clonic spasm in the presence of severe toxicity. (1)
+ Analgesics: see Quaalude + Sedatives
+ Barbiturates: Quaalude potentiates barbiturates. (1, 51)
+ CNS Depressants: Quaalude is not recommended for use with other central nervous system depressants. (1)
+ Codeine: Quaalude potentiates or acts synergistically with codeine as an antitussive, and it potentiates the analgesic action of codeine and codeine with acetylsalicylic acid. (1, 51)
+ Doriden: see Quaalude + Thorazine
+ Drugs: Quaalude may potentiate other drugs or itself be potentiated; adjust dosage with care when used with other drugs. (1)
+ Psychotherapeutic Agents: see Quaalude + Sedatives
+ Reserpine: see Quaalude + Thorazine
+ Sedatives: Care should be used during administration with other sedative, analgesic, or psychotherapeutic drugs or with alcohol because of possible potentiation of the effects. (1, 51)
+ Thorazine: Thorazine, Doriden, reserpine, and alcohol potentiate Quaalude. (1, 51)

QUANTRIL + Reserpine: see Reserpine + Quantril

QUELICIN CHLORIDE INJECTION: see Succinylcholine

QUELICIN CHLORIDE INJECTION
+ Antibiotics: Since neomycin, streptomycin, and certain other antibiotics have neuromuscular blocking effects and may cause respiratory depression, these antibiotics should not be instilled or used to irrigate either the peritoneal or the thoracic cavity when muscle relaxants such as succinylcholine have been administered. Under some circumstances, neomycin given orally may be absorbed in appreciable quantities. The possibility exists that subsequent administration of muscle relaxants may be a potential anesthetic hazard. (1)
+ Antimalarial Drugs: see Quelicin Chloride Injection + Insecticides, Neurotoxic
+ Cholinesterase Inhibitors: see Quelicin Chloride Injection + Neostigmine

QUELICIN CHLORIDE INJECTION (continued)

+ Cyclopropane: Bradycardia, arrhythmias, and even transient sinus arrest have occurred after the intravenous injection of succinylcholine. The incidence of such responses increases with repeated doses of the drug. These effects, which are thought to be the result of vagal stimulation, are reportedly enhanced by cyclopropane and halothane and tend to be less pronounced in patients premedicated with atropine or scopolamine. Prior administration of morphine may increase the likelihood of both bradycardia and arrest. (1)

+ Depressant Drugs: The most serious side effect of succinylcholine is prolonged postoperative apnea commonly caused by excessive doses of the drug. This may be due to a combination of several factors including depressant drugs employed in premedication or anesthetic induction. (1)

+ Digitalis: Unless absolutely necessary, succinylcholine should not be administered to those patients who are digitalized. Arrhythmias and cardiac arrest have occurred from a potentiation by succinylcholine of the effects of digitalis on cardiac conduction in digitalized patients. (1)

+ Halothane: see Quelicin Chloride Injection + Cyclopropane

+ Hypothermia: Hypothermia increases sensitivity to the effects of succinylcholine. This is probably caused by decreased enzymatic activity so that breakdown of succinylcholine and of acetylcholine at the myoneural junction is slowed. (1)

+ Insecticides, Neurotoxic: A low level of plasma pseudocholinesterase may be associated with prolonged paralysis of respiration following the use of succinylcholine. Low levels of this enzyme are often found in patients after exposure to neurotoxic insecticides and those receiving antimalarial drugs. (1) Mechanism: Prolonged apnea may result from succinylcholine administration for reasons associated with its breakdown. It is hydrolyzed in two stages. The first stage is a rapid one during which one molecule is quickly split off leaving succinylmonocholine which also has neuromuscular blocking activity. The second stage takes place more slowly, as the succinylmonocholine is hydrolyzed to succinic acid and choline. The enzyme responsible for these hydrolytic reactions is plasma pseudocholinesterase.

+ Morphine: see Quelicin Chloride Injection + Cyclopropane

+ Muscle Relaxants: If other muscle relaxants are to be used during the same procedure, the possibility of a synergistic or antagonistic effect should be considered. (1)

QUELICIN CHLORIDE INJECTION (continued)

+ Neomycin: see Quelicin Chloride Injection + Antibiotics
+ Neostigmine: Drugs which inhibit plasma pseudocholinesterase such as neostigmine and Tensilon may be associated with a prolonged paralysis of respiration following the use of succinylcholine. The usual depolarized block caused by succinylcholine is not antagonized by the anticholinesterases such as Tensilon or neostigmine. Instead, these agents may prolong depolarized block and the resulting apnea by preventing enzymatic hydrolysis of succinylcholine at the motor end plate.

  When succinylcholine is given over a prolonged period, the characteristic depolarization block of the myoneural junction may change to a non-depolarizing block which results in prolonged respiratory depression or apnea. Under such circumstances, small repeated doses of neostigmine or Tensilon may act as antagonists. These drugs should be preceded by 0.4 to 0.6 mg of atropine intravenously to guard against bradycardia. (1) Mechanism: see mechanism under Quelicin Chloride Injection + Insecticides, Neurotoxic.
+ Procaine: Drugs which compete with succinylcholine for the enzyme (plasma pseudocholinesterase) should not be given concurrently with succinylcholine, as this may cause a prolonged paralysis of respiration. (1) Mechanism: see mechanism under Quelicin Chloride Injection + Insecticides, Neurotoxic.
+ Streptomycin: see Quelicin Chloride Injection + Antibiotics
+ Tensilon: see Quelicin Chloride Injection + Neostigmine
+ Tubocurarine: Since recovery from succinylcholine is rapid and motor end plate returns to its resting state, tubocurarine may be administered subsequently without modification of the usual dosage. However, prolonged administration of succinylcholine may lead to a gradual change from a depolarized block to a partial non-depolarized block, thus increasing the sensitivity of the motor end plate to a non-depolarizing relaxant. This does not contraindicate the use of tubocurarine, but the dosage should be adjusted accordingly. The effect of a single dose of tubocurarine is additive to the effect of subsequent doses of that drug or to other non-depolarizing relaxants but is antagonistic to that of succinylcholine and other depolarizing agents. This antagonism is so great that at a time when there is no discernable residual effect of the non-depolarizing relaxant, three or four times more depolarizing relaxant is required for the production of neuromuscular block than under ordinary circumstances. Therefore some investigators advise against the use of succinylcholine for abdominal closure if tubocurarine has been employed

QUELICIN CHLORIDE INJECTION (continued)

+ Tubocurarine (continued): throughout the procedure. They feel that an effective dose of succinylcholine will be unnecessarily large and may lead to prolonged apnea following completion of surgery. (1)

QUESTRAN: see also Cuemid

QUESTRAN

+ Anticoagulants: Increased bleeding tendencies may develop in some patients using Questran chronically, due to hypoprothrombinemia associated with a nutritional deficiency of vitamin K. Prompt response to parenteral vitamin K_1 may be anticipated, and recurrence can be prevented by the oral administration of this vitamin (as in Mephyton tablets). (1) See Questran + Warfarin
+ Butazolidin: The absorption of Butazolidin may be delayed (but not decreased) when taken with Questran, as suggested by studies in the rat. Patients for whom Butazolidin is also prescribed should ingest this drug one hour before Questran. (1)
+ Chlorothiazide: Patients for whom chlorothiazide is also prescribed should ingest this drug one hour before Questran. (1)
+ Drugs: Since Questran is an anion exchange resin, Questran has a strong affinity for acidic materials. It may also absorb neutral or, less likely, basic materials to some extent. As a precautionary measure, it may seem wise to direct patients to ingest other drugs 30 minutes to one hour before Questran. (1)
+ Thyroid Hormone: see Questran + Thyroxine
+ Thyroxine: Cholestyramine administered concurrently with thyroid hormone produces malabsorption of thyroxine by binding the drug in the intestinal lumen. The malabsorption can be avoided by allowing a time interval of four to five hours between ingestion of the two drugs. In the hypothyroid patient, malabsorption of thyroid hormone can increase the severity of the hypothyroid state with a resultant rise in serum cholesterol to pretreatment levels; therefore, cholestyramine may be of limited efficiency in lowing serum cholesterol concentrations if allowed to interfere with the replacement therapy for the correction of the hypothyroid state. (136)
+ Warfarin: In rats, the anticoagulant activity of a large single dose of warfarin was unaffected by the administration of Questran, whether warfarin was given 30 minutes before or simultaneously with the resin. Plasma warfarin levels were lower, however, when the two drugs were given together. Patients for whom warfarin is also prescribed should ingest this drug one hour before Questran. (1) See Questran + Anticoagulants.

QUIACTIN

+ Alcohol: Patients should avoid the concomitant use of alcohol, since the effects may be additive. (2)
+ MAO Inhibitors: see Quiactin + Psychotropic Agents
+ Phenothiazines: see Quiactin + Psychotropic Agents
+ Psychotropic Agents: Other psychotropic agents, particularly phenothiazines or monoamine oxidase inhibitors, that are known to potentiate the action of other drugs should not be given with Quiactin. (2)

QUIBRON

+ Aminophylline: see Quibron + Theophylline
+ Theophylline: Quibron should not be given within 12 hours after rectal administration of any preparation containing theophylline or aminophylline. (1)
+ Xanthine Derivatives: Other formulations containing xanthine derivatives should not be given concurrently with Quibron. (1)

QUICK TEST + Thrombolysin: see Thrombolysin + Quick Test

QUIDE: see Phenothiazines

QUIDE

+ Alcohol: Concomitant use with alcohol should be avoided due to the potential additive effect. (1)
+ Analgesics: see Quide + CNS Depressants
+ Antihistamines: see Quide + CNS Depressants
+ Atropine: Phenothiazines potentiate heat, atropine, and phosphorous insecticides. (1)
+ Barbiturates: see Quide + CNS Depressants
+ Bone Marrow Depressants: Quide is contraindicated in patients with bone marrow depression. (1)
+ CNS Depressants: Quide should be used with caution in patients taking central nervous system depressants (e.g., opiates, analgesics, antihistamines, barbiturates, narcotics) because of the possibility of potentiation. (1)
+ Drugs: Quide's potent antiemetic action may mask the diagnosis of drug intoxication. (1)
+ Epinephrine: Epinephrine should not be used to treat hypotension, since it may aggravate rather than improve the hypotension (reverse epinephrine effect). (1)
+ Heat: see Quide + Atropine
+ Metrazol: see Quide + Picrotoxin
+ Narcotics: see Quide + CNS Depressants
+ Opiates: see Quide + CNS Depressants
+ Phosphorous Insecticides: see Quide + Atropine

QUIDE (continued)

+ Picrotoxin: If a stimulant is desired for the treatment of central nervous system depression, amphetamine, dextroamphetamine, and caffeine-sodium benzoate are suitable. Stimulants capable of inducing convulsions (like picrotoxin or Metrazol) should be avoided.(1)
+ Trisoralen: see Trisoralen

QUINAGLUTE DURA-TABS: see Quinidine

QUINAGLUTE DURA-TABS + Digitalis: Extreme caution should be exercised in using the drug in patients with digitalis intoxication. (1)

QUINIDEX: see Quinidine

QUINIDEX + Digitalis: Use with care in patients with digitalis intoxication. (1)

QUINIDINE: see also Quinidine Sulfate and various trademarked name preparations

QUINIDINE: see Alkaline Agents + Acidic Media

QUINIDINE: see listed under the following agents:

Anticoagulants
Coumadin
Dicumarol
Mylaxen
Panwarfin
Phenothiazine
Protalba
Provell Maleate
Raudixin
Rauwiloid
Rauwiloid plus Veriloid
Reserpine
Serpasil
Singoserp
Succinylcholine
Tubocurarine
Tubocurarine Chloride Inj.
Veriloid

QUINIDINE SULFATE: see Quinidine

QUINIDINE SULFATE + Digitalis: Occasionally arrhythmias may be produced by digitalis intoxication, and the use of quinidine in this situation is extremely dangerous because the cardiac glycoside may already have caused serious impairment of the intracardiac conduction system. (1)

QUININE: see Alkaline Agents + Acidifying Agents (Urinary)

QUININE: see listed under the following agents:

Anticoagulants
Coumadin
Dicumarol
Panwarfin
Sintrom
Tromexan

QUINORA: see Quinidine

QUINORA + Digitalis: Quinora is contraindicated when arrhythmia is due to digitalis intoxication. (1)

RABIES ANTISERUM + Rabies Vaccine (Duck Embryo) Dried Killed Virus (Lilly): see Rabies Vaccine (Duck Embryo) Dried Killed Virus (Lilly) + Rabies Antiserum

RABIES VACCINE (Duck Embryo) DRIED KILLED VIRUS (Lilly)

+ Adrenal Corticosteroids: see Rabies Vaccine (Duck Embryo) Dried Killed Virus (Lilly) + Adrenocorticotrophin

+ Adrenocorticotrophin: Adrenocorticotrophin and adrenal corticosteroids may reduce host resistance to certain infectious agents either through suppression of antibody response or through other and as yet poorly understood mechanisms. Therefore they should not be administered following exposure to infectious agents (such as rabies) for which no satisfactory antimicrobial therapy is available. To do so may alter the host-parasite relationship sufficiently to cause severe or fatal illness in spite of prophylactic administration of a vaccine. Under these circumstances, the occurrence of disease, actually due to the altered pattern of resistance, might be attributed to a vaccine failure. (1)

+ Rabies Antiserum: Although the most recent report of the WHO Expert Committee on Rabies recommends the use of Rabies Antiserum in certain circumstances, there is evidence that the serum administered concurrently with vaccine seriously interferes with the development of active immunity. If serum is used, the Committee recommends supplementary doses of vaccine 10 and 20 days after the last usual dose. (1)

RADIATION THERAPY: see also Irradiation and X-Ray Therapy

RADIATION THERAPY: see listed under the following agents:

Alkeran	Leukeran
Alkylating Agents	Matulane
Antimetabolic Agents	Methotrexate
Cosmegen	Thio-Tepa
Fluorouracil	Velban

RADIOACTIVE COMPOUNDS: see listed under the following agents:
Dicumarol
Panwarfin

RADIOCAPS-131 + Antithyroid Drugs: Radioiodine I-131 in any amount is considered by most authorities to be contraindicated in patients under treatment with antithyroid drugs (except in the presence of thyroid carcinoma). (1)

RAUDIXIN: see also Reserpine

RAUDIXIN
+ Aldomet: see Raudixin + Ismelin
+ Anesthesia: Since some patients on rauwolfia preparations have experienced marked hypotension under surgical anesthesia, it may be advisable to discontinue therapy for a period of about two weeks prior to elective surgery. Emergency surgery may be done using anticholinergic or adrenergic drugs if necessary to prevent vagal circulatory responses; other supportive measures may be used as indicated. (1)
+ Apresoline: see Raudixin + Ismelin
+ Digitalis: Use Raudixin cautiously with digitalis and quinidine, since cardiovascular reactions including angina-like symptoms and cardiac arrhythmias have occurred with reserpine preparations. (1)
+ Ganglionic Blocking Agents: see Raudixin + Ismelin
+ Hygroton: see Raudixin + Ismelin
+ Ismelin: Concomitant use of Raudixin and Ismelin, ganglionic blocking agents, Veratrum, Apresoline, Aldomet, Hygroton, or thiazides necessitates an immediate reduction in dosage of these latter preparations by at least 50% to avoid a precipitous drop in blood pressure particularly of the orthostatic type. A similar reduction is necessary when one or more of these agents is added to an established Raudixin regimen. Adjustments in dosage may subsequently be made in accordance with the patient's response. (1)
+ Quinidine: see Raudixin + Digitalis
+ Thiazides: see Raudixin + Ismelin
+ Veratrum: see Raudixin + Ismelin

RAU-SED: see Reserpine

RAU-SED
+ Electroshock: Electroshock therapy should not be given within seven days after reserpine therapy. (1)
+ Surgery: Patients on reserpine therapy should not receive the drug for two weeks prior to surgery, if possible. Emergency surgery may be carried out by using vagal blocking agents to prevent and treat vagal circulatory response. (1)

RAUTENSIN: see Reserpine

RAUTRAX-N: see Raudixin and Naturetin

RAUWILOID: see Reserpine

RAUWILOID (continued)

+ Aldomet: see Rauwiloid + Ismelin
+ Apresoline: see Rauwiloid + Ismelin
+ Digitalis: Use Rauwiloid cautiously with digitalis or quinidine, since cardiac arrhythmias have occurred with rauwolfia alkaloids. (1)
+ Electroshock Therapy: Electroshock therapy should not be given to patients taking rauwolfia alkaloids, since severe and even fatal reactions have been reported. The drug should be discontinued for two weeks before giving electroshock therapy. (1)
+ Ganglionic Blocking Agents: see Rauwiloid + Ismelin
+ Hygroton: see Rauwiloid + Ismelin
+ Ismelin: Concomitant use of Rauwiloid with Ismelin, ganglionic blocking agents, Apresoline, Aldomet, Hygroton, or thiazides necessitates an immediate dosage reduction of both agents by about 50%. Careful observation for changes in blood pressure must be made when combined therapy is used. (1)
+ Quinidine: see Rauwiloid + Digitalis
+ Surgery: Since some patients receiving rauwolfia preparations have experienced marked hypotension when undergoing surgery, it may be advisable to discontinue therapy for a period of about two weeks prior to elective surgery. Emergency surgery may be carried out by using, if necessary, anticholinergic or adrenergic drugs to prevent circulatory responses; other supportive measures may be used as indicated. (1)
+ Thiazides: see Rauwiloid + Ismelin

RAUWILOID + VERILOID: see also Reserpine

RAUWILOID + VERILOID

+ Digitalis: Rauwiloid + Veriloid should be used with caution in patients with digitalis intoxication. (1)
+ Quinidine: Co-administration with quinidine is a relative contraindication. (1)

RAUWOLFIA ALKALOIDS: see Reserpine

RAUWOLFIA ALKALOIDS: see listed under the following agents:

Anesthetics, General	Naqua
Bristuron	Naturetin
Hydromox	Renese
Metahydrin	

RAUWOLFIA ALKALOIDS

+ Anesthesia: Marked hypotension has been reported when used concomitantly.

RAUWOLFIA ALKALOIDS (continued)

+ Anticonvulsants: It may necessitate (in epileptic patients) dosage adjustment of the anticonvulsant. Rauwolfia alkaloids lower convulsive threshold and shorten seizure latency.

RAUWOLFIA COMPOUNDS: see listed under the following agents:
MAO Inhibitors
Niamid

RAUWOLFIA DERIVATIVES: see listed under the following agents:
Esidrix
Ismelin

RAUWOLFIA DRUGS + Regitine: see Regitine + Rauwolfia Drugs

RAUZIDE: see Raudixin and Naturetin

REDISOL + Ascorbic Acid: When Redisol tablets are dissolved in a medium containing ascorbic acid, such as orange juice, the solution should be administered at once, since ascorbic acid will have a deleterious effect on the stability of vitamin B_{12}. (1)

REGITINE: see listed under the following agents:
MAO Inhibitor
Parnate

REGITINE

+ Analgesics: see Regitine + Sedatives
+ Anesthetics: see Regitine + Sedatives
+ Antihypertensive Agents: Since antihypertensive agents may interfere with response to Regitine, withdraw them at least two weeks before the test is given. (1) See also Regitine + Sedatives and Rauwolfia Drugs.
+ Cardiac Glycosides: It is preferable not to give cardiac glycosides following any untoward response to Regitine until the patient has fully recovered from the reaction. (1)
+ Digitalis: see Regitine + Sedatives and Cardiac Glycosides
+ Epinephrine: Do not use epinephrine in cases of Regitine overdosage, since it may cause a fall in blood pressure instead of a rise. (1)
+ Insulin: see Regitine + Sedatives
+ Narcotics: see Regitine + Sedatives
+ Phenobarbital: see Regitine + Sedatives
+ Rauwolfia Drugs: Withhold antihypertensives until blood pressure returns to the untreated, hypertensive level. This may necessitate withdrawal of rauwolfia drugs for as long as one month before the test. (1) See Regitine + Antihypertensive Agents.

REGITINE (continued)

+ Sedatives: Withhold sedatives (such as phenobarbital), analgesics, narcotics, anesthetics, antihypertensives, thiocyanates, and all other drugs not deemed absolutely essential (such as digitalis and insulin) for at least 24 hours (preferably 48 to 72 hours) prior to the test. Thiocyanates should be withheld for four to six days preceding the test. Other drugs administered prior to the test with Regitine may show a false-positive reaction. (1)
+ Thiocyanates: see Regitine + Sedatives

REGROTON: see Hygroton and Reserpine

RENASUL A + Dolonil: see Dolonil + Renasul A

RENCAL

+ Aluminum Hydroxide: Because the absorption and urinary excretion of phosphate may be important in producing the desired effect, the concomitant administration of aluminum hydroxide, which reacts with phosphate, probably should be avoided. (1, 2)
+ Iron: Rencal also binds other metals such as magnesium and zinc, and there is some evidence that it interferes with the absorption of iron. No clinical evidence has indicated that these actions of the drug have caused any adverse effects, but the possibility that the long-term use of Rencal might lead to deficiencies of these metals should be borne in mind. (2)
+ Vitamin D: Patients being treated with Rencal should avoid taking vitamin D, since this increases the absorption of calcium. (1, 2)

RENESE: see Diuretics, Thiazide

RENESE

+ Adrenocortical Steroids: Since all diuretic agents may reduce serum levels of sodium, chloride, and potassium, especially with brisk diuresis or when used concomitantly with steroids, patients should be observed regularly for early signs of fluid or electrolyte imbalance, and serum electrolyte studies should be performed periodically when feasible. (1) See also Renese + Digitalis.
+ Antihypertensive Agents: Like other benzothiadiazines, the action of Renese may be used with other antihypertensives (e.g., rauwolfia alkaloids and ganglionic blocking agents). If used with another antihypertensive agent, lower than usual doses of both drugs should be considered. (1, 2)
+ Corticotrophin: see Renese + Digitalis
+ Diabetes: Thiazide diuretics are known to disturb glucose tolerance in some individuals even when there is no history of glucose tolerance

RENESE (continued)

+ Diabetes (continued): or diabetes in the individual or his family. (1, 2)
+ Digitalis: Patients in whom potassium depletion may occur, such as those receiving digitalis, adrenocortical steroids, or corticotrophin concomitantly, should take some food daily that is high in potassium content (e.g., orange juice) or potassium salt. Hypokalemia predisposes to digitalis toxicity and should be avoided when digitalis is part of the therapy. (1, 2)
+ Ganglionic Blocking Agents: see Renese + Antihypertensive Agents
+ Levophed: see Renese + Pressor Amines
+ Pressor Amines: Recent reports indicate that the thiazides decrease arterial responsiveness to pressor amines (e.g., Levophed). (1, 2)
+ Rauwolfia Alkaloids: see Renese + Antihypertensive Agents
+ Trisoralen: see Trisoralen
+ Tubocurarine: Extra precaution may be necessary in patients who require tubocurarine or its derivatives, since thiazides may augment the paralyzing action of tubocurarine. (1, 2)

RENESE-R: see Renese and Reserpine

REPOISE: see Phenothiazines

REPOISE

+ Alcohol: Patients should be advised against the use of alcohol during treatment with Repoise or any other phenothiazine, since potentiation of the effects of alcohol may occur. (1, 94) See also Repoise + CNS Depressants.
+ Analgesics: see Repoise + CNS Depressants
+ Anesthetics: Repoise should be used with caution in patients who are about to receive general anesthesia. (1, 94) See also Repoise + CNS Depressants.
+ Antihistamines: see Repoise + CNS Depressants
+ Antihypertensive Agents: Care should be exercised when antihypertensive agents are used concomitantly because the actions of these agents may be potentiated by Repoise. (94)
+ Atropine: Exaggerated responses to atropine, environmental heat, and phosphorous insecticides have been associated with phenothiazine therapy. (1)
+ Barbiturates: see Repoise + CNS Depressants
+ Bone Marrow Depressants: Do not use Repoise in the presence of bone marrow depression. (1)
+ CNS Depressants: When Repoise treatment is combined with barbiturates, analgesics, narcotics, antihistamines, hypnotics, or other central nervous system acting drugs, the dosage of both should be

REPOISE (continued)

+ CNS Depressants (continued): reduced to one-half the conventional dosage. The phenothiazine drugs, as a class, may potentiate the effects of other drugs, particularly central nervous system depressants. Repoise is generally contraindicated in comatose patients, especially those whose depression is due to barbiturates, narcotics, alcohol, analgesics, antihistamines, or other drugs. (1, 94)
+ Drugs, Overdosage: Because of its anti-emetic effect, Repoise may mask symptoms of drug overdosage. (1)
+ Epinephrine: If hypotension should occur with Repoise, epinephrine is contraindicated because of its tendency to decrease blood pressure in the presence of phenothiazines. Levophed or Neosynephrine should be used if a vasopressor is indicated. (1, 94)
+ Hypnotics: see Repoise + CNS Depressants
+ Narcotics: see Repoise + CNS Depressants
+ Phosphorous Insecticides: see Repoise + Atropine

RESERPINE: see individual trade name preparations.

RESERPINE: see listed under the following agents:

Anhydron
Ansolysen
Aquatag
Cogentin
Digitalis
Dopar
Esidrix
Eutonyl
Exna
Fluothane
Inderal
Inversine
Larodopa
Levodopa
Levoprome
MAO Inhibitor
Metahydrin
Naqua
Organophosphate Pesticides
Ostensin
Parnate
Phenothiazines
Protopan
Quaalude
Unitensin

RESERPINE

+ Amphetamine: see Reserpine + Sympathomimetics, Indirect-Acting
+ Anesthesia, General: Reserpine depletes the body stores of catecholamines (epinephrine and norepinephrine) and thus may cause severe hypotension, particularly during general anesthesia. (5, 26, 55, 126) See also Ismelin + Anesthesia.
+ Digitalis: The concomitant administration of digitalis and reserpine should be used cautiously, as atrial arrhythmias, ectopic ventricular activity, and varying degrees of heart block and electrolyte imbalance may occur. (25) See also Digitalis + Reserpine.

RESERPINE (continued)

+ Electroshock Therapy: Use lower milliamperage and shorter duration of stimulus initially, since more prolonged and more severe convulsions, as well as apnea, have been reported with previously well tolerated stimulation.
+ Ganglionic Blocking Agents: Concomitant administration with ganglionic blocking agents necessitates a dosage reduction of at least 50% of the more toxic agents, thus minimizing the incidence and severity of their side effects.
+ Ismelin: see Ismelin + Rauwolfia Derivatives
+ Levophed: Direct-acting vasopressors such as Levophed given to counteract severe hypotension in a patient who has been on reserpine will induce a much greater response than would be expected under normal circumstances. These antihypertensives work to prevent the uptake of the Levophed into inactivation sites and therefore potentiate the effect of the vasopressor. (55, 80)
+ Metaraminol (Aramine, Pressonex, Pressorol): see Reserpine + Sympathomimetics, Indirect-Acting
+ Methamphetamine: see Reserpine + Sympathomimetics, Indirect-Acting
+ Nitoman: Nitoman is able to prevent the action of reserpine. Mechanism: Nitoman displaces the reserpine from its site of action. (118)
+ Phenethylamine: see Reserpine + Sympathomimetics, Indirect-Acting
+ Quantril: Quantril prevents the action of reserpine. Mechanism: Quantril blocks the reserpine from its receptor site. (118)
+ Quinidine: The concomitant use of quinidine and reserpine should be used cautiously since arrhythmias have occurred with rauwolfia preparations.
+ Sympathomimetics, Indirect-Acting: Indirect-acting pressor amines (e.g., Tyramine, Wyamine, amphetamine, phenethylamine, metaraminol, methamphetamine) are ineffectual in patients who are being effectively treated with reserpine. Mechanism: Reserpine and rauwolfia alkaloids reduce the norepinephrine content at adrenergic neuron terminals and also prevent the storage of norepinephrine in the neurosecretory granules and by this action impairs adrenergic transmission and lowers blood pressure. The indirect-acting sympathomimetics work by releasing the norepinephrine. (34, 50, 55)
+ Tyramine: see Reserpine + Sympathomimetics, Indirect-Acting
+ Wyamine: see Reserpine + Sympathomimetics, Indirect-Acting

RESISTOPEN

+ Food: Oral doses of Resistopen should be taken one or two hours before

RESISTOPEN (continued)

+ Food (continued): meals to minimize retention of the drug and its destruction by gastric acidity. (2)
+ Neutrapen: Since Resistopen is relatively unaffected by Neutrapen, it is unlikely that the use of Neutrapen would be of value in the management of allergic reactions. (1)

RESPIRATORY DEPRESSANTS: see listed under the following agents

Analgesics
Nembutal Sodium Inj.
Talwin

RETET: see Tetracycline

RETET

+ Foods, Calcium-containing: see Retet + Milk Formulas
+ Meals: Oral forms of tetracycline should be given one hour before or two hours after meals. (1) This aids in the absorption of the tetracycline.
+ Milk Formulas: Pediatric dosage forms (oral) should not be given with milk formulas or other calcium-containing foods and should be given at least one hour before feeding. (1) Mechanism: Tetracyclines are chelating agents that form insoluble chelates with multivalent cations (Ca^{++}, Mg^{++}, Al^{+++}), and gastrointestinal absorption is inhibited.

RITALIN: see listed under the following agents:

Hypertensin
Ismelin
MAO Inhibitors

RITALIN

+ Alkaline Solution: see Ritalin + Barbiturates
+ Anticonvulsants: see Ritalin + Tranquilizers
+ Barbiturates: Do not inject Ritalin Parenteral Solution through tubing or a syringe which contains a barbiturate or strongly alkaline solution, since a heavy precipitate is formed.
  Laboratory and clinical studies show a clearcut antagonism between the analeptic effect of Ritalin and the sedation produced by barbiturates. (1,2)
+ Diphenylhydantoin: Ritalin may increase the serum levels of diphenylhydantoin if used concomitantly in a patient, causing diphenylhydantoin toxicity. Mechanism: It is believed that Ritalin will inhibit the microsomal drug metabolizing enzymes in the liver, so that diphenylhydantoin serum levels will reach toxic proportions. (125)

RITALIN (continued)

+ Epinephrine: see Ritalin + Pressor Agents
+ Hypertensin: see Ritalin + Pressor Agents
+ Ismelin: Ritalin may decrease the hypotensive effect of Ismelin. (1) Mechanism: It is postulated that Ritalin acts at the same site as Ismelin and blocks the uptake of Ismelin at sympathetic nerve endings and perhaps displaces Ismelin from these sites. (72)
+ MAO Inhibitors: Use Ritalin cautiously with monoamine oxidase inhibitors as it may cause additive central stimulation. (1) Mechanism: For drugs passing the blood-brain barrier, central excitation is possible because MAO inhibitors will increase the catecholamine concentration in the brain and be additive with Ritalin. (115)
+ Norepinephrine: see Ritalin + Pressor Agents
+ Pressor Agents: Ritalin has potentiated the effects of pressor agents in pharmacological experiments. Use cautiously with pressor agents such as norepinephrine, epinephrine, or Hypertensin. (1)
+ Tranquilizers: Laboratory and clinical studies show a clearcut antagonism between the analeptic effect of Ritalin and the sedation produced by tranquilizers and anticonvulsants. (1)
+ Tromexan: Ritalin may increase the anticoagulant effect of Tromexan if used concomitantly in a patient. Mechanism: It is believed that Ritalin will inhibit the microsomal drug metabolizing enzymes in the liver, causing excessive anticoagulation. (125)

RITONIC: see Ritalin

RITONIC + MAO Inhibitors: Do not give monoamine oxidase inhibitors to patients taking Ritonic. (1)

ROCYTE + Antihypertensive Agents: Cobalt may produce dilatation of blood vessels and lowering of blood pressure by action on the sympathetic nervous system. (1)

RONDOMYCIN

+ Aluminum Hydroxide Gel: Aluminum hydroxide gel given with antibiotics has been shown to decrease their absorption and is contraindicated. (1) Mechanism: see Rondomycin + Milk.
+ Foods, Calcium-containing: see Rondomycin + Milk
+ Meals: To aid absorption of the drug, it should be given at least one hour before or two hours after eating. (1)
+ Milk: Pediatric dosage forms of Rondomycin should not be given with milk formulas or calcium-containing foods. (1) Mechanism: Tetracyclines are chelating agents that form insoluble chelates with

RONDOMYCIN (continued)

+ Milk (continued): multivalent cations (Ca^{++}, Mg^{++}, Al^{+++}), and gastrointestinal absorption is inhibited.
+ Trisoralen: see Trisoralen

RUBBER: see listed under the following agents:

Mucomyst
Penthrane

RUBEOVAX, LYOVAC

+ Alkylating Agents: see Rubeovax, Lyovac + Corticosteroids
+ Antimetabolites: see Rubeovax, Lyovac + Corticosteroids
+ Blood Transfusion: Vaccination should be deferred in children who have had either a blood transfusion of more than 0.01 cc of gamma globulin per pound of body weight during the preceding six weeks. (1)
+ Corticosteroids: Measles vaccination is contraindicated in children who are receiving therapy with corticosteroids, irradiation, alkylating agents, or antimetabolites. (1)
+ Gamma Globulin: see Rubeovax, Lyovac + Blood Transfusion
+ Immunization Agents: In general, it is good practice to separate the administration of live measles virus vaccine by a space of one month from other elective immunization procedures. (1)
+ Irradiation: see Rubeovax, Lyovac + Corticosteroids

S

SALCORT + Diabetes: Corticosteroids should be used with caution in patients with diabetes. (1)

SALICYLATES: see listed under the following agents:

Anticoagulants
Anturane
Benemid
Coumadin
Dianabese
Dicumarol
Hedulin
Indocin
Lasix
Liquamar
Methotrexate
Orinase
Panwarfin
Persistin
Ponstel
Sintrom
Sulfonamides, Long-acting
Tolinase
Tromexan

SALICYLIC ACID: see Acidic Agents + Alkaline Media

SALICYLIC ACID: see Acidic Agents + Alkanizing Agents (Urinary)

SALICYLIC ACID: see listed under the following agents:

Antacids
Liquamar
Methotrexate

SALINE SOLUTION: see listed under the following agents:

Ilotycin Gluceptate I.V.
Levophed
Surgicel

SALT DEPLETION: see listed under the following agents:

Ansolysen
Inversine

SALURETIC AGENTS: see listed under the following agents:

Inversine
Unitensin

SALURON: see Diuretics, Thiazide

SALURON

+ Anesthesia, General: see Saluron + Pressor Amines and Tubocurarine
+ Antihypertensive Agents: Saluron is capable of enhancing the effects of other antihypertensive agents (in particular the ganglionic blocking

SALURON (continued)

+ Antihypertensive Agents (continued): agents and Apresoline). Combination therapy with these drugs may necessitate a reduction in the dose of the latter. (1)
+ Apresoline: see Saluron + Antihypertensive Agents
+ Corticotrophin: see Saluron + Steroids
+ Curare and Derivatives: see Saluron + Tubocurarine
+ Diabetes: Hyperglycemia and glycosuria have occurred in patients receiving thiazides. Patients who have diabetes mellitus or who are suspected of being prediabetic should be kept under close observation if treated with Saluron. Adjustment of diabetic control is frequently indicated. (1, 2)
+ Digitalis: Patients taking digitalis should be observed for signs of digitalis toxicity, since depletion of the body stores of potassium, by Saluron, may lessen digitalis requirements. Potentiation of digitalis effects upon the heart may cause atrial or ventricular arrhythmias. (1, 2)
+ Ganglionic Blocking Agents: see Saluron + Antihypertensive Agents
+ Pressor Amines: Since thiazide compounds decrease the response of the arterial system to pressor amines, great caution should be exercised in administering this drug to patients who undergo general anesthesia and surgery. It is advisable to discontinue therapy for one week or so before elective surgery. (1, 2)
+ Steroids: Decreased serum potassium is more likely to be met in cases associated with concomitant thiazide and steroid or corticotrophin administration. (1, 2)
+ Trisoralen: see Trisoralen
+ Tubocurarine: Since thiazide compounds enhance the effects of tubocurarine, great caution should be exercised in administering this drug to patients who undergo general anesthesia and surgery or who are treated with curare or its derivatives. It is advisable to discontinue therapy for one week or so before elective surgery. (1, 2)

SALUTENSIN: see Saluron, Reserpine, Proveratrine

SALUTENSIN + Digitalis: The use of Salutensin together with digitalis may increase the risk of digitalis-like intoxication (due perhaps to the diuretic effect of Saluron, since depletion of the stores of potassium may lessen digitalis requirements and the vagotonic effect of proveratrine and reserpine). If there is evidence of myocardial irritability (extrasystoles, bigeminy or AV block), dosage of Salutensin should be reduced or discontinued regardless of whether the patient is on digitalis. (1)

SANDRIL: see Reserpine

SANDRIL INJECTION + Anesthesia: Sandril should be discontinued for approximately one week before elective surgery and anesthesia because it may produce excessive hypotension. (1)

SCOPOLAMINE + Levoprome: see Levoprome + Scopolamine

SEBUCARE + Hair Preparations: Do not use additional hair grooming aids. (1)

SECONAL SODIUM
+ Alcohol: see Seconal Sodium + Analgesics
+ Analgesics: Special care is needed when any barbiturate is given to patients who have received analgesics, other sedatives or hypnotics, ether, curare-like drugs, alcohol, or tranquilizers because an additive respiratory-depressant effect may ensue. (1)
+ Curare-like Drugs: see Seconal Sodium + Analgesics
+ Ether: see Seconal Sodium + Analgesics
+ Hypnotics: see Seconal Sodium + Analgesics
+ Sedatives: see Seconal Sodium + Analgesics
+ Tranquilizers: see Seconal Sodium + Analgesics

SEDATIVES: see listed under the following agents:

Ambodryl
Anti-Nausea Suprettes
Apodol
Benadryl
Carbocaine
Diabinese
Disophrol
Doriden
Dymelor
Eutonyl
Forhistal
Furoxone
Hypoglycemic Drugs, Oral
Leritine
Levoprome
Marplan
Narcotic Antagonist
Niamid
Orinase
Parest
Parnate
Pentothal
Phenothiazines
Placidyl
Probanthine with Dartal
Pyribenzamine
Quaalude
Regitine
Seconal Sodium
Somnofac
Sopor
Sparine
Stelazine
Tacaryl

SEDORZYL + Digitalis: Cardiac arrhythmias may occur if given to digitalized patients. (1)

SELSUN
+ Mercuric Ointments: see Selsun + Metallic Compounds
+ Metallic Compounds: Selsun should not be used until all traces of any ointment containing metallic compounds (e.g., mercuric ointments) have been removed from the scalp. (1)

SEMOPEN + Food: Absorption of Semopen is greater if the drug is given when the stomach is empty. (2)

SER-AP-ES: see listed under the components, Reserpine, Apresoline, and Esidrix

SERAX: see Benzodiazepines

SERAX
+ Alcohol: Care should be taken to warn patients taking Serax that their tolerance to alcohol may be lowered. (1, 2, 55)
+ Antidepressants, Tricyclic: There may be an additive effect when used concomitantly. (55)
+ Barbiturates: There may be an additive effect when used concomitantly. (55)
+ MAO Inhibitors: Concomitant usage causes enhanced sedation. (55)
+ Phenothiazines: There may be an additive effect when used concomitantly. (55)
+ Psychotropic Agents: The concomitant administration of Serax and other psychotropic agents should be avoided. (2)

SERC + Antihistamines: In the absence of studies determining whether there is an interaction between this histamine-like drug and antihistamines, it is recommended that Serc not be used concurrently with antihistamine agents. (1, 95)

SERENTIL: see Phenothiazines

SERENTIL
+ Alcohol: see Serentil + CNS Depressants
+ Anesthetics: see Serentil + CNS Depressants
+ Atropine: see Serentil + CNS Depressants
+ CNS Depressants: Attention should be paid to the fact that phenothiazines are capable of potentiating central nervous system depressants (e.g., anesthetics, opiates, alcohol, etc.) as well as atropine and phosphorous insecticides. (1)
+ Insecticides, Phosphorous: see Serentil + CNS Depressants
+ Opiates: see Serentil + CNS Depressants

SEROMYCIN

+ Alcohol: The risk of convulsions due to Seromycin is increased by the ingestion of ethyl alcohol. (137)
+ Antituberculosis Agents: The administration of Seromycin with other antituberculosis drugs has been associated in a few instances with vitamin B_{12} and/or folic acid deficiency, megaloblastic anemia, and sideroblastic anemia. (1)

SERPASIL: see also Reserpine

SERPASIL

+ Anesthesia: Discontinue the drug at least two weeks prior to surgery, if possible, since an unexpected degree of hypotension and bradycardia has been reported in patients undergoing anesthesia while under treatment with rauwolfia alkaloids. For emergency surgical procedures, give vagal blocking agents parenterally to prevent or reverse hypotension and/or bradycardia. (1)
+ Digitalis: Use reserpine cautiously with digitalis, since arrhythmias have occurred with rauwolfia preparations. Angina-like syndrome, ectopic cardiac rhythms may occur particularly if Serpasil is used concurrently with digitalis. (1)
+ Electroshock Therapy: When patients on reserpine receive electroshock therapy, use lower milliamperage and a shorter duration of stimulus initially, since more prolonged and severe convulsions as well as apnea have been reported with previously well tolerated stimulation. Shock therapy within seven days after giving the drug is hazardous; discontinue the drug for two weeks before giving electroshock therapy. (1)
+ Ismelin: Concurrent use of Ismelin and rauwolfia derivatives may cause bradycardia, mental depression, and postural hypotension. (1)
+ Quinidine: Use reserpine cautiously with quinidine, since arrhythmias have occurred with rauwolfia preparations. (1)

SERPASIL-APRESOLINE: see listed under components, Serpasil and Apresoline

SERUM + Hypertensin: see Hypertensin + Serum

SIGNEMYCIN

+ Aluminum Hydroxide Gel: Aluminum hydroxide gel given with antibiotics has shown to decrease their absorption and is contraindicated. (1) Mechanism: Tetracyclines are chelating agents that form insoluble chelates with multivalent cations (Ca^{++}, Mg^{++}, Al^{+++}), and gastrointestinal absorption is inhibited.

SIGNEMYCIN (continued)
+ Foods, Calcium-containing: see Signemycin + Meals
+ Meals: To aid absorption of the drug, it should be given one hour before or two hours after eating. Pediatric dosage forms of tetracycline should not be given with milk formulas or calcium-containing foods. (1) Mechanism: see Signemycin + Aluminum Hydroxide Gel.
+ Milk: see Signemycin + Meals

SILVER NITRATE + Oxycel: see Oxycel + Silver Nitrate

SILVER PREPARATIONS: see listed under the following agents:
Gantrisin Opthalmic Sol. and Oint.
Sulamyd Sodium Opthalmic Sol.
Thiosulfil Opthalmic with Methylcellulose

SINEQUAN
+ Alcohol: Patients should be cautioned that their response to alcohol may be potentiated. (1)
+ Ismelin: Other structurally related psychotherapeutic agents (e.g., iminodibenzyls and dibenzocycloheptenes) are capable of blocking the effects of Ismelin in both animal and man. Sinequan, however, does not show this effect in animals. At the usual clinical dosage, 75 to 150 mg per day, Sinequan can be given concomitantly with Ismelin and related compounds without blocking the antihypertensive effect. At doses of 300 mg per day or above, Sinequan does exert a significant blocking effect. In addition, Sinequan was similar to the other structurally related psychotherapeutic agents as regards its ability to potentiate norepinephrine response in animals. However, in the human this effect was not seen. This is in agreement with the low incidence of the side effect of tachycardia seen clinically. (1)
+ MAO Inhibitors: Serious side effects and even death have been reported following the concomitant administration of certain drugs with MAO inhibitors. Therefore MAO inhibitors should be discontinued at least two weeks before the cautious initiation of therapy with Sinequan. The exact length of time may vary and is dependent upon the particular monoamine oxidase inhibitor being used, the length of time it has been administered, and the dosage involved. (1)
+ Norepinephrine: see Sinequan + Ismelin
+ Trisoralen: see Trisoralen

SINGOSERP: see also Reserpine

SINGOSERP
+ Anesthesia: If possible, discontinue therapy at least two weeks before

SINGOSERP (continued)

+ Anesthesia (continued): surgery, since an unexpected degree of hypotension and bradycardia has been reported in patients undergoing anesthesia while receiving rauwolfia alkaloids. For emergency surgery, give vagal blocking agents parenterally to prevent or reverse hypotension and/or bradycardia. (1, 2)
+ Anticonvulsants: Singoserp should be used cautiously in patients with epilepsy, and the dosage of anticonvulsant drugs may have to be adjusted. (2)
+ Antihypertensive Agents: Thiazides (e.g., hydrochlorothiazide), Apresoline, or Ismelin may be used in combination with Singoserp in lower doses than when they are used alone. Since these agents may have a more potent antihypertensive effect in combination than when used alone, watch patient carefully during period of dosage adjustment. (1, 2) See Singoserp + Ismelin.
+ Apresoline: see Singoserp + Antihypertensive Agents
+ Digitalis: Use Singoserp cautiously with digitalis and quinidine, since cardiac arrhythmias have occurred with rauwolfia alkaloids. Singoserp is contraindicated in cases of digitalis intoxication. (1)
+ Hydrochlorothiazide: see Singoserp + Antihypertensive Agents
+ Ismelin: Concurrent use of Ismelin and rauwolfia derivatives may cause bradycardia, mental depression, and postural hypotension. (1) See also Singoserp + Antihypertensive Agents
+ Quinidine: see Singoserp + Digitalis
+ Thiazides: see Singoserp + Antihypertensive Agents

SINGOSERP-ESIDRIX: see listed under components, Singoserp and Esidrix

SINGOSERP-ESIDRIX

+ Antihypertensive Agents: Because Singoserp-Esidrix potentiates the action of other antihypertensive agents, such additions to the regimen should be gradual and effects observed carefully. Dosages of ganglionic blockers in particular should be reduced by 50%. (1)
+ Ganglionic Blockers: see Singoserp-Esidrix + Antihypertensive Agents

SINTROM: see Anticoagulants

SINTROM: The following factors may be responsible for increased prothrombin time (potentiation of hypoprothrombinemic effect): (1)

Alcohol
Anesthetics
Antibiotics (penicillin, Chloromycetin, Aureomycin)
Butazolidin
Drugs affecting blood elements

SINTROM (continued, factors that may be responsible for increased prothrombin time):

Drugs with hepatotoxic action
Low Choline
Low Cystine
Low Protein diet, dietary deficiency
Low Vitamin C
Prolonged Narcotics
Quinine
Salicylates (excess of 1 gm/day)
Sulfonamides

SINTROM: The following factors may be responsible for decreased prothrombin time (reversal of hypoprothrombinemic effect): (1)

Antihistamines
Barbiturates (The concurrent administration of barbiturates may antagonize the effects of coumarin-like anticoagulants.)
Corticosteroids
Diet high in Vitamin K (vegetables, fish, fish oils)
Mineral Oil
Vitamin K

SINTROM

+ Blood, Whole: Phytonadione (vitamin K_1) and fresh whole blood will counteract the effect of Sintrom. (2)
+ Phytonadione (vitamin K_1): see Sintrom + Blood, Whole

SKELAXIN

+ Benedict's Test: see Skelaxin + Diabetes
+ Diabetes: In several nondiabetic patients who received Skelaxin, urinalysis revealed the presence of a reducing agent thought to be a metabolic product of the drug. False-positive Benedict's tests due to this reducing substance in the urine were noted in nine patients. (1, 2)

SKELETAL MUSCLE RELAXANTS + Cholinesterase Inhibitors: see Cholinesterase Inhibitors + Skeletal Muscle Relaxants

SKF SPECIAL RESIN No. 648 + Digitalis: Because lowering potassium levels may increase the toxicity of digitalis, SKF Special Resin No. 648 should be used cautiously in digitalized patients. (1)

SMALLPOX VACCINE: see listed under the following agents:

Adrenocorticosteroids
Lirugen

SMALLPOX VACCINE (continued)
Measles Virus Vaccine, Live, Attenuated
M-Vac Measles Virus Vaccine, Live, Attenuated
Pfizer-Vax Measles-L
Sterneedle

SMALLPOX VACCINE DRIED (Dryvax)
+ ACTH: Smallpox Vaccine Dried is contraindicated for persons of any age receiving therapy with x-ray, ACTH, corticosteroids, or immunosuppressive drugs. (1,2)
+ Corticosteroids: see Smallpox Vaccine Dried (Dryvax) + ACTH
+ Drugs, Immunosuppressive: see Smallpox Vaccine Dried (Dryvax) + ACTH
+ X-Ray Therapy: see Smallpox Vaccine Dried (Dryvax) + ACTH

SOAP: see listed under the following agents:
Akrinol
Benasept Vaginal Jelly
Sterisil
Virac
Zephiran

SODA LIME + Trilene: see Trilene + Soda Lime

SODIUM + Aldactazide: see Aldactazide + Sodium

SODIUM BICARBONATE + Benemid: see Benemid + Sodium Bicarbonate

SODIUM BISULFITE + Cardio-Green: see Cardio-Green + Sodium Bisulfite

SODIUM CHLORIDE: see listed under the following agents:
Ostensin
Percorten

SODIUM CHLORIDE SOLUTION + Fungizone: see Fungizone + Sodium Chloride Solution

SODIUM IODIDE I-131 + Antithyroid Drugs: Treatment with antithyroid drugs (except in the presence of cancer of the thyroid) represents a contraindication to the use of radioiodine. (1)

SOFTRAN
+ MAO Inhibitors: see Softran + Psychotropic Agents
+ Phenothiazines: see Softran + Psychotropic Agents
+ Psychotropic Agents: Other psychotropic agents, particularly phenothiazines or monoamine oxidase inhibitors, that are known to potentiate the action of other drugs should not be given with Softran. (2)

SOLACEN
+ Alcohol: Simultaneous administration of Solacen with alcohol or with other psychotropic agents, particularly phenothiazines or monoamine oxidase inhibitors, which are known to potentiate the action of other drugs, may result in additive actions. (1, 2)
+ CNS Depressants: Simultaneous administration to psychotic patients of Solacen and other central nervous system depressants has in a few instances been associated with the occurrence of grand mal seizures. Seizures have been reported with administration of phenothiazines alone, but not with Solacen alone; nevertheless, use Solacen cautiously in individuals receiving other central nervous system depressants or having a history of convulsive seizures. (1, 2)
+ MAO Inhibitors: see Solacen + Alcohol
+ Phenothiazines: see Solacen + CNS Depressants
+ Psychotropic Agents: see Solacen + Alcohol

SOLU-CORTEF: see Cortef

SOLU-MEDROL: see Medrol

SOLUTIONS + Depo-Medrol: see Depo-Medrol + Solutions

SOMACORT: see Prednisolone

SOMACORT + Diabetes: Corticosteroid therapy in diabetic patients may necessitate increased dosage of antidiabetic medications. (1)

SOMNOFAC
+ Alcohol: see Somnofac + Sedatives
+ Analgesics: see Somnofac + Sedatives
+ CNS Depressants: Somnofac is not recommended for use with psychotropic drugs or other central nervous system depressants. (1)
+ Psychotherapeutic Drugs: see Somnofac + Sedatives
+ Psychotropic Agents: see Somnofac + CNS Depressants
+ Sedatives: Care should be used during administration with other sedatives, analgesics or psychotherapeutic drugs, or with alcohol because of possible potentiation of the effects. (1)

SOMNOS: see Chloral Hydrate

SOMNOS + Alcohol: Chloral hydrate should not be given to patients who are taking or have recently taken alcohol. (1)

SOPOR
+ Alcohol: see Sopor + Sedatives
+ Analgesics: see Sopor + Sedatives
+ Psychotherapeutic Agents: see Sopor + Sedatives

SOPOR (continued)

+ Sedatives: Care should be used during administration with other sedatives, analgesics or psychotherapeutic drugs, or with alcohol because of possible potentiation of the effects. (1)

SORBITRATE

+ Acetylcholine: see Sorbitrate + Norepinephrine
+ Alcohol: An occasional patient exhibits marked sensitivity to the hypotensive effects of nitrates, and severe responses (nausea, vomiting, weakness, restlessness, pallor, perspiration, and collapse) can occur with the usual therapeutic dose. Alcohol may enhance this effect. (1)
+ Histamine: see Sorbitrate + Norepinephrine
+ Norepinephrine: Sorbitrate can act as a physiological antagonist to norepinephrine, acetylcholine, histamine, and many other agents. (1)

SPAN-RD + MAO Inhibitors: Do not use in patients taking monoamine oxidase inhibitors. (1) Mechanism: see MAO Inhibitor + Sympathomimetics, Indirect-Acting

SPARINE: see Phenothiazines

SPARINE

+ Alcohol: see Sparine + CNS Depressants
+ Analgesics: see Sparine + CNS Depressants. In surgical and obstetrical sedation, Sparine is administered intramuscularly usually in 50 mg doses and in conjunction with reduced doses of analgesics such as meperidine. In general, doses of analgesics and sedatives may be reduced by one-third to one-half when Sparine is used as an adjunct. (1)
+ Anticonvulsants: Patients with a history of epilepsy should be treated with phenothiazine compounds only when such therapy is absolutely necessary. In such cases adequate anticonvulsant therapy should be given concomitantly. (1)
+ Barbiturates: see Sparine + CNS Depressants and Analgesics
+ CNS Depressants: Sparine potentiates the action of analgesics and central nervous system depressants. Such agents should therefore be given in reduced dosage to patients receiving Sparine.
  Sparine should not be used in comatose states due to central nervous system depressants (alcohol, barbiturates, opiates, etc.). (1)
+ Epinephrine: If it is desirable to administer a vasopressor drug, Levophed appears to be the most suitable. Epinephrine should not be used because Sparine may reverse its action, causing a further lowering of blood pressure instead of the usual elevating effects. (1)

SPARINE (continued)
+ Meperidine: see Sparine + Analgesics
+ Opiates: see Sparine + CNS Depressants and Analgesics
+ Phosphorous Insecticides: It has been reported that Sparine may potentiate the effects of organic phosphates found in certain insecticides. (1)
+ Sedatives: see Sparine + Analgesics

SPARTOCIN
+ Oxytocic Agents: If any other oxytocic agent is to be administered, it is advisable to allow a lapse of two or more hours after the last dose of Spartocin before initiating such treatment. (1)
+ Oxytocin: If a patient fails to respond to Spartocin treatment and it is decided to initiate an oxytocin infusion, the patient must be carefully observed for unusual sensitivity due to an apparent synergistic action of Spartocin and oxytocin which may result in tetanic uterine contractions. Therefore it is advisable to let several hours elapse after the last dose of Spartocin before starting this treatment. (1)

STAPHCILLIN
+ Antimicrobial Agents: Other agents should not be physically mixed with Staphcillin but should be administered separately in accordance with its recommended route of administration and dosage schedule. (1)
+ Kanmycin Solution: The two antibiotics should not be mixed, as they rapidly inactivate each other. (138)
+ Neutrapen: Because of the resistance of this agent to destruction by penicillinase, parenteral B. cereus penicillinase may not be effective for the treatment of allergic reactions. (1)

STAPHYLOCOCCUS-STREPTOCOCCUS UBA + Antiseptic: If any of the antiseptic used to cleanse the rubber stopper is carried into the vial by the needle, denaturation of the proteins may occur and thus the special advantage of the antigens will be lost. (1)

STECLIN
+ Food: Oral forms of tetracycline should be given one hour before or two hours following meals. (1)
+ Foods, Calcium-containing: see Steclin + Milk Formulas
+ Milk Formulas: Pediatric dosage forms (oral) should not be given with milk formulas and should be given at least one hour prior to feeding. (1) Mechanism: Tetracyclines are chelating agents that form insoluble chelates with multivalent cations (Ca^{++}, Mg^{++}, Al^{+++}), and gastrointestinal absorption is inhibited.
+ Trisoralen: see Trisoralen

STECLIN PARENTERAL + Hemastix: see Hemastix + Steclin Parenteral

STELAZINE: see Phenothiazines

STELAZINE
+ Alcohol: see Stelazine + CNS Depressants
+ Anesthetics: see Stelazine + CNS Depressants
+ Antihypertensive Agents: Stelazine may potentiate the action of antihypertensive agents, and care should be exercised when used concomitantly. (2)
+ Bone Marrow Depressants: Stelazine is contraindicated in cases of bone marrow depression. (1)
+ CNS Depressants: Stelazine has no clinically significant potentiating action. However, if agents such as sedatives, narcotics, anesthetics, tranquilizers, or alcohol are used either simultaneously or successively with the drug, the possibility of an undesirable additive depressant effect should be considered. Stelazine is contraindicated in greatly depressed states due to central nervous system depressants. (1, 2)
+ Drugs, Overdosage: The antiemetic effect of Stelazine may mask signs of overdosage of toxic drugs. (1)
+ Epinephrine: see Stelazine + Epinephrine
+ Metrazol: see Stelazine + Picrotoxin
+ Narcotics: see Stelazine + CNS Depressants
+ Picrotoxin: In the treatment of Stelazine overdosage, stimulants that may cause convulsions (e.g., picrotoxin or Metrazol) should be avoided. Amphetamine, methamphetamine, or caffeine and sodium benzoate is recommended. (1)
+ Pressor Agents: If hypotension occurs with Stelazine, the standard measures for managing circulatory shock should be initiated. If it is desirable to administer a vasoconstrictor, Levophed or Neosynephrine are the most suitable. Other pressor agents, including epinephrine, are not recommended because phenothiazines may reverse the usual elevating action of these agents and cause a further lowering of blood pressure. (1, 2)
+ Sedatives: see Stelazine + CNS Depressants
+ Tranquilizers: see Stelazine + CNS Depressants

STEMEX + Diabetes: Stemex, like other glucocorticoids, may aggravate diabetes mellitus so that higher insulin dosage may become necessary or latent diabetes mellitus may be precipitated. (1)

STERANE + Diabetes: Diabetic patients should be kept under close observation with careful adjustment of diet and insulin, since the gluconeo-

STERANE + Diabetes (continued): genesis produced by corticosteroids will often cause an increase in insulin requirements. (1)

STERAPRED + Diabetes: Physicians should weigh possible undesirable effects against anticipated clinical improvement in patients with diabetes mellitus. Prednisone, particularly in large doses, may cause hyperglycemia and glycosuria. (1)

STERASAL-K + Diabetes: Physicians should weigh possible undesirable effects against anticipated clinical improvement in patients with diabetes mellitus. Prednisone, particularly in large doses, may cause hyperglycemia and glycosuria. (1)

STERAZOLIDIN: see Butazolidin and Prednisone

STERILIZING SOLUTION: see listed under the following agents:
Insulin (Iletin, Lente)
Insulin, Protamine Zinc

STERISIL + Soap: A douche may be used prior to application, but soapy solutions should not be used for douching since the drug is inactivated by soap. (2)

STERNEEDLE
+ Measles Vaccine: see Sterneedle + Smallpox Vaccine
+ Smallpox Vaccine: Recent smallpox or measles vaccination are contraindications for tuberculin testing. (1)

STEROID HORMONES + Anticoagulants: see Anticoagulants + Steroid Hormones

STEROIDS: see listed under the following agents:
Brevital
Bristuron
Cendevax
Diuril
Esidrix
HydroDIURIL
Indocin
Lasix
Lirugen
Measles Virus Vaccine, Live, Attenuated
Measles Virus Vaccine, Live, Attenuated (Edmonston B and Schwarz)
Metahydrin
M-Vac Measles Virus Vaccine, Live, Attenuated
Naturetin
Pfizer-Vax Measles-L
Saluron
Stoxil
Tetanus Toxoids

STIMULANTS: see listed under the following agents:
Hypertensin
Ismelin
MAO Inhibitor
Niamid

STOXIL
+ Adrenocorticosteroids: see Adrenocorticosteroids + Stoxil
+ Boric Acid: Boric acid should not be administered during the course of therapy since it may cause irritation in the presence of Stoxil. (1)
+ Drugs: Stoxil Opthalmic Solution should not be mixed with other medications. (1)
+ Steroids: In superficial infections (dendritic keratitis and geographic ulcers of the corneal epithelium), Stoxil should not be used in combination with steroids. In deep infections, if such combined therapy is judged necessary, it must be employed with caution and the patient must be observed closely. Although there is evidence that Stoxil Opthalmic Solution and especially the Opthalmic Ointment have been used successfully in combination with steroids in cases complicated by iritis or corneal edema, the physician should bear in mind that steroids can accelerate the spread of viral infections and are normally contraindicated in herpes simplex of the eye. (1)

STREPTOMYCIN: see listed under the following agents:
Anectine
Antiemetics, Phenothiazine-type
Coly-Mycin M Inj.
Dramamine
Flaxedil
Garamycin
Kantrex Inj.
Magnesium Sulfate Inj.
Quelicin Chloride Inj.
Succinylcholine
Sucostrin
Tubocurarine (d-Tubocurarine)
Tubocurarine Chloride Inj.
Viocin Sulfate

STREPTOMYCIN + Ototoxic Antibiotics: Do not give concurrently or in series with other antibiotics capable of causing eighth nerve damage because the ototoxicity is frequently additive. (1)

STREPTOMYCIN CALCIUM CHLORIDE + Sulamyd Opth. Sol.: see Sulamyd Opth. Sol. + Streptomycin Calcium Chloride

STRIATRAN
+ Alcohol: Patients should avoid the concomitant use of Striatran and alcohol, since the effects may be additive. (2)
+ MAO Inhibitors: see Striatran + Psychotropic Agents
+ Phenothiazines: see Striatran + Psychotropic Agents
+ Psychotropic Agents: Other psychotropic agents, particularly phenothiazines or monoamine oxidase inhibitors, that are known to potentiate the action of other drugs should not be given with Striatran. (2)

STRYCHNINE + Nembutal Sod. Sol.: see Nembutal Sod. Sol. + Strychnine

SUBLIMAZE

+ Barbiturates: see Sublimaze + CNS Depressants
+ CNS Depressants: Other central nervous system depressants (e.g., barbiturates, major tranquilizers, and other narcotics), given so their actions overlap those of Sublimaze, must be used in reduced doses (as low as one-fourth to one-third the dose usually recommended) because of additive or potentiating effects. (1)
+ Dextromethorphan: A pharmacologic study in dogs suggests that the possibility of muscular rigidity may be increased when Sublimaze is given to patients receiving medications containing d-methorphan, although there have been no reports of this in clinical studies. (1)
+ MAO Inhibitors: There is no evidence that Sublimaze is potentiated by monoamine oxidase inhibitors, but since potentiation is found with other narcotic analgesics, the use of Sublimaze with MAO inhibitors is not recommended. (1)
+ Narcotics: see Sublimaze + CNS Depressants
+ Tranquilizers, Major: see Sublimaze + CNS Depressants

SUCCINYLCHOLINE: see also Anectine, Quelicin, and Sucostrin

SUCCINYLCHOLINE: see listed under the following agents:

Anticholinesterase Agents	Motilyn
Brevital	Mylaxen
Cholinesterase Inhibitors	Organophosphate Pesticides
Cozyme	Phospholine Iodide
Floropryl	Protopam
Ilopan	Surbex-T Solution
Levoprome	Surital
Magnesium Sulfate Inj.	Tensilon

SUCCINYLCHOLINE

+ Alkaline Solutions: Succinylcholine is rapidly hydrolyzed by alkaline solutions and therefore loses potency rapidly if mixed with alkaline preparations such as Pentothal Sodium, Surital, or Brevital. Such mixtures, if used at all, must be used within a few minutes of preparation. Separate injections of the drug are preferable.
+ Antibiotics: The following antibiotics: neomycin, streptomycin, dihydrostreptomycin, polymixin B, Coly-Mycin, viomycin, Kantrex, and possibly bacitracin, have been shown to have neuromuscular blocking effects and may cause respiratory depression. They may on occasion act as weak depolarizing agents and cause a transmission defect at the myoneural junction; therefore they may potentiate the action of muscle relaxants and prolong recovery time following induced muscle paralysis. (131,139,140)

SUCCINYLCHOLINE (continued)

+ Antimalarial Drugs: A low level of pseudocholinesterase may be associated with a prolonged paralysis of respiration following the use of succinylcholine. Low levels of this enzyme are found in patients receiving antimalarial drugs. (See also Quelicin + Antimalarial Drugs.)
+ Bacitracin: see Succinylcholine + Antibiotics
+ Brevital: see Succinylcholine + Alkaline Solutions
+ Cholinergic Agents: Succinylcholine and decamethonium (Syncurine) produce paralysis by not only displacing acetylcholine from receptor sites, but produce persistent depolarization. Acetylcholine or acetylcholine-like substances would aggravate the effects of succinylcholine and decamethonium due to an additive depolarizing effect. (118)
+ Cholinesterase Inhibitors: Anticholinesterase agents such as neostigmine, physostigmine, Tensilon, Humorsol, Mytelase and organic phosphates such as Floropryl, Phospholine, and organophosphate insecticides usually prolong the action of succinylcholine rather than antagonize its action. When succinylcholine is given over a prolonged period of time, the characteristic depolarization block of the myoneural junction may change to a non-depolarizing block which results in prolonged respiratory depression or apnea. Under such circumstances small repeated doses of neostigmine or Tensilon may act as antagonists. These drugs should be preceded by 0.4 to 0.6 mg of atropine to guard against bradycardia. Mechanism: A low level of pseudocholinesterase in the plasma may be associated with prolonged paralysis of respiration following the use of succinylcholine. The anticholinesterases may prolong the block and the resulting apnea by preventing enzymatic hydrolysis of succinylcholine at the motor end plates. Cholinesterase inhibitors by also preventing the destruction of acetylcholine, which acts as a depolarizing agent at the motor end plates, augments the depolarizing action of succinylcholine. (131, 141, 142)
+ Cinchona Alkaloids: Quinidine and related alkaloids are known to have a curare-like action on skeletal muscles. Caution is suggested in the use of these compounds concomitantly with, or in the immediate post-recovery period following the use of muscle relaxants.
+ Coly-Mycin: see Succinylcholine + Antibiotics
+ Curare-like Muscle Relaxants: see Succinylcholine + Nondepolarizing Muscle Relaxants
+ Cyclopropane: see Succinylcholine + Halothane

SUCCINYLCHOLINE (continued)

+ Digitalis: Unless absolutely necessary, succinylcholine should not be administered to those patients who are digitalized. Arrhythmias and cardiac arrest have occurred from a potentiation by succinylcholine of the effects of digitalis on cardiac conduction in digitalized patients. (See also Quelicin + Digitalis.) (131)
+ Dihydrostreptomycin: see Succinylcholine + Antibiotics
+ Electrolyte Imbalance: Succinylcholine should be used cautiously in those patients who have a disturbed electrolyte balance. Patients who have electrolyte imbalance will require less succinylcholine. (131)
+ Ether: see Succinylcholine + Nitrous Oxide
+ Floropryl: see Succinylcholine + Cholinesterase Inhibitors
+ Halothane: Bradycardia, arrhythmias, and even transient sinus arrest have occurred after the intravenous injection of succinylcholine. These effects, which are thought to be the result of vagal stimulation, have been reported to be enhanced by cyclopropane and halothane and tend to be less pronounced in patients premedicated with atropine or scopolamine. Prior administration of morphine may increase the likelihood of bradycardia and arrest. (14, 131, 144) See also Succinylcholine + Nitrous Oxide.
+ Humorsol: see Succinylcholine + Cholinesterase Inhibitors
+ Hypothermia: Hypothermia increases sensitivity to the effects of succinylcholine. This is probably caused by decreased enzymatic activity so that the breakdown of succinylcholine and of acetylcholine at the myoneural junction is slowed.
+ Irradiation: Sensitivity to the effects of succinylcholine occurs in patients whose pseudocholinesterase levels are low as in post-irradiation patients. (See also Sucostrin + Irradiation.)
+ Kantrex: see Succinylcholine + Antibiotics
+ Morphine: see Succinylcholine + Halothane
+ Mylaxen: The administration of Mylaxen markedly potentiates and prolongs the neuromuscular effects of succinylcholine. <u>Mechanism</u>: Mylaxen inhibits plasma cholinesterase and also produces non-depolarization and neuromuscular blockade. (127, 143)
+ Mytelase: see Succinylcholine + Cholinesterase Inhibitors
+ Neomycin: see Succinylcholine + Antibiotics
+ Neostigmine: see Succinylcholine + Cholinesterase Inhibitors
+ Nitrous Oxide: Bradycardia, sinus arrest, and supraventricular and ventricular arrhythmias may be found after injection of succinylcholine in patients who have been anesthetized with nitrous oxide, Trilene, ether, halothane, or cyclopropane. (14, 144)

SUCCINYLCHOLINE (continued)

+ Nondepolarizing Muscle Relaxants: Prolonged administration of succinylcholine may lead to a gradual change from a depolarized block to a nondepolarized block, thus increasing the sensitivity of the motor end plates to a nondepolarizing relaxant. This does not contraindicate the use of curare-like muscle relaxants (e.g., tubocurarine), but the dosage should be adjusted accordingly.
  The effect of a single dose of tubocurarine is additive to the effect of subsequent doses of that drug or to other nondepolarizing relaxants but is antagonistic to that of succinylcholine and other depolarizing agents. This antagonism is so great that at a time when there is no discernable residual effect of the nondepolarizing relaxant, three or four times more depolarizing relaxant is required for the production of neuromuscular block than under ordinary circumstances. Therefore some investigators advise against the use of succinylcholine for abdominal closure if tubocurarine has been employed throughout the procedure. They feel that an effective dose of succinylcholine will be unnecessarily large and may lead to prolonged apnea following completion of surgery.
+ Organic Phosphates: see Succinylcholine + Cholinesterase Inhibitors
+ Organophosphate Insecticides: Succinylcholine may produce prolonged apnea if used in conjunction with or in patients exposed to the organophosphate insecticides. This effect is due to the inhibition of pseudocholinesterase by the insecticide, permitting accumulation of acetylcholine. Mechanism: The organophosphate combines with the cholinesterases causing inhibition of their action. (142) See also Succinylcholine + Cholinesterase Inhibitors
+ Pentothal Sodium: see Succinylcholine + Alkaline Solutions
+ Phospholine: see Succinylcholine + Cholinesterase Inhibitors
+ Physostigmine: see Succinylcholine + Cholinesterase Inhibitors
+ Polymixin B: see Succinylcholine + Antibiotics
+ Procaine: Drugs which compete with succinylcholine for the enzyme (plasma pseudocholinesterase) should not be given concurrently with succinylcholine as it may intensify the action of succinylcholine.
+ Quinidine: see Succinylcholine + Cinchona Alkaloids
+ Streptomycin: see Succinylcholine + Antibiotics
+ Surital: see Succinylcholine + Alkaline Solutions
+ Tensilon: see Succinylcholine + Cholinesterase Inhibitors
+ Trilene: see Succinylcholine + Nitrous Oxide
+ Tubocurarine: see Succinylcholine + Nondepolarizing Muscle Relaxants.

SUCCINYLCHOLINE (continued)
+ Viomycin: see Succinylcholine + Antibiotics

SUCOSTRIN: see Succinylcholine

SUCOSTRIN
+ Alkaline Solutions: Because succinylcholine is rapidly hydrolyzed in alkaline solutions, it is incompatible with Pentothal Sodium; separate injections are therefore preferable. (1)
+ Anticholinesterase Agents: Neostigmine or other anticholinesterases as well as Tensilon are contraindicated as antidotes for succinylcholine, since they prolong its effect rather than antagonize its action. (1) Mechanism: see Succinylcholine + Cholinesterase Inhibitors.
+ Cholinesterase Inhibitors: Succinylcholine should be used cautiously in patients exposed to cholinesterase inhibitors such as diisopropylfluorophosphate (DPF) or diethyl-p-nitrophenylthiophosphate (Parathion). In such cases, fresh whole blood or plasma may be indicated to replenish cholinesterase levels. (1) See Sucostrin + Anticholinesterase Agents. See also Succinylcholine + Cholinesterase Inhibitors, and Organophosphate Insecticides.
+ Curare-Type Muscle Relaxants: If a curare-type muscle relaxant is used in conjunction with succinylcholine, its effect should be permitted to wear off before succinylcholine is administered.

  Prolonged apnea may be due to a dual block or a mixed block. A dual block may occur with succinylcholine alone, usually after excessive doses or prolonged administration. In a dual block the muscle fiber responds first by depolarization but eventually changes to show all the characteristics of a non-depolarizing block. A mixed block may occur when a depolarizing and a non-depolarizing muscle relaxant are used together. It is important to determine which type of block is responsible for prolonged apnea, since a depolarizing block is potentiated and a non-depolarizing block antagonized by neostigmine or Tensilon. (1)
+ Dihydrostreptomycin: see Sucostrin + Procaine
+ DPF: see Sucostrin + Cholinesterase Inhibitors
+ Electrolyte Imbalance: Succinylcholine should be used cautiously in those patients who have disturbed electrolyte balance. Patients who have electrolyte imbalance will require less succinylcholine. (1)
+ Irradiation: Sensitivity to the effects of succinylcholine occurs in patients whose pseudocholinesterase levels are low as in post-irradiation patients. (1)
+ Neomycin: see Sucostrin + Procaine

SUCOSTRIN (continued)
+ Neostigmine: see Sucostrin + Anticholinesterase Agents
+ Parathion: see Sucostrin + Cholinesterase Inhibitors
+ Pentothal Sodium: see Sucostrin + Alkaline Solutions
+ Procaine: The intravenous injection of procaine or parenteral use of neomycin, streptomycin, or dihydrostreptomycin may intensify the action of succinylcholine. (1) Mechanism: see Succinylcholine + Antibiotics, and Procaine.
+ Streptomycin: see Sucostrin + Procaine
+ Tensilon: see Sucostrin + Anticholinesterase Agents

SUGRACILLIN + Meals: Oral penicillin should be given on a fasting stomach not less than one hour before meals or at least two to three hours after meals. (1)

SULADYNE + Dolonil: see Dolonil + Suladyne

SULAMYD SODIUM OPTH. SOL.
+ Silver Preparations: The solutions of Sulamyd are incompatible with silver preparations. (1)
+ Streptomycin Calc. Chloride: The solutions are incompatible. (1)

SULFA DRUGS: see listed under the following agents:
Anturane
Benemid

SULFABID + Orinase: Diabetic patients taking Orinase should be observed closely because of recently reported potentiation of the hypoglycemic action of Orinase by sulfonamide compounds. (1, 50, 58) Mechanism: Studies have shown a pronounced delay in the plasma disappearance rate of tolbutamide after Sulfabid was administered. Diminished hepatic metabolism or renal clearance could be responsible for the prolonged Orinase effect, but it is possible that the Orinase is displaced from a plasma protein to a tissue protein that releases it more slowly. By such means Orinase disappearance would be delayed. (50)

SULFADIAZINE + Anturane: see Anturane + Sulfadiazine

SULFAS, ANTIMICROBIAL + Dymelor: see Dymelor + Sulfas, Antimicrobial

SULFASUXIDINE
+ Anticoagulants: Sulfasuxidine may reduce fecal output of thiamine and decrease synthesis of vitamin K, requiring supplementation. Hypoprothrombinemia and possibly hemorrhage might result from deficiency of vitamin K. (1)

SULFASUXIDINE (continued)

+ Enemas: Cleansing enemas should not be used during the 48 hours preceding operation. If a cleansing enema is desirable, however, it may be given on the morning of operation and should contain 6 gm of sulfasuxidine per liter to avoid too great a reduction of the drug in the colon. (1)
+ Mineral Oil: Mineral oil interferes with the action of sulfasuxidine. (1)
+ Purgatives: The use of purgatives should be avoided. (1)

SULFATES + Metycaine Hydrochloride: see Methycaine Hydrochloride + Sulfates

SULFATHALIDINE

+ Anticoagulants: Sulfathalidine may reduce fecal output of thiamine and decrease synthesis of vitamin K, requiring supplementation. Hypoprothrombinemia and possibly hemorrhage might result from deficiency of vitamin K. (1)
+ Enemas: Cleansing enemas should not be used during the 48 hours preceding operation. If a cleansing enema is desirable, however, it may be given on the morning of operation and should contain 6 gm of Sulfathalidine per liter to avoid too great a reduction of the drug in the colon. (1)
+ Mineral Oil: Mineral oil interferes with the action of Sulfathalidine. (1)
+ Purgatives: The concomitant use of purgatives should be avoided. (1)

SULFATHIAZOLE + Methenamine: Since sulfathiazole may form an insoluble precipitate with formaldehyde in the urine, it is probable that no methenamine compound should be given during sulfonamide therapy. (2)

SULFONAMIDES: see Acidic Agents + Alkaline Media

SULFONAMIDES: see Acidic Agents + Alkanizing Agents (Urinary)

SULFONAMIDES: see listed under the following agents:

Antacids
Benemid
Coumadin
Inversine
Methotrexate
Orinase
Pabalate
Pabirin Buffered Tablets
Para-Amino Benzoic Acid (PABA)
P-B-Sal-C
Potaba
Sintrom
Thio-Tepa
Tolinase
Tromexan

SULFONAMIDES + Methenamine: Since other sulfonamides may form an insoluble precipitate with formaldehyde in the urine, it is probable that no methenamine compound should be given during sulfonamide therapy.(2)

SULFONAMIDES, ANTIBACTERIAL: see listed under the following agents:
Diabinese
Methotrexate

SULFONAMIDES, BACTERIOSTATIC: see listed under the following agents:
Diabinese
Dymelor
Hypoglycemic Drugs, Oral
Orinase

SULFONAMIDES, DIURETIC + Methotrexate: see Methotrexate + Sulfonamides, Diuretic

SULFONAMIDES, HYPOGLYCEMIC + Methotrexate: see Methotrexate + Sulfonamides, Hypoglycemic

SULFONAMIDES, LONG-ACTING
+ Anticoagulants: see Sulfonamides, Long-Acting + Butazolidin
+ Anturane: see Sulfonamides, Long-Acting + Butazolidin
+ Butazolidin: Highly bound acidic agents such as Butazolidin, Tandearil, Anturane, Salicylic Acid, and anticoagulants such as dicumarol and Tromexan are able to displace the long-acting, albumin-bound sulfonamides from plasma proteins. Since these sulfonamides are not rapidly metabolized or excreted, the displaced molecules diffuse into the tissues. As a result, the total concentration of the sulfonamides decline in plasma but rise in skeletal muscle, cerebral spinal fluid, and brain, and their antibacterial activities are enhanced. (55, 118) Mechanism: Displacement from protein binding sites of one drug by another, thereby increasing the concentration of unbound drug.
+ Dicumarol: see Sulfonamides, Long-Acting + Butazolidin
+ Salicylic Acid: see Sulfonamides, Long-Acting + Butazolidin
+ Tandearil: see Sulfonamides, Long-Acting + Butazolidin
+ Tromexan: see Sulfonamides, Long-Acting + Butazolidin

SULFONAMIDE-TYPE AGENTS: see listed under the following agents:
Butazolidin
Tandearil

SULFONYLUREA AGENTS: see listed under the following agents:
Acusul
Anturane
Butazolidin
Eskadiazine
Tandearil

SULFONYLUREA, HYPOGLYCEMIC: see Hypoglycemic Drugs, Oral

SULFONYLUREAS, HYPOGLYCEMIC (continued): see listed under the following agents:

Alcohol
Diuretics, Thiazide
Sulla
Sul-Spantab
Sultacof
Sultussin
Trisulfaminic Tablets and Suspension

SULFONYLUREAS, HYPOGLYCEMIC

+ Alcohol: A reaction to alcohol has been reported in some patients receiving sulfonylurea compounds. This reaction may occur with any of the sulfonylurea agents now available. The reaction signs are flushing, a feeling of facial warmth, and occasionally a pounding headache, nausea, giddiness, tachycardia, and dyspnea. In severe cases, there may be blurring of vision. The reaction comes on three to ten minutes after drinking small amounts of alcohol and may last an hour or longer. This reaction resembles the Antabuse-alcohol reaction and is called a disulfiram reaction. (55, 76) Mechanism: see Alcohol + Sulfonylurea, Hypoglycemic.

+ Anticoagulants, Coumarin: Coumarin anticoagulants have been known to potentiate the action of the sulfonylureas, resulting in prolonged hypoglycemia. (50, 55)

+ Butazolidin: Butazolidin has been known to potentiate the action of the sulfonylureas, resulting in prolonged hypoglycemia. (55)

+ Insulin: There have been isolated reports of the potentiation of exogenous insulin by sulfonylureas in juvenile diabetes. This may be due to the release of the insulin from its binding to antibody by the sulfonylurea derivatives. (76)

+ Sulfonamides: Sulfonylureas such as Orinase are bound avidly to protein. When sulfonamides are given in addition to sulfonylureas, the latter will be displaced and hypoglycemia may ensue. (50, 55)

+ Thiazide Diuretics: Thiazide diuretics may aggravate the diabetic state, particularly when given with a sulfonylurea, and the patient may require an increased dosage of oral hypoglycemics. (55)

SULLA

+ Sulfonylurea, Hypoglycemic: Because of the danger of severe hypoglycemic reactions, Sulla should be administered with caution to patients receiving oral hypoglycemic agents of the sulfonylurea type. (1) Mechanism: Displacement of the oral hypoglycemic agent from its protein binding site. (50)

+ Trisoralen: see Trisoralen

SUL-SPANSION + Sulfonylureas, Hypoglycemic: Sulfonamides may potentiate the hypoglycemic action of sulfonylureas; therefore caution should

SUL-SPANSION + Sulfonylureas, Hypoglycemic (continued): be observed. (1) Mechanism: Displacement of the hypoglycemic agent from its protein binding site. (50)

SUL-SPANTAB
+ Butazolidin: see Butazolidin + Sul-Spantab
+ Sulfonylureas, Hypoglycemic: Sulfonamides may potentiate the hypoglycemic action of sulfonylureas; therefore caution should be observed. (1) Mechanism: Displacement of the hypoglycemic agent from its protein binding site. (50)

SULTACOF + Sulfonylureas, Hypoglycemic: Sulfonamide therapy may potentiate the hypoglycemic action of sulfonylureas. (1) Mechanism: Displacement of the hypoglycemic agent from its protein binding sites. (50)

SULTUSSIN + Sulfonylureas, Hypoglycemic: Sulfonamide therapy may potentiate the hypoglycemic action of sulfonylureas. (1) Mechanism: Displacement of the hypoglycemic agent from its protein binding sites. (50)

SUMYCIN
+ Foods, Calcium-containing: see Sumycin + Milk
+ Meals: Oral forms of tetracycline should be given one hour before or two hours following meals. (1)
+ Milk: Pediatric dosage forms of tetracycline (oral) should not be given with milk formulas or other calcium-containing food and should be given at least one hour prior to feeding. (1) Mechanism: Tetracyclines are chelating agents that form insoluble chelates with multivalent cations (Ca^{++}, Mg^{++}, Al^{+++}), and gastrointestinal absorption is inhibited.
+ Trisoralen: see Trisoralen

SUPRARENIN: see Epinephrine

SUPRARENIN
+ Chloroform: Suprarenin is contraindicated during chloroform anesthesia, as it may produce ventricular contractions. (1) Mechanism: see Chloroform + Epinephrine.
+ Cyclopropane: Suprarenin is contraindicated during cyclopropane anesthesia, as it may produce premature ventricular contractions. (1) Mechanism: see Cyclopropane + Epinephrine.

SURBEX-T SOLUTION
+ Neostigmine: see Surbex-T Solution + Parasympathetic Drugs

SURBEX-T SOLUTION (continued)

+ Parasympathetic Drugs: In order to prevent exaggeration of the effects of parasympathetic drugs such as neostigmine, an interval of 12 hours should be allowed between administration of such drugs and Surbex-T Solution. (1) Mechanism: Pantothenyl alcohol is converted to pantothenic acid and thus claimed to produce more acetylcholine, thus there may be an additive effect.
+ Succinylcholine: Respiratory embarrassment has been observed when the Surbex-T component, d-pantothenyl alcohol, was given shortly after succinylcholine. (1) Mechanism: d-Pantothenyl alcohol acts as a precursor for acetylcholine and thereby potentiates the effects of muscle relaxants.

SURGERY: see listed under the following agents:

Antimetabolic Agents	Pertofrane
Choloxin	Rauwiloid
Euthroid	Thyrolar
Fluorouracil	Tofranil

SURGICEL ABSORBABLE HEMOSTAT

+ Anti-Infective Agents: see Surgicel Absorbable Hemostat + Hemostatic Agents
+ Hemostatic Agents: Do not impregnate with other hemostatic agents or penicillin or other anti-infective agents. The hemostatic effect of Surgicel is not enhanced by the addition of thrombin, the activity of which is destroyed by the low pH of the gauze. (1)
+ Penicillin: see Surgicel Absorbable Hemostat + Hemostatic Agents
+ Saline: see Surgicel Absorbable Hemostat + Water
+ Thrombin: see Surgicel Absorbable Hemostat + Hemostatic Agents
+ Water: Do not moisten Surgicel with water or saline. Its hemostatic effect is greater when applied dry. (1)

SURITAL

+ Air: Injections of air from the syringe into the solution should be avoided, since this may hasten development of cloudiness. (1)
+ Buffering Agents: see Surital + Diluent
+ Diluent: It is preferable that solvents employed for making solutions (as with other barbiturates and thiobarbiturates) should not contain any substance such as phenol or buffer agents, since they may tend to cause precipitation. (1)
+ Diluents, Acid: Clear solutions prepared with Dextrose 5% may be employed; this solvent occasionally is sufficiently acid to cause precipitation. (1)

SURITAL (continued)

+ Phenol: see Surital + Diluent
+ Succinylcholine: Solutions of Surital and succinylcholine may be given concurrently, but they should not be mixed prior to administration.
  (1) Mechanism: see Succinylcholine + Surital

SYMMETREL

+ Alcohol: Although no evidence of clinical intolerance has been reported for the use of Symmetrel following the usage of alcoholic beverages, attention should be paid to any evidence of intolerance. (1)
+ CNS Stimulants: see Symmetrel + Psychopharmacologic Agents
+ Psychopharmacologic Agents: Because of the limited long-term experience with this medication in patients on psychopharmacologic therapeutic agents or central nervous system stimulants, if these patients are given Symmetrel, careful observation is required. (1)

SYMPATHOMIMETIC AGENTS: see listed under the following agents:

Aventyl
Dilor
Dopram
Eutonyl
MAO Inhibitor
Matulane
Nardil
Norpramin
Pertofrane
Vivactil

SYMPATHOMIMETIC AMINES: see listed under the following agents:

Bronkometer
Bronkosol
Eutonyl
Marplan
Nardil
Niamid
Parnate

SYMPATHOMIMETIC AMINES, INDIRECT-ACTING: see listed under the following agents:

Eutonyl
Furoxone
Ismelin
Reserpine

SYMPATHOMIMETICS: see also under the following listings: Sympathomimetic Agents; Sympathomimetic Amines; Sympathomimetics, Direct-Acting; Sympathomimetics, Indirect-Acting; and Sympathomimetic Amines, Indirect-Acting

SYMPATHOMIMETICS: see listed under the following agents:

Nebair
Parnate
Vivactil

SYMPATHOMIMETICS, DIRECT-ACTING: see listed under the following agents:

Aldomet
Ismelin
MAO Inhibitor

SYMPATHOMIMETICS, DIRECT-ACTING

+ Aldomet: see Sympathomimetics, Direct-Acting + Ismelin
+ Ismelin: If a direct-acting sympathomimetic such as Levophed is given to counteract severe hypotension in a patient on reserpine, Ismelin, or Aldomet, there is a much greater pressor response than would be expected under normal circumstances. These antihypertensives work to prevent the uptake of Levophed into inactivation sites and therefore potentiates the effect of the vasopressor. (55)
+ Reserpine: see Sympathomimetics, Direct-Acting + Ismelin

SYMPATHOMIMETICS, INDIRECT-ACTING: see listed under the following agents:

Aldomet
Ismelin
MAO Inhibitor

SYMPATHOMIMETICS, INDIRECT-ACTING + Reserpine: Indirect-acting sympathomimetics such as metaraminol (Aramine, Pressonex, Pressorol) or Wyamine depend on the release of catecholamines for its pressor effect. If an indirect-acting sympathomimetic is given to counteract hypotension in a patient on reserpine, the pressor response would be less than expected. If the patient has been on high doses of the antihypertensive for a long period of time, the vasopressor may be completely ineffective. (55)

SYNCILLIN + Food: Absorption of Syncillin is greater if the drug is given when the stomach is empty. (2)

SYNCURINE: see listed under the following agents:

Cholinesterase Inhibitors
Magnesium Sulfate Inj.
Mylaxen
Tensilon

SYNCURINE + Cholinergic Agents: Syncurine produces paralysis by not only displacing acetylcholine from receptor sites but produce persistent depolarization. Acetylcholine or acetylcholine-like substances

SYNCURINE + Cholinergic Agents (continued): would aggravate the effects of Syncurine due to an additive depolarizing effect. (118)

SYNKAVITE + Anticoagulant: Synkavite will reverse the hypoprothrombinemic effects of anticoagulants. (1)

SYNTETRIN I.M.: see listed under the following agents:

Hemastix
Trisoralen

SYNTETRIN I.V.: see listed under the following agents:

Hemastix
Trisoralen

SYNTHROID

+ Cytomel: see Cytomel + Synthroid
+ Diabetes: In patients with diabetes mellitus, look for possible changes in metabolic activity which may decrease insulin or other antidiabetic drug dosage requirements. (1)

SYNTOCINON + Actospar: see Actospar + Syntocinon

T

TACARYL: see Phenothiazines

TACARYL

+ Alcohol: see Tacaryl + Barbiturates
+ Analgesics: see Tacaryl + Barbiturates
+ Barbiturates: Because of possible potentiating properties, the product should be used with caution in patients receiving alcohol, analgesics, or sedatives (particularly barbiturates). (1)
+ Phenothiazines: When Tacaryl is given concomitantly with other phenothiazines, the likelihood of serious side effects associated with these phenothiazines may be increased. In these circumstances, physicians should be alert to possible occurrence of such reactions as agranulocytosis, jaundice, extrapyramidal symptoms, seizures, orthostatic hypotension, autonomic reactions, skin disorders, peripheral edema, and hyperpyrexia. (1)
+ Sedatives: see Tacaryl + Barbiturates

TALWIN

+ Barbiturates: Talwin should not be mixed in the same syringe with soluble barbiturates because precipitation will occur. (1)
+ Narcotics: Talwin is a narcotic-antagonist. Patients dependent on narcotics and receiving Talwin may occasionally experience certain withdrawal symptoms. Talwin should be given with special caution to such patients. It has been observed that some patients previously given narcotic-analgesics for one month or longer had mild withdrawal symptoms when the drug was replaced with the analgesic, Talwin. Symptoms believed to be indicative of antagonism to the opiates may be observed rarely with administration of Talwin to patients receiving opiates for a short time. Intolerance or untoward reactions have been observed after administration of Talwin to patients who have received single doses or who have had limited exposure to narcotics. (1)
+ Respiratory Depressants: Talwin should be administered only with caution and in low dosage to patients with respiratory depression from other medication. (1)

TANDEARIL: see listed under the following agents:

Anticoagulants	Dianabol	Panwarfin
Coumadin	Dicumarol	Sulfonamides, Long-acting
Diabinese	Orinase	Tolinase

TANDEARIL (continued)

+ Anticoagulants, Coumarin-type: Coumarin-type anticoagulants depress prothrombin activity. This is accentuated in some cases when Tandearil is simultaneously employed in treatment; occational instances of severe bleeding have been reported. Patients receiving coumarin-type anticoagulants should be very carefully followed for evidence of excessive increase of prothrombin time when Tandearil is added to the regimen. Anticoagulant therapy can then be properly adjusted, if necessary. When prescribed alone, Tandearil has not been shown to influence prothrombin activity. (1, 2)
+ Drugs: Patients receiving other potent chemotherapeutic agents which may increase possibility of toxic reactions are contraindicated. (1)
+ Insulin: Recent reports have indicated that pyrazole compounds may potentiate the pharmacologic action of sulfonylurea and sulfonamide-type agents and insulin. Patients receiving such concomitant therapy, therefore, should be carefully observed for this effect. (1, 2)
+ Sulfonamide-Type Agents: see Tandearil + Insulin
+ Sulfonylurea Agents: see Tandearil + Insulin

TAR DISTILLATE 'DOAK' + Trisoralen: see Trisoralen

TARACTAN: Since Taractan is structurally similar to the phenothiazines, all known serious side effects or toxicity associated with phenothiazines should be borne in mind when patients receive this agent for prolonged periods. See also Phenothiazines.

TARACTAN

+ Alcohol: see Taractan + CNS Depressants
+ Analgesics: see Taractan + CNS Depressants
+ Anesthetics: see Taractan + CNS Depressants
+ Antihypertensive Agents: Taractan may potentiate the action of antihypertensive agents, and care should be exercised when used concomitantly. (2)
+ CNS Depressants: Taractan should be used with care when employed in conjunction with anesthetics, hypnotics, opiates, analgesics, and other central nervous system depressants, since Taractan intensifies the central action of these drugs. (1, 2)
+ Epinephrine: Epinephrine should not be used concomitantly, since Taractan may reverse its action and cause profound hypotension. If a vasopressor is indicated, vasopressors such as Levophed may be used if the resulting hypotension is severe or prolonged. (1, 2)
+ Hypnotics: see Taractan + CNS Depressants
+ Opiates: see Taractan + CNS Depressants
+ Trisoralen: see Trisoralen

TEGETROL
+ Drugs: Where feasible, Tegetrol should not be used in conjunction with any potent drug which may increase the possibility of toxic reactions. (1)
+ MAO Inhibitors: On theoretical grounds the use of Tegetrol with MAO inhibitors is not recommended. When it seems desirable to administer Tegetrol to patients receiving MAO inhibitors, the MAO inhibitor should be discontinued and as long a drug-free interval should elapse as the clinical situation permits, with a minimum of seven days. Initial dosage of Tegetrol in these situations should be low with gradual increments given under close observation. (1)
+ Trisoralen: see Trisoralen

TEGOPEN + Food: The presence of food in the stomach or small intestine reduces the absorption of Tegopen and decreases the ultimate plasma level obtainable from a given dose of the drug. Furthermore, although it is more resistant than many penicillins, Tegopen is degraded in the stomach. Therefore this antibiotic must be given one or two hours before a meal, both to ensure optimal absorption and to avoid its excessive destruction by increased gastric acidity. (1, 2)

TEMARIL: see Phenothiazines

TEMARIL
+ Alcohol: see Temaril + CNS Depressants
+ Analgesics: see Temaril + CNS Depressants
+ Antihistamines: see Temaril + CNS Depressants
+ Barbiturates: see Temaril + CNS Depressants
+ CNS Depressants: Temaril should be used with caution where there is a risk of potentiating central nervous system depressants (opiates, analgesics, antihistamines, barbiturates, alcohol).
Temaril is contraindicated in states of central nervous system depression from agents such as barbiturates, alcohol, narcotics, or analgesics. (1)
+ Epinephrine: If a pressor agent is indicated, in Temaril overdosage, Neosynephrine or Levophed may be used. Epinephrine is contraindicated in these circumstances because it may paradoxically enhance any hypotensive effect. (1)
+ Narcotics: see Temaril + CNS Depressants
+ Opiates: see Temaril + CNS Depressants

TENSILON: see listed under the following agents:

Anectine
Phospholine Iodide
Quelicin Chloride Inj.
Succinylcholine
Sucostrin
Tubocurarine

TENSILON (continued)

+ Curare: Because of its brief effect, Tensilon should not be given prior to the administration of curare, tubocurarine, Flaxedil, or Metubine; it should be used at the time when its effect is needed. (1)
+ Flaxedil: see Tensilon + Curare
+ Metubine: see Tensilon + Curare
+ Succinylcholine: see Tensilon + Syncurine
+ Syncurine: Tensilon should not be used as an antidote for Syncurine and succinylcholine because it may prolong rather than antagonize their relaxant action on skeletal muscle. (1) Mechanism: see Succinylcholine + Cholinesterase Inhibitors
+ Tubocurarine: see Tensilon + Curare

TENUATE + MAO Inhibitors: Tenuate should not be given concurrently with monoamine oxidase inhibitors. (1) Mechanism: see MAO Inhibitor + Sympathomimetics, Indirect-Acting

TEPANIL + MAO Inhibitors: Tepanil should not be given concurrently with monoamine oxidase inhibitors. (1) Mechanism: see MAO Inhibitor + Sympathomimetics, Indirect-Acting

TERGEMIST

+ Polymixin-B Sulfate: Tergemist is incompatible with Polymixin-B Sulfate. (1)
+ Zephiran: Tergemist is incompatible with Zephiran. (1)

TERRAMYCIN

+ Aluminum Hydroxide Gel: Aluminum Hydroxide Gel given with antibiotics has been shown to decrease absorption and is contraindicated. (1) Mechanism: Tetracyclines are chelating agents that form insoluble chelates with multivalent cations (Ca^{++}, Mg^{++}, Al^{+++}), and gastrointestinal absorption is inhibited.
+ Foods, Calcium-containing: see Terramycin + Milk
+ Meals: To aid absorption of the drug, it should be given at least one hour before or two hours after eating. (1)
+ Milk: Pediatric dosage forms should not be given with milk formulas or calcium-containing foods; this is to aid absorption of the drug. (1) Mechanism: see Terramycin + Aluminum Hydroxide Gel.
+ Trisoralen: see Trisoralen

TERRAMYCIN PARENTERAL + Hemastix: see Hemastix + Terramycin Parenteral

TETANUS ANTITOXIN + Tetanus Toxoid, Aluminum Precipitated: see Tetanus Toxoid, Aluminum Precipitated + Tetanus Antitoxin

TETANUS IMMUNE GLOBULIN (HUMAN): see Pro-Tet

TETANUS IMMUNE GLOBULIN (HUMAN) + Tetanus Toxoid: All authorities agree that active immunization with tetanus toxoid should be instituted in those patients who receive passive immunization with tetanus immune globulin. However, some feel that the initiation of such a program should be delayed for one month after the administration of tetanus immune globulin, whereas other authorities believe that interference with antibody response from the passively acquired material is not sufficiently great to outweigh the advantages of beginning active immunization when the patient is at hand. (2)

TETANUS TOXOID: see listed under the following agents:
Pro-Tet
Tetanus Immune Globulin (Human)

TETANUS TOXOID
+ Steroids: Individuals receiving therapy with steroids may not respond optimally to active immunization procedures. Administration of immunizing agents should be deferred in such individuals or repeated thereafter. (1)
+ Tetanus Antitoxin: When Tetanus Toxoid and antitoxin are both used, the two should not be mixed in the same syringe and they should be administered in different extremities. (1)

TETANUS TOXOID, ALUM PRECIPITATED
+ Adrenal Corticosteroid: see Tetanus Toxoid, Alum Precipitated + Adrenocorticotrophin
+ Adrenocorticotrophin: By mechanisms incompletely understood, adrenocorticotrophin and adrenal corticosteroids have been shown to reduce host resistance to disease. Therefore they should not be administered following exposure to infectious agents (mumps, rabies, tetanus) for which no satisfactory antimicrobial therapy is available. To do so may alter the host-parasite relationship to such an extent that severe or fatal illness may result in spite of the prophylactic administration of a vaccine. Under these circumstances, the occurrence of disease, actually due to the altered pattern of resistance, might be attributed to a vaccine failure. (1)
+ Tetanus Antitoxin: If antitoxin and toxoid are injected at the same time, they should be given in different extremities. (1)

TETRACHEL
+ Foods, Calcium-containing: see Tetrachel + Milk Formulas
+ Meals: Oral forms of tetracycline should be given one hour before or two hours after meals; this is to aid the absorption of the drug. (1)

TETRACHEL (continued)

+ Milk Formulas: Pediatric dosage forms (oral) should not be given with milk formulas or other calcium-containing foods and should be given at least one hour prior to feeding. (1) Mechanism: Tetracyclines are chelating agents that form insoluble chelates with multivalent cations (Ca^{++}, Mg^{++}, Al^{+++}), and gastrointestinal absorption is inhibited.
+ Trisoralen: see Trisoralen

TETRACHEL INTRAVENOUS + Diluents, Calcium-containing: Solutions containing calcium may cause a precipitate; they are to be avoided. (1) Mechanism: see Tetrachel + Milk Formulas.

TETRACYCLINE: see also various trade-named preparations.

TETRACYCLINE: see listed under the following agents:
Anticoagulants (see Anticoagulants + Antibiotics)
Liquaemin Sodium
Mucomyst
Penicillin

TETRACYCLINE

+ Antacids: Tetracyclines are chelating agents that form insoluble chelates with multivalent cations (Ca^{++}, Mg^{++}, Al^{+++}); gastrointestinal absorption of these antibiotics are therefore inhibited. (55, 81)
+ Meals: As an aid to absorption of the drug, tetracycline should be given one hour before or two hours after meals.
+ Penicillin: There is some evidence that the use of a bacteriostatic agent such as tetracycline may prevent the bactericidal action of a drug such as penicillin. (55)
+ Trisoralen: see Trisoralen

TETRACYN

+ Aluminum Hydroxide Gel: Aluminum hydroxide gel given with antibiotics has been shown to decrease absorption and is contraindicated. (1) Mechanism: Tetracyclines are chelating agents that form insoluble chelates with multivalent cations (Ca^{++}, Mg^{++}, Al^{+++}), and gastrointestinal absorption is inhibited.
+ Foods, Calcium-containing: see Tetracyn + Milk Formulas
+ Meals: To aid absorption of the drug, it should be given at least one hour before or two hours after eating. (1)
+ Milk Formulas: Pediatric dosage forms of tetracycline should not be given with milk formulas or calcium-containing foods. (1) Mechanism: see Tetracyn + Aluminum Hydroxide Gel.
+ Trisoralen: see Trisoralen

TETRACYN PARENTERAL + Hemastix: see Hemastix + Tetracyn Parenteral

TETRAMAX
+ Food: It is believed that higher blood levels are attained when taking tetracycline one or two hours after eating or on an empty stomach. (1)
+ Foods, Calcium-containing: Pediatric dosage forms of TetraMAX should not be given with milk formulas and should be given at least one hour prior to feeding or meals. (1) Mechanism: Tetracyclines are chelating agents that form insoluble chelates with multivalent cations (Ca^{++}, Mg^{++}, Al^{+++}), and gastrointestinal absorption is inhibited.
+ Milk Formulas: Pediatric dosage forms of TetraMAX should not be given with milk formulas. (1) Mechanism: see TetraMAX + Foods, Calcium-containing.
+ Trisoralen: see Trisoralen

TETRA-SOLGEN
+ Adrenal Corticosteroids: see Tetra-Solgen + Adrenocorticotrophin
+ Adrenocorticosteroids: see Adrenocorticosteroids + Tetra-Solgen
+ Adrenocorticotrophin: Adrenocorticotrophin (ACTH) and adrenal corticosteroids may suppress the antibody response to the vaccine. Therefore, if possible, it would seem advisable to avoid administration of the vaccine concomitantly with these hormones. (1)

TETREX
+ Foods, Calcium-containing: Pediatric dosage forms should not be given with calcium-containing foods and should be given at least one hour prior to feeding or meals. (1) Mechanism: Tetracyclines are chelating agents that form insoluble chelates with multivalent cations (Ca^{++}, Mg^{++}, Al^{+++}), and gastrointestinal absorption is inhibited.
+ Meals: Oral dosage forms of tetracycline should be given one hour before or two hours after meals for more effective absorption. (1)
+ Milk: Pediatric dosage forms of tetracycline should not be given with milk formulas and should be given at least one hour prior to feeding. (1) Mechanism: see Tetrex + Foods, Calcium-containing.
+ Trisoralen: see Trisoralen

THEOKON ELIXIR + Xanthine Preparations: Avoid concurrent use of additional xanthines contained in other medications. (1)

THEO-ORGANIDIN
+ Aminophylline: see Theo-Organidin + Theophylline

THEO-ORGANIDIN (continued)

+ Theophylline: Theo-Organidin should not be taken within twelve hours after rectal administration of any preparation containing theophylline or aminophylline. (1)
+ Xanthine Derivatives: Other formulations containing xanthine derivatives should not be given concurrently with Theo-Organidin. (1)

THEOPHYLLINE: see Alkaline Agents + Acidic Media

THEOPHYLLINE: see Alkaline Agents + Acidifying Agents (Urinary)

THEOPHYLLINE: see listed under the following agents:

Aqualin Suprettes
Optiphyllin
Organophosphate Pesticides
Protopam
Quibron
Theo-Organidin

THEOPHYLLINE PREPARATIONS + Elixophylline: see Elixophylline + Theophylline Preparations

THEPTINE + MAO Inhibitors: Do not use Theptine in patients taking monoamine oxidase inhibitors. (1) Mechanism: see MAO Inhibitor + Sympathomimetics, Indirect-Acting.

THERATUSS + Barbiturates: Because of its chemical relationship to the phenothiazines, the possibility that Theratuss may potentiate barbiturate activity should be borne in mind. (2)

THIAZIDES: see also Diuretics, Thiazide

THIAZIDES: see listed under the following agents:

Aldomet
Inversine
Ismelin
Levophed
Raudixin
Rauwiloid
Singoserp
Unitensin

THIOCYANATES + Regitine: see Regitine + Thiocyanates

THIOGUANINE

+ Cytotoxic Chemotherapy: see Thioguanine + Irradiation
+ Irradiation: The antimetabolic agents are contraindicated for patients in a poor nutritional state or those with minimal leukocyte and thrombocyte counts because the bone marrow may become further depressed. Such conditions are likely to occur when the patient has had recent surgery, radiation therapy, or treatment with other cytotic chemotherapeutics. (2)

THIOMERIN + Ammonium Chloride: To potentiate diuresis, an acidifying agent such as ammonium chloride may be administered in oral tablets of 2 to 3 gm three or four times daily for two days before injection of a mercurial. (1)

THIOSULFIL + Methenamine: Since Thiosulfil may form an insoluble precipitate with formaldehyde in the urine, it is probable that no methenamine compound should be given during sulfonamide therapy. (2)

THIOSULFIL DUO-PAK + Dolonil: see Dolonil + Thiosulfil Duo-Pak

THIOSULFIL-A FORTE + Dolonil: see Dolonil + Thiosulfil-A Forte

THIOSULFIL OPTH. with METHYLCELLULOSE + Silver Preparations: These preparations should not be used concomitantly. (1)

THIO-TEPA

+ Antimicrobial Agents: Antimicrobials such as Chloromycetin (chloramphenicol) or a sulfonamide derivative should be avoided inasmuch as these agents may depress the bone marrow. (1, 2)
+ Chloromycetin: see Thio-Tepa + Antimicrobial Agents
+ Neoplastic Chemotherapy: Smaller dosage and greater caution should be employed in the presence of leukopenia presisting from previous therapy with other agents. (1)
+ Radiation Therapy: It should be remembered that lesions in a heavily radiated area may show a slow or lessened response, while toxicity can be enhanced because Thio-Tepa is radiometric. Combined therapy with radiation should be approached with care, as there may be greater than an additive effect on the bone marrow. (1, 2)
+ Sulfonamides: see Thio-Tepa + Antimicrobial Agents

THORA-DEX: see Thorazine and Dexedrine

THORAZINE: see Phenothiazines

THORAZINE: see listed under the following agents:

Alcohol	Pyrilgin
Dipyrone	Quaalude
Fluothane	Trisoralen
Narone	Wyamine

THORAZINE

+ Alcohol: see Thorazine + CNS Depressants. The use of alcohol with Thorazine should be avoided due to possible additive effects and hypotension. (1)
+ Analgesics: see Thorazine + CNS Depressants

THORAZINE (continued)

+ Anesthetics: see Thorazine + CNS Depressants
+ Antihistamines: see Thorazine + CNS Depressants
+ Atropine: Phenothiazines may potentiate the action of atropine or related drugs. (1)
+ Barbiturates: see Thorazine + CNS Depressants. Thorazine does not potentiate the anticonvulsant action of barbiturates. Therefore dosage of anticonvulsants, including barbiturates, should not be reduced if Thorazine is started. Instead, start Thorazine at low doses and increase as needed. (1)
+ Bone Marrow Depressants: Thorazine is contraindicated in the presence of bone marrow depression. (1)
+ CNS Depressants: Thorazine prolongs and intensifies the action of central nervous system depressants such as anesthetics, barbiturates, narcotics, analgesics, antihistamines, opiates, and alcohol. When Thorazine is administered concomitantly, about one-fourth to one-half the usual dosage of such agents is required. When Thorazine is not being administered for potentiation, it is best to stop such depressants before starting Thorazine treatment. These agents may subsequently be reinstated at low doses and increased as needed.
  Thorazine is contraindicated in the presence of large amounts of central nervous system depressants (alcohol, barbiturates, narcotics, etc.) (1, 2)
+ Drugs, Overdosage: Thorazine, due to its antiemetic effect, may mask signs of overdosage of toxic drugs. (1)
+ Epinephrine: see Thorazine + Pressor Agents
+ Heat: Phenothiazines may potentiate the action of heat. (1)
+ MAO Inhibitors: Thorazine is contraindicated in patients taking monoamine oxidase inhibitors. (1) Mechanism: It is postulated that MAO inhibitors inhibit the hepatic microsomal enzymes which metabolize phenothiazines thereby potentiating the effects of the phenothiazine and causing extrapyramidal symptoms or hypotension. (115)
+ Metrazol: see Thorazine + Picrotoxin
+ Narcotics: see Thorazine + CNS Depressants
+ Opiates: see Thorazine + CNS Depressants
+ Phosphorous Insecticides: Phenothiazines may potentiate the action of phosphorous insecticides. (1)
+ Picrotoxin: In the advent of Thorazine overdosage, if a stimulant is desired, use amphetamine, dextroamphetamine, or caffeine and sodium benzoate. Avoid stimulants that may cause convulsions (e.g., picrotoxin and Metrazol). (1)

THORAZINE (continued)

+ Pressor Agents: To control hypotension due to Thorazine, pressor agents, including epinephrine, should not be used as they may cause further lowering of blood pressure. If a vasoconstrictor is required, Levophed and Neosynephrine are the most suitable. (1)

THROMBIN

+ Acid: see Thrombin + Antacids
+ Antacids: When Thrombin is used in cases of gastroduodenal hemorrhage, one teaspoonful of an antacid every three hours is advised to neutralize gastric acidity to prevent inactivation of Thrombin and to protect the formed blood clot from digestion. (1)
+ Surgicel Absorbable Hemostat: see Surgicel Absorbable Hemostat + Thrombin

THROMBIN, TOPICAL

+ Acids, Dilute: Dilute acids, alkalies, heat, and salts of heavy metals are detrimental to Thrombin activity. (1)
+ Alkalies: see Thrombin, Topical + Acids, Dilute
+ Heavy Metal Salts: see Thrombin, Topical + Acids, Dilute

THROMBOLYSIN

+ Anesthesia: Thrombolysin should be administered with caution immediately after anesthesia or surgery, since spontaneous fibrinolytic activity is usually increased for a short period of time after either process. (1)
+ Anticoagulants: Thrombolysin has been used as an adjunct to conventional anticoagulant therapy; however, it must be borne in mind that simultaneous use with anticoagulants may be hazardous. (2)
+ Heparin: Large doses of Thrombolysin may temporarily affect the amount of Heparin required to attain the desired prolongation of the clotting time. Smaller doses probably do not have any significant effect in this respect. (1)
+ Quick Test: Large doses of Thrombolysin may increase the clotting time and the (Quick) one-stage prothrombin time. (2)

THYROGLOBULIN + Cytomel: see Cytomel + Thyroglobulin

THYROID: see also Thyroid Hormone, Thyroid Medications, and Thyroid Preparations as well as Thyroglobulin (various trade-named products) and Thyroxin (various trade-named products)

THYROID: see listed under the following agents:

Cytomel
Norpramin
Pertofrane
Tofranil

THYROID HORMONE + Questran: see Questran + Thyroid Hormone

THYROID MEDICATION + Choloxin: see Choloxin + Thyroid Medication

THYROID PREPARATIONS: see listed under the following agents:

Aventyl | Pertofrane
Norpramin | Vivactil

THYROLAR

+ Anticoagulants: The institution of thyroid replacement therapy may potentiate anticoagulant effect with agents such as warfarin or bishydroxycoumarin (Dicumarol) and reduction of one-third in anticoagulant dosage should be undertaken upon initiation of Thyrolar-therapy. Subsequent anticoagulant dosage adjustment should be made on the basis of frequent prothrombin determinations. (1)
+ Catecholamines: see Thyrolar + Epinephrine
+ Diabetes: In patients with diabetes mellitus addition of thyroid hormone therapy may cause an increase in the required dosage of insulin or oral hypoglycemic agents. Conversely, decreasing the dose of thyroid hormone may possibly cause hypoglycemic reactions if the dosage of insulin or oral agents is not adjusted. (1)
+ Dicumarol: see Thyrolar + Anticoagulants
+ Epinephrine: Injection of epinephrine in patients with coronary artery disease may precipitate an episode of coronary insufficiency. This may be enhanced in patients receiving thyroid preparations. Careful observation is required if catecholamines are administered to patients in this category. (1)
+ Surgery: Thyrolar-treated patients with concomitant coronary artery disease should be carefully observed during surgery, since the possibility of precipitating cardiac arrhythmias may be greater in patients treated with thyroid hormone. (1)
+ Warfarin: see Thyrolar + Anticoagulants

THYROXIN: see Levothyroxin as well as various trade-named preparations

THYROXIN + Questran: see Questran + Thyroxin

TIGAN

+ Barbiturates: see Tigan + CNS-Acting Agents
+ Belladonna Derivatives: see Tigan + CNS-Acting Agents
+ CNS-Acting Agents: During acute febrile illness, encephalitides, gastroenteritis, dehydration, and electrolyte imbalance, especially in children, the elderly, or debilitated, central nervous system reactions (e.g., opisthotonos, convulsions, coma, and extrapyramidal

TIGAN (continued)

+ CNS-Acting Agents (continued): symptoms) have been reported with or without use of Tigan or other antiemetic agents. In such disorders, exercise caution in administering Tigan, particularly in patients recently receiving other central nervous system acting agents (phenothiazines, barbiturates, belladonna derivatives). (1)
+ Drugs, Overdosage: The antiemetic effects of Tigan may obscure signs of toxicity due to overdosage of other drugs. (1)
+ Phenothiazines: see Tigan + CNS-Acting Agents

TINDAL: see Phenothiazines

TINDAL

+ Alcohol: see Tindal + CNS Depressants
+ Analgesics: see Tindal + CNS Depressants
+ Anesthetics, General: see Tindal + CNS Depressants
+ Antihistamines: see Tindal + CNS Depressants
+ Antihypertensive Agents: Tindal may potentiate the action of antihypertensive agents, and care should be exercised when used concomitantly. (2)
+ Atropine: Atropine, heat, and phosphorous insecticides may increase the toxicity of phenothiazines. (1)
+ Barbiturates: see Tindal + CNS Depressants
+ Bone Marrow Depressants: As with all phenothiazines, any evidence of bone marrow depression are contraindications to the use of Tindal. (1)
+ CNS Depressants: As with all phenothiazines, any depression of the central nervous system associated with drugs (barbiturates, general anesthetics, alcohol, narcotics, analgesics, antihistamines) are contraindications to the use of Tindal. Because of its sedative action, it should be assumed that Tindal may cause further depression of the central nervous system in comatose or depressed states whatever the etiology, including that caused by central nervous system depressants. (1)
+ Epinephrine: As with other phenothiazines, epinephrine should not be employed to correct the hypotension, since its action may be blocked and partially reversed by Tindal. Pressor agents such as Levophed or Neo-Synephrine are preferable. (1, 2)
+ Heat: see Tindal + Atropine
+ Metrazol: see Tindal + Picrotoxin
+ Narcotics: see Tindal + CNS Depressants
+ Phosphorous Insecticides: see Tindal + Atropine

TINDAL (continued)

+ Picrotoxin: In cases of Tindal overdosage, amphetamine, dextroamphetamine, or caffeine and sodium benzoate can be administered as stimulants and are preferable to agents which might cause convulsions, such as picrotoxin or Metrazol. (1)
+ Trisoralen: see Trisoralen

TITROID + Cytomel: see Cytomel + Titroid

TOFRANIL: see also Antidepressants, Tricyclic

TOFRANIL: see Alkaline Agents + Acidifying Agents (Urinary)

TOFRANIL: see listed under the following agents:

Eutonyl	Matulane
Ismelin	Nardil
MAO Inhibitors	Niamid
Marplan	Parnate

TOFRANIL

+ Adrenergic Neuron-Blocking Agents: see Tofranil + Ismelin
+ Alcohol: Patients should be warned that the concomitant use of alcoholic beverages may be associated with exaggerated effects. (1)
+ Anticholinergics: In occasional susceptible patients or in patients receiving anticholinergic drugs (including antiparkinsonism agents), the atropine-like effects may become more pronounced (e.g., paralytic ileus). (1, 22, 59) Close supervision and careful adjustment of dosage is required when Tofranil is administered concomitantly with anticholinergics. Mechanism: Tofranil has anticholinergic properties which is additive to other anticholinergic drugs.
+ Antiparkinsonism Agents: see Tofranil + Anticholinergics
+ Diabetes: Glucose tolerance may be inconsistently affected by Tofranil, as reflected in altered blood glucose levels. (1)
+ Ismelin: Tofranil may block the pharmacologic action of the antihypertensive Ismelin and related adrenergic neuron-blocking agents. (1) Mechanism: see Ismelin + Antidepressants, Tricyclic.
+ Levophed: Tofranil enhances rather than diminishes pressor response to Levophed. (2)
+ MAO Inhibitors: The concomitant administration of monoamine oxidase inhibiting compounds is contraindicated. Hyperpyretic crises or severe convulsive seizures may occur in patients receiving such combinations. The potentiation of adverse effects can be serious, even fatal. When it is desired to substitute Tofranil in patients receiving a monoamine oxidase inhibitor, as long an interval should elapse as the clinical situation will allow, with a minimum of 14

TOFRANIL (continued)

+ MAO Inhibitors (continued): days. Initial dosage should be low and increases should be gradual and cautiously prescribed. (1, 2, 59) Mechanism: see MAO Inhibitor + Antidepressants, Tricyclic.
+ Surgery: Prior to elective surgery, Tofranil should be discontinued for as long as the clinical situation will allow. (1)
+ Sympathomimetic Agents: Close supervision and careful adjustment of dosage is required when Tofranil is administered concomitantly with sympathomimetic agents. (1)
+ Thyroid: It is advised that caution be observed in prescribing Tofranil in hyperthyroid patients or in patients receiving thyroid medication conjointly. Transient cardiac arrhythmias have occurred in rare instances. (1)
+ Trisoralen: see Trisoralen

TOLERAN + Meals: Although the absorption of iron is best when the iron is taken between meals, gastrointestinal disturbances may be controlled by reducing the dose and giving the preparation shortly after meals. (2)

TOLINASE: see also Sulfonylurea, Hypoglycemic and Hypoglycemic Drugs, Oral

TOLINASE

+ Alcohol: Although disulfiram (Antabuse)-like alcoholic flushes have been reported with other sulfonylureas, these have not been reported to date with Tolinase. Nevertheless, the possibility of their occurrence should be kept in mind. (1) Mechanism: see Alcohol + Sulfonylurea, Hypoglycemic.
+ Benemid: see Tolinase + Butazolidin
+ Butazolidin: Drugs which may prolong or enhance the action of Tolinase and thereby increase the risk of hypoglycemia include: insulin, DBI, sulfonamides, Tandearil, Butazolidin, salicylates, Benemid, and MAO Inhibitors. (1) Mechanism: see Hypoglycemic Drugs, Oral + Butazolidin.
+ DBI: see Tolinase + Butazolidin
+ Diabinese: Patients should be carefully observed for hypoglycemia during transition due to prolonged retention of Diabinese and subsequent overlapping effect. (1)
+ Diuretics, Thiazide: Caution should be observed in administering the thiazide-type diuretics to diabetic patients on Tolinase therapy because the thiazides have been reported to aggravate diabetes mellitus and to result in increased sulfonylurea requirements. (1) Mechanism: see Diuretics, Thiazide + Sulfonylureas, Hypoglycemic.

TOLINASE (continued)

+ Insulin: see Tolinase + Butazolidin
+ MAO Inhibitors: see Tolinase + Butazolidin
+ Salicylates: see Tolinase + Butazolidin. Mechanism: Displacement of the Tolinase from its protein binding site, thereby increasing the concentration of free unbound Tolinase and inducing a further hypoglycemic effect. (118)
+ Sulfonamides: see Tolinase + Butazolidin. Mechanism: see Hypoglycemic Drugs, Oral + Sulfonamides.
+ Tandearil: see Tolinase + Butazolidin

TORECAN: see Phenothiazines

TORECAN

+ Alcohol: see Torecan + CNS Depressants
+ Anesthesia: see Torecan + CNS Depressants. When used in the treatment of nausea and/or vomiting associated with anesthesia and surgery, it is recommended that Torecan be administered at or shortly before termination of anesthesia.
  As with other phenothiazine antiemetics, restlessness and postoperative CNS depression during anesthesia recovery may occur but have not been of a serious degree in Torecan-treated patients. However, the possibility of more severe degree or any of the other known reactions of this class of drug must be weighed. (1, 2)
+ Atropine: see Torecan + CNS Depressants
+ CNS Depressants: Attention should be paid to the fact that phenothiazines are capable of potentiating central nervous system depressants (e.g., opiates, alcohol, anesthetics, etc.) as well as atropine and phosphorous insecticides. (1, 2)
+ Epinephrine: The administration of epinephrine should be avoided in the treatment of drug-induced hypotension in view of the fact that phenothiazines may induce a reversed epinephrine effect on occasion. Should a vasoconstrictor be required, the most suitable are Levophed and Neosyneprine. (1)
+ Opiates: see Torecan + CNS Depressants
+ Phosphorous Insecticides: see Torecan + CNS Depressants

TOSCAMINE + Oxytocin: If the patient fails to respond to Toscamine and it is decided to initiate an oxytocic infusion, several hours should elapse after the last dose of Toscamine before starting this form of treatment. The patient must be carefully observed for sensitivity due to the known synergistic action of Toscamine and oxytocin (Pitocin, Syntocinon) which may result in tetanic uterine convulsions. (1)

TRANCOPAL
+ Alcohol: Patients should avoid the concomitant use of Trancopal and alcohol, since the effects may be additive. (2)
+ Monoamine Oxidase Inhibitors: see Trancopal + Psychotropic Agents
+ Phenothiazines: see Trancopal + Psychotropic Agents
+ Psychotropic Agents: Other psychotropic agents, particularly phenothiazines or monoamine oxidase inhibitors, that are known to potentiate the action of other drugs should not be given with Trancopal. (2)

TRANQUILIZERS: see listed under the following agents:

Alvodine
Ambodryl
Amytal Sodium Sterile
Analgesics
Benadryl
Carbocaine
Disophrol
Eutonyl
Flurothyl
Furoxone
Medomin
Methadone
Phenothiazines
Placidyl
Protopam
Ritalin, Parenteral
Seconal Sodium
Solacen
Stelazine

TRANQUILIZERS, MAJOR: see listed under the following agents:

Inapsine
Sublimaze

TRAVERT + Blood: Do not administer Travert simultaneously with blood. (1)

TRECATOR
+ Diabetes: The management of patients with diabetes mellitus may become more difficult when these same patients are treated with Trecator. (137)
+ Drugs: Trecator may intensify the adverse effects of other drugs administered concomitantly. (2)

TREPIDONE
+ Alcohol: Patients should avoid the concomitant use of Trepidone and alcohol, since the effects may be additive. (2)
+ MAO Inhibitors: see Trepidone + Psychotropic Agents
+ Phenothiazines: see Trepidone + Psychotropic Agents
+ Psychotropic Agents: Other psychotropic agents, particularly phenothiazines or monoamine oxidase inhibitors, that are known to potentiate the action of other drugs should not be given with Trepidone. (2)

TRIAVIL: see Elavil and also Trilafon

TRIBURON OINTMENT + Plastics: The base of the ointment may affect certain plastics. (1)

TRIDIONE
+ Drugs: Concurrent administration of other drugs known to cause toxic effects should be avoided or used only with extreme caution. (1)
+ Methyl Phenyl Hydantoin (Nuvarone): Concomitant administration is not advised, since development of aplastic anemia has been reported. (1)

TRIETHYLENEMELAMINE (TEM)
+ Acid: see Triethylenemelamine + Food. Absorption may be enhanced and made more predictable if transformation in the acidic environment of the stomach is prevented by the simultaneous administration of sodium bicarbonate. (145)
+ Food: TEM should be taken in the morning with plain water on an empty stomach. Food should be withheld for one hour afterward, since TEM tends to be inactivated in an acid medium and is reactive with organic materials. (1) (See Triethylenemelamine + Acid.)
+ Nitrogen Mustard: It is important that TEM should not follow nitrogen mustard therapy until the full effect of such treatment has been observed; probably the white count is the best index of such effects and, accordingly, TEM should not be administered until the white count begins to increase following the depression usually observed after bone-marrow depressed therapy. (1)
+ X-Ray Therapy: It is important that TEM should not follow x-ray therapy until the full effect of such treatment has been observed; probably the white count is the best index of such effects and, accordingly, TEM should not be administered until the white count begins to increase following the depression usually observed after bone-marrow depressed therapy. (1)

TRILAFON: see Phenothiazines

TRILAFON
+ Alcohol: see Trilafon + CNS Depressants
+ Analgesics: see Trilafon + CNS Depressants
+ Anesthetics: see Trilafon + CNS Depressants
+ Anticonvulsant Drugs: As with other phenothiazine drugs, Trilafon may lower convulsant threshold in susceptible individuals and should be used with great caution in patients with a history of convulsive disorders. An increase in dosage of anticonvulsant drugs

TRILAFON (continued)

+ Anticonvulsant Drugs (continued): may be required if the patient is treated concomitantly with Trilafon. (1)
+ Antihistamines: see Trilafon + CNS Depressants
+ Antihypertensive Agents: Trilafon may potentiate the action of antihypertensive agents, and care should be exercised when used concomitantly. (2)
+ Atropine: Atropine, heat, and phosphorous insecticides may increase the toxicity of phenothiazines. (1)
+ Barbiturates: see Trilafon + CNS Depressants
+ Bone Marrow Depressants: Trilafon is contraindicated in the presence of bone marrow depression. (1)
+ CNS Depressants: Trilafon is contraindicated in patients who have depression of the central nervous system from drugs (barbiturates, anesthetics, alcohol, narcotics, analgesics, hypnotics, antihistamines). Potentiation of central nervous system depressants can occur with phenothiazines. (1, 2) Trilafon may exert a slight potentiating effect on central nervous system depressants such as barbiturates, alcohol, narcotics, and similar drugs. (1)
+ Drugs: The antiemetic effect of Trilafon may obscure signs of toxicity due to overdosage of other drugs. (1)
+ Epinephrine: Epinephrine should not be employed to correct hypotension that may occur with Trilafon, since its action may be blocked and partially reversed by Trilafon. Levophed should be employed to counteract hypotensive effects. (1, 2)
+ Heat: see Trilafon + Atropine
+ Hypnotics: see Trilafon + CNS Depressants
+ Narcotics: see Trilafon + CNS Depressants
+ Phosphorous Insecticides: see Trilafon + Atropine
+ Trisoralen: see Trisoralen

TRILENE: see under the following agents:

Epinephrine
Succinylcholine
Vasopressors

TRILENE

+ Epinephrine: Epinephrine is contraindicated when Trilene is employed. When indicated, any vasoconstrictor other than epinephrine may be employed. (1) Mechanism: Trilene sensitizes the heart to exogenous catecholamines. (123)
+ Soda Lime: Under no circumstances should Trilene be used in a closed circuit with soda lime. Contact with soda lime in a closed system may result in the formation of toxic decomposition products which may cause cranial nerve palsies. (1) Mechanism: If Trilene

TRILENE (continued)

+ Soda Lime (continued): is employed in a closed system with carbon dioxide absorbers, impurities are formed from contact of Trilene with soda lime. One of these decomposition products, dichloroacetylene, is toxic, causing cranial nerve palsies, including anesthesia over the cutaneous distribution of the trigeminal nerve and is also spontaneously explosive. (123)

TRINALIS SUPRETTES

+ Corticosteroids: Concurrent therapy with corticosteroids should be avoided. (1)
+ Mercurials: Concurrent therapy with mercurials should be avoided. (1)

TRI-SOLGEN

+ Adrenal Corticosteroids: see Tri-Solgen + Adrenocorticotrophin
+ Adrenocorticosteroids: see Adrenocorticosteroids + Tri-Solgen
+ Adrenocortocotrophin: It has been demonstrated in animals that adrenocorticotrophin and adrenal corticosteroids suppress the antibody response following immunization. In man such suppression is not evident with usual therapeutic doses of these hormones. Until these species' differences have been reconciled, however, it would seem reasonable not to administer the hormones and immunizing agents concurrently. Apart from their effect on antibody synthesis, corticosteroids have been shown to reduce host resistance to disease. Therefore they should not be administered following exposure to infectious agents (mumps, rabies, tetanus) for which no satisfactory antimicrobial therapy is available. To do so may alter the host-parasite relationship to such an extent that severe or fatal illness may occur in spite of the prophylactic administration of a vaccine. Under these circumstances, the occurrence of disease, actually due to the altered pattern of resistance, might be attributed to a vaccine failure. (1)

TRISORALEN + Photosensitizing Agents: No other preparation that has photosensitizing properties should be given with Trisoralen. (1, 2) The following agents have been reported to have photosensitizing properties:

Achromycin
Acridine (101)
Aldactazide
Aminobenzoic Acid (101)
Anhydron
Anthracene (101)

TRISORALEN + Photosensitizing Agents (continued)

- Aquatag
- Aureomycin
- Aventyl
- Biosphenol (bithionol) (100, 101)
- Caplaril
- Capsebon
- Carbo-Cort Creme and Lotion
- Carbo-Cort Forte Creme
- Coal Tar Derivatives (101)
  - Anthracene
  - Acridine
  - Phenanthrene
  - Pyridium
- Compazine
- Contraceptives, Oral (102)
- Darbid
- Declomycin (101)
- Diabinese (101)
- Diethylstilbesterol (102)
- Diuretics, Thiazide (100)
- Diuril
- Dyazide
- Dymelor
- Dyrenium
- Ecolid
- Enduron
- Enovid (102)
- Epidol
- Esidrix
- Eskatrol
- Etrafon
- Exna
- Fulvicin
- Furocoumarins (found in fruits and perfumes) (101)
- Garamycin
- Grifulvin
- Grisactin
- Griseofulvin (100, 101)
- Haldol
- Hydromox (99)
- HydroDIURIL

TRISORALEN + Photosensitizing Agents (continued)

Hygroton
Impregon (tetrachlorsalicylanilide) (33, 101)
Jadit (n-butyl-4 chlorosalicylamide) (33, 101)
Kesso Tetra
Kynex
Levoprome
Macrodantin
Matulane
Mellaril
Meloxine (99)
Mepergan
Metahydrin
Methatar Creme
Midicel
Murel Inj.
Mysteclin-F
Naqua
Nardil
Navane
NegGram
Neo-A-Fil (digalloyl trioleate) (33, 101)
Norpramin
Oracon (102)
Oretic
Orinase
Ornade
Ortho-Novum (102)
Oxsoralen (99)
Panmycin
Permitil
Pertofrane
Phenanthrene (101)
Phenergan
Phenothiazines
Proketazine
Prolixin
Pyridium (101)
Quide
Renese
Repoise
Retet

TRISORALEN + Photosensitizing Agents (continued)

- Rondomycin
- Saluron
- Sandalwood Oil (Sandela) (82)
- Syntetrin I.M. and I.V.
- Signemycin
- Sinequan
- Sumycin
- Sulfonamides (33, 100, 101)
- Sulfonylureas (100)
- Sulla
- Tar Distillate 'Doak'
- Taractan
- Tegetrol
- Terramycin
- Tetrachel
- Tetrachlorosalicylanilide (Impregon) (100, 101)
- Tetracycline
- Tetracyn
- Tetrex
- Thiazides and related Sulfonamides (33, 100, 101)
- Thorazine
- Tindal
- Tofranil
- Trilafon
- Tri-Sulfameth
- Tri-Sulfanyl
- Ulcotar
- Unguentum Bossi
- Urobiotic
- Vesprin
- Vibramycin
- Vivactil
- Zetar Emulsion
- Zetar Shampoo
- Ze-Tar-Quin
- Zetone Cream and Lotion

TRI-SULFAMETH + Trisoralen: see Trisoralen

TRISULFAMINIC TABLETS and SUSPENSION + Sulfonylureas, Hypoglycemic: Sulfonamide therapy may potentiate the hypoglycemic action of sulfonylureas. (1) Mechanism: see Sulfonylureas, Hypoglycemic + Sulfonamides.

TRI-SULFNYL + Trisoralen: see Trisoralen

TROMEXAN: see Anticoagulants

TROMEXAN: see listed under the following agents:
Ritalin
Sulfonamides, Long-Acting

TROMEXAN: The following factors may be responsible for increased prothrombin time (increased hypoprothrombinemia): (1)
Alcohol
Anesthetics
Antibiotics (penicillin, Chloromycetin, Aureomycin)
Butazolidin
Drugs affecting blood elements, quinine
Drugs with hepatotoxic action
Low Choline
Low Cystine
Low protein diet, dietary deficiencies
Low Vitamin C
Prolonged Narcotics
Salicylates (excess of 1 gm/day)

TROMEXAN: The following agents may be responsible for decreased prothrombin time (decreased hypoprothrombinemia): (1)
Antihistamines
Corticosteroids
Diet high in Vitamin K (vegetables, fish, and fish oils)
Mineral Oil
Vitamin K

TROMEXAN + Diabetes: Carbohydrate metabolism is stated to be generally unaffected, but increased resistance to insulin has been noted in isolated instances of severe diabetes. (1)

TRYPTOPHANE (or Trypotophan): see listed under the following agents:
Nardil
Parnate
Velban

TUAZOLE: see other methaqualone preparations (Parest, Quaalude, Sopor, Somnofac)

TUBERCULIN TEST: see listed under the following agents:
Attenuvax, Lyovac
Biavax

TUBOCURARINE: see listed under the following agents:

Anhydron
Aquatag
Bristuron
Diuretics, Thiazide
Diuril
Enduron
Esidrix
Ether
Fluoromar
Fluothane
HydroDIURIL
Hydromox
Lasix
Magnesium Sulfate Inj.
Metahydrin
Naqua
Naturetin
Oretic
Penthrane
Pentothal Sodium
Quelicin Chloride Inj.
Renese
Saluron
Succinylcholine
Tensilon
Vinethene

TUBOCURARINE

+ Antibiotics: Tubocurarine (d-tubocurarine) acts by polarizing the motor end plates and are potentiated by certain antibiotics with similar activity (neomycin, streptomycin, polymixin). (55) Mechanism: These antibiotics in sufficiently high doses produce neuromuscular blockade which combines the features of the competitive block with a reduction in the release of acetylcholine by the motor nerve impulse. (131)
+ Cholinesterase Inhibitors: The effects of tubocurarine (d-tubocurarine) can be reversed by chlinesterase inhibitors. (55, 131) (See Tubocurarine + Neostigmine.)
+ Cinchona Alkaloids: Quinidine may potentiate the action of non-depolarizing neuromuscular blocking agents such as tubocurarine. The cinchona alkaloids, of which quinidine is one, have neuromuscular blocking effects. This blockade seems to be related to a curariform activity at the myoneural junction, as well as depression of muscle action potential. This potentiation may be a dose related response. (146)
+ Diuretics, Thiazide: see Tubocurarine + Drugs, Potassium-Depleting
+ Drugs, Potassium-Depleting: Tubocurarine (d-tubocurarine) is potentiated by drugs that cause potassium depletion, such as thiazide diuretics and mineralocorticosteroids. (55, 131)
+ Ether: Ether will prolong the action of tubocurarine (d-tubocurarine). (126, 131, 144)
+ Mineralocorticosteroids: see Tubocurarine + Drugs, Potassium-Depleting
+ Mylaxen: The concomitant use of these agents always incrased the neuromuscular effects of each other. (143)

TUBOCURARINE (continued)

+ Neomycin: see Tubocurarine + Antibiotics
+ Neostigmine: Tubocurarine (d-tubocurarine) produces paralysis by occupying receptors at the motor end plates normally reserved for acetylcholine. This effect of the drug may be reduced competitively by acetylcholine or neostigmine which inhibits cholinesterase thereby increasing the level of acetylcholine. In addition, tubocurarine is antagonized by Tensilon which acts like acetylcholine. (118) Mechanism: Displacement of the tubocurarine from drug receptors.
+ Polymixin: see Tubocurarine + Antibiotics
+ Quinidine: see Tubocurarine + Cinchona Alkaloids
+ Streptomycin: see Tubocurarine + Antibiotics
+ Tensilon: see Tubocurarine + Neostigmine

TUBOCURARINE CHLORIDE INJECTION: see Tubocurarine

TUBOCURARINE CHLORIDE INJECTION

+ Antibiotics: Certain antibiotics such as neomycin and streptomycin have neuromuscular blocking effects and may themselves cause respiratory depression. They should not be added to solutions for instillation or irrigation of either the peritoneal or thoracic cavity when muscle relaxants have been administered. Subsequent administration of muscle relaxants may be a potential anesthetic hazard. (1)
+ Cinchona Alkaloids: Quinidine and related cinchona alkaloids are known to have a curare-like action on skeletal muscle. Accordingly, caution is suggested in the use of these compounds concomitantly with, or in the immediate post-recovery period following the use of muscle relaxants. (1)
+ Drugs, Electrolyte-Depleting: Some patients may exhibit unduly intense and prolonged response to tubocurarine chloride even in moderate dosages. Many factors are known to be involved and any one or any combination may be responsible in a given patient. Among the many factors responsible for apnea when overdose has not been a possible cause are: reduced blood flow to skeletal muscles causing delayed removal of the drug, increased body temperature, poor renal excretion, and electrolyte imbalance. (1)
+ Ether: Ether and Penthrane act as non-depolarizing relaxants. When tubocurarine is used during ether or Penthrane anesthesia, dosage of tubocurarine should be reduced. If ether or Penthrane is used, only one-third or one-half as much tubocurarine is

TUBOCURARINE CHLORIDE INJECTION (continued)

+ Ether (continued): required as with other inhalation anesthetic agents such as cyclopropane or nitrous oxide. (1)
+ Muscle Relaxants: Considerable caution is necessary in administering tubocurarine after another muscle relaxant has been given during the same surgical operation, since there is a possibility of a synergistic or antagonistic effect. After a single dose of succinylcholine, recovery is quick and the endplate returns to its resting state. Tubocurarine may be given without modification of the usual dose. However, prolonged administration of succinylcholine may produce a non-depolarized block (phase II block), thus increasing sensitivity to subsequent tubocurarine injections. This may not contraindicate the use of tubocurarine, but dosage should be reduced. (1)
+ Narcotics: Narcotics should be used with care since they are central respiratory depressants and large amounts administered to patients given muscle relaxants may further depress respiration. (1)
+ Neomycin: see Tubocurarine Chloride Injection + Antibiotics
+ Neostigmine: Neostigmine antagonizes only the skeletal muscular blocking action of tubocurarine and may aggravate such side effects as hypotension or bronchospasm. If the response to small doses of neostigmine lasts only a few minutes, more of the drug may be given, but care should be taken to avoid overdosage, since this drug can itself produce neuromuscular block and has caused bradycardia and cardiac arrest. The prior administration of 0.4 to 0.6 mg of atropine to minimize the possibility of muscarinic side effects associated with neostigmine is recommended. (1)
+ Penthrane: see Tubocurarine Chloride Injection + Ether
+ Quinidine: see Tubocurarine Chloride Injection + Cinchona Alkaloids
+ Streptomycin: see Tubocurarine Chloride Injection + Antibiotics
+ Succinylcholine: see Tubocurarine Chloride Injection + Muscle Relaxants

TYBATRAN

+ Alcohol: see Tybatran + Psychotropic Agents
+ CNS Depressants: Simultaneous administration to psychotic patients of Tybatran with phenothiazines and other central nervous system depressants has in a few instances been associated with the occurrence of grand mal or petit mal seizures. Seizures have been reported with administration of phenothiazines alone, but not with the administration of Tybatran alone; nevertheless, Tybatran should be used cautiously in individuals receiving other central

TYBATRAN (continued)

+ CNS Depressants (continued): nervous system depressants or having a history of convulsive seizures. (1) (See also Tybatran + Psychotropic Agents.)
+ MAO Inhibitors: see Tybatran + Psychotropic Agents
+ Phenothiazines: see Tybatran + CNS Depressants and also with Psychotropic Agents
+ Psychotropic Agents: The simultaneous administration of Tybatran with alcohol or with other psychotropic agents, particularly phenothiazines or monoamine oxidase inhibitors, which are known to potentiate the action of other drugs, may result in additive actions. (1)

TYPHOID VACCINE

+ Adrenal Corticosteroids: see Typhoid Vaccine + Adrenocorticotrophin
+ Adrenocorticotrophin: It has been demonstrated in animals that adrenocorticotrophin and adrenal corticosteroids suppress the antibody response following immunization. In man, such suppression is not evident with usual therapeutic doses of these hormones. Until these species' differences have been reconciled, however, it would seem reasonable not to administer the hormones and immunizing agents concomitantly. (1)

THPHUS VACCINE

+ Adrenal Corticosteroids: see Typhus Vaccine + Adrenocorticotrophin
+ Adrenocorticotrophin: It has been demonstrated in animals that adrenocorticotrophin and adrenal corticosteroids suppress the antibody response following immunization. In man, such suppression is not evident with usual therapeutic doses of these hormones. Until these species' differences have been reconciled, however, it would seem reasonable not to administer the hormones and immunizing agents concomitantly. (1)

TYRAMINE: see listed under the following agents:

MAO Inhibitor
Reserpine

U

ULCOTAR + Trisoralen: see Trisoralen

ULO

+ CNS Depressants: Since ULO is a centrally acting drug, it should be used with caution in patients taking drugs that depress or stimulate the central nervous system. (1, 2)
+ CNS Stimulants: see ULO + CNS Depressants

ULTRAN

+ Alcohol: Patients should avoid the concomitant use of Ultran and alcohol, since the effects may be additive. (1, 2)
+ CNS Depressants: The administration of Ultran with other central nervous system depressants may result in additive effects. (1)
+ MAO Inhibitors: see Ultran + Psychotropic Agents
+ Phenothiazines: see Ultran + Psychotropic Agents
+ Psychotropic Agents: Other psychotropic agents, particularly phenothiazines and monoamine oxidase inhibitors, that are known to potentiate the action of other drugs should not be given with Ultran. (2)

UNGUENTUM BOSSI + Trisoralen: see Trisoralen

UNIPEN

+ Meals: The oral dosage forms should be taken in the fasting state, preferably one to two hours before meals. (1)
+ Neutrapen: Neutrapen would probably be ineffective for the treatment of allergic reactions. (1)

UNITENSIN

+ Aldomet: see Unitensin + Antihypertensive Agents
+ Anesthetic Agents: Caution must be exercised when either of these preparations (Unitensin preparations) is used in conjunction with surgery because of the additive hypotensive effects of anesthetic and pre-anesthetic agents. (1)
+ Antihypertensive Agents: Concomitant use of Unitensin with ganglionic blocking agents, Apresoline, Ismelin, Aldomet, reserpine, Hygroton, or thiazides necessitates immediate reduction in dosage of both agents by about 50%. Patients receiving combined therapy with Unitensin or its formulations and any other antihypertensive agent must be carefully followed for changes in blood pressure. (1)
+ Apresoline: see Unitensin + Antihypertensive Agents

UNITENSIN (continued)

+ Digitalis: Veratrum alkaloids may increase the heightened cardiac irritability produced by digitalis and concurrent use of Unitensin formulations with digitalis may lead to ectopic rhythms. (1)
+ Ganglionic Blocking Agents: see Unitensin + Antihypertensive Agents
+ Heat: During periods of excessively hot weather, the dosage may need to be reduced in some patients. (1)
+ Hygroton: see Unitensin + Antihypertensive Agents
+ Ismelin: see Unitensin + Antihypertensive Agents
+ Morphine: The bradycrotic effects of veratrum alkaloids is additive to, but not synergistic with, that produced by morphine and related drugs. (1)
+ Pre-Anesthetic Agents: see Unitensin + Anesthetics
+ Reserpine: see Unitensin + Antihypertensive Agents
+ Saluretic Agents: Clinical studies indicate that concomitant treatment of hypertensive patients with Unitensin or its formulations and a saluretic agent can result in a greater reduction of blood pressure than does treatment with either agent alone. This additive effect should be borne in mind whenever a saluretic agent is added to the regimen of patients previously or currently receiving Unitensin, and lower doses of both agents should be given. (1)
+ Thiazides: see Unitensin + Antihypertensive Agents

URACIL MUSTARD

+ Cytotoxic Drugs: Uracil Mustard should not be administered until several weeks after completion of a course of treatment with another cytotoxic drug, since it is essential to allow for recovery of the bone marrow to normal. It should be borne in mind that cytotoxic drugs may markedly reduce tolerance to Uracil Mustard. (2)
+ X-Ray Therapy: Uracil Mustard should not be administered until several weeks after completion of a course of treatment with x-ray, since it is essential to allow the bone marrow to return to normal. It should be borne in mind that irradiation may markedly reduce tolerance to Uracil Mustard. (2)

URECHOLINE + Ganglionic Blocking Agents: Special care is required when Urecholine is given to patients receiving ganglionic blocking compounds because a critical fall in blood pressure may occur. Usually, severe abdominal symptoms appear before there is such a fall in the blood pressure. (1)

UREMIDE + Dolonil: see Dolonil + Uremide

URICOSURIC AGENTS + Zyloprim: see Zyloprim + Uricosuric Agents

URIPLEX + Dolonil: see Dolonil + Uriplex

UROBIOTIC: see listed under the following agents:
Dolonil
Trisoralen

UROBIOTIC

+ Aluminum Hydroxide Gel: Aluminum Hydroxide Gel given with antibiotics has been shown to decrease absorption and is contraindicated. (1) Mechanism: Tetracyclines are chelating agents that form insoluble chelates with multivalent cations (Ca^{++}, Mg^{++}, Al^{+++}), and gastrointestinal absorption is inhibited.
+ Meals: To aid absorption of Urobiotic, it should be given at least one hour before or two hours after eating. (1)

UROKON + Benemid: see Benemid + Urokon

UROPEUTIC + Dolonil: see Dolonil + Uropeutic

V

VACCINES + Mumpsvax, Lyovac: see Mumpsvax, Lyovac + Vaccines

VACCINES, LIVE: see listed under the following agents:

Measles Virus Vaccine, Live, Attenuated
M-Vac Measles Virus Vaccine, Live, Attenuated
Orimune Poliovirus Vaccine, Live, Oral, Trivalent

VACCINES, LIVE VIRUS: see listed under the following agents:

Cendevax
Lirugen
Orimune Poliovirus Vaccine, Live, Oral, Trivalent
Pfizer-Vax Measles-L

VACUETTS SUPPOSITORIES

+ Mineral Oil: Do not use mineral oil or petroleum jelly as a lubricant. (1)
+ Petroleum Jelly: see Vacuetts Suppositories + Mineral Oil

VALIUM: see Benzodiazepines

VALIUM

+ Alcohol: see Valium + CNS Depressants
+ Anticonvulsants: When using oral Valium adjunctively in convulsive disorders, possibility of increase in frequency and/or severity of grand mal seizures may require increase in dosage of standard anticonvulsant medication. (1)
+ Antidepressants: see Valium + Psychotropic Agents
+ Barbiturates: see Valium + Psychotropic Agents
+ CNS Depressants: Since Valium has a central nervous system depressant effect, patients should be advised against simultaneous ingestion of alcohol or other CNS depressant drugs during Valium therapy. (1, 55) See also Valium + Psychotropic Agents
+ Narcotics: see Valium + Psychotropic Agents
+ Phenothiazines: see Valium + Psychotropic Agents
+ Psychotropic Agents: The concurrent administration of Valium and other psychotropic agents is not recommended. If such combination therapy is used, careful consideration should be given to the pharmacology of the agents to be employed with Valium, particularly with known compounds which may potentiate the action of Valium, such as phenothiazines, narcotics, barbiturates, monoamine oxidase inhibitors, and other antidepressants. (1, 2, 55)

VALIUM INJECTION
+ Narcotics: Injectable Valium has produced transient hypotension and respiratory depression when used with narcotic drugs. Since Valium may have an additive effect with narcotics, appropriate reduction in narcotic dosage is possible. (1)
+ Parenteral Fluids: Injectable Valium should not be added to parenteral fluids or be further diluted. (1)

VALMID
+ Alcohol: see Valmid + CNS Depressants
+ CNS Depressants: The concurrent ingestion of Valmid with alcohol or other CNS depressants, especially in overdosage, will increase the potential hazards of these agents. (1)

VANCOMYCIN + Magnesium Sulfate Injection: see Magnesium Sulfate Injection + Vancomycin

VANOXIDE LOTION
+ Cleansers, Abrasive: Comcomitant use of Ultraviolet, Cold Quartz and harsh abrasive cleansers is not recommended. (1)
+ Cold Quartz: see Vanoxide Lotion + Cleansers, Abrasive
+ Ultraviolet: see Vanoxide Lotion + Cleansers, Abrasive

VAPO-N-ISO METERMATIC + Epinephrine: Isoproterenol and epinephrine may be used interchangeably if the patient becomes unresponsive to one or the other but should not be used concomitantly. (1)

VARIDASE + Anticoagulants: It should be emphasized that Varidase should not be injected intramuscularly when there is evidence of a defect in blood coagulation or where the liver function is depressed and blood coagulation may be prolonged thereby. (1)

VASOCONSTRICTORS: see listed under the following agents:
Compazine
Ergotrate Maleate Injection
Metycaine Hydrochloride Injection
Nardil
Parnate

VASOCONSTRICTORS
+ Chloroform: see Vasoconstrictors + Cyclopropane
+ Cyclopropane: Avoid injection of solutions of vasoconstrictors during first 15 minutes after administration of cyclopropane, chloroform, halothane, or related drugs (may cause cardiac arrhythmias). (1)
+ Halothane: see Vasoconstrictors + Cyclopropane

VASOCONSTRICTORS, NASAL + MAO Inhibitors: see MAO Inhibitors + Vasoconstrictors, Nasal

VASODILATORS: see listed under the following agents:

Aventyl
Norpramin
Pertofrane
Vivactil

VASODILATORS, CORONARY: see listed under the following agents:

Gerilid
Geriliquid
Nicotron

VASOPRESSOR AGENTS: see listed under the following agents:

Enduron
Esimil
Eutonyl
Hydeltrasol Injection
Ismelin
Levoprome
Phenothiazines

VASOPRESSORS
+ Chloroform: see Vasopressors + Trilene
+ Cyclopropane: see Vasopressors + Trilene
+ Fluothane: see Vasopressors + Trilene
+ MAO Inhibitors: Monoamine oxidase inhibitors should be used very cautiously, if at all, with the vasopressors, since excessively high blood pressure may ensue. (2)
+ Oxytocics: Oxytocics should be used very cautiously, if at all, since excessively high blood pressure may ensue and even rupture of a cerebral blood vessel. (1, 2)
+ Trilene: The administration of epinephrine, Levophed, or other vasopressors with Trilene, Fluothane, chloroform, or cyclopropane may be dangerous, for these drug combinations tend to increase the likelihood of cardiac arrhythmias. (2)

VASOXYL + Ergot Alkaloids: Caution should be observed when used closely following the parenteral injection of ergot alkaloids to prevent an excessive rise in blood pressure. (1)

VEHICLES + Feosol Elix.: see Feosol Elixir + Vehicles

VELBAN
+ Aspartic Acid: see Velban + Glutamic Acid
+ Drugs, Oncolytic: see Velban + Radiation
+ Glutamic Acid: Reversal of the antitumor effect of Velban by glutamic acid and tryptophan has been observed. In addition, glutamic acid and aspartic acid have protected mice from lethal doses of Velban.

VELBAN (continued)

+ Glutamic Acid (continued): Aspartic acid was relatively ineffective in reversing the antitumor effect. (1, 145)
+ Radiation: Although the thrombocyte count usually is not significantly lowered by therapy with Velban, patients whose bone marrow has been recently impaired by prior therapy with radiation or other oncolytic drugs may show marked thrombocytopenia (less than 200,000 platelets per cu mm). (1)
+ Tryptophan: see Velban + Glutamic Acid

VERACILLIN + Meals: Veracillin is best absorbed when taken on an empty stomach (one or two hours before meals). (1)

VERATRUM + Naturetin: see Naturetin + Veratrum

VERATRUM ALKALOIDS: see listed under the following agents:

Anhydron
Benuron
Esidrix
HydroDIURIL
Hydromox
Metahydrin
Naqua

VERATRUM DERIVATIVES: see listed under the following agents:

Bristuron
Naturetin

VERATRUM VIRIDE + Raudixin: see Raudixin + Veratrum Viride

VERCYTE

+ Cytotoxic Chemotherapy: Since Vercyte frequently causes bone marrow depression, it should not be administered to patients with bone marrow depression resulting from cytotoxic chemotherapy. (1, 96)
+ X-Ray: Since Vercyte frequently causes bone marrow depression, it should not be administered to patients with bone marrow depression resulting from x-ray. (1, 96)

VERILOID

+ Digitalis: Veriloid should be used with caution in digitalis intoxication. (1)
+ Quinidine: Co-administration with quinidine is a relative contraindication. (1)

VESPRIN: see Phenothiazines

VESPRIN

+ Alcohol: see Vesprin + CNS Depressants
+ Analgesics: see Vesprin + CNS Depressants

VESPRIN (continued)

+ Anesthesia, Spinal: Although some investigators have used Vesprin prior to spinal anesthesia without any untoward effects, the drug is generally not recommended when spinal anesthesia is contemplated. (1)
+ Anesthetics: Vesprin exerts a definite potentiating effect on general anesthetics. Patients on Vesprin therapy who are undergoing surgery should be watched carefully for possible hypotensive phenomena. Moreover, it should be remembered that dosages of anesthetics and central nervous system depressants should be reduced. (1)
+ Antihistamines: see Vesprin + CNS Depressants
+ Atropine: As with other phenothiazines, potentiation of Atropine occurs with Vesprin. (1)
+ Barbiturates: see Vesprin + CNS Depressants
+ CNS Depressants: As with other phenothiazines, potentiation of CNS depressants (opiates, analgesics, antihistamines, barbiturates, narcotics, alcohol) occurs with Vesprin. (1, 2) See also Vesprin + Anesthetics.
+ Epinephrine: If severe hypotension should occur with Vesprin, supportive measures including the use of intravenous pressor agents should be instituted immediately. Levophed is the most suitable drug for this purpose; epinephrine should not be used since phenothiazine derivatives have been found to reverse its action, resulting in further lowering of blood pressure. (1, 2)
+ Hypnotics: Phenothiazine compounds should not be used in patients receiving large doses of hypnotics. (1)
+ Narcotics: see Vesprin + CNS Depressants
+ Opiates: see Vesprin + CNS Depressants
+ Phosphorous Insecticides: Phenothiazines have been reported to potentiate phosphorous insecticides. (1)
+ Trisoralen: see Trisoralen

VIBRAMYCIN

+ Aluminum Hydroxide Gel: Simultaneous administration of Aluminum Hydroxide Gel with tetracycline antibiotics including Vibramycin has been shown to decrease absorption. (1) Mechanism: Tetracyclines are chelating agents that form insoluble chelates with multivalent cations (Ca^{++}, Mg^{++}, Al^{+++}), and gastrointestinal absorption is inhibited.
+ Trisoralen: see Trisoralen

VI-DEXEMIN + MAO Inhibitors: Do not use in patients taking MAO Inhibitors. (1) Mechanism: See MAO Inhibitor + Sympathomimetics, Indirect-Acting.

VINETHENE
+ Antibiotics: see Antibiotics + Vinethene
+ Tubocurarine: Although much of the relaxing effect of Vinethene on muscle is due to the depression of reflex activity on the spinal cord, it has a peripheral neuromuscular blocking effect, being synergistic with tubocurarine. (110)

VIOCIN SULFATE: see also Viomycin

VIOCIN SULFATE
+ Drugs, Electrolyte-Depleting: Viocin may cause disturbances in serum electrolyte pattern. These may be alleviated readily by the administration of supplemental potassium chloride. (1) Mechanism: Disturbances in electrolyte balance due to urinary loss of calcium, potassium, and chloride, together with an increase in bicarbonate in the blood, has been observed in a number of instances. (137)
+ Streptomycin: The similar nature of the toxic manifestations of streptomycin and viomycin sulfate would suggest that these two drugs should not be used concomitantly except when the potentialities of other drug combinations have been exhausted. In such instances, careful laboratory control is essential in order to prevent or minimize possible toxic reactions. Streptomycin and viomycin are ototoxic; these agents must never be given together. (1, 137)

VIOMYCIN + Succinylcholine: see Succinylcholine + Viomycin

VIRAC + Soap: Soap interferes with the action of Virac and therefore should not be used concomitantly. (1, 2)

VISTARIL
+ Alcohol: Patients should avoid the concomitant use of Vistaril and alcohol, since the effects may be additive. (2)
+ Anticoagulants: It has been observed that, in some patients receiving anticoagulant therapy concurrently with Vistaril, the requirements for anticoagulant dosage may be decreased. Patients receiving both drugs should be followed closely, and appropriate laboratory studies should be performed regularly. (1, 112)
+ Barbiturates: see Vistaril + CNS Depressants

VISTARIL (continued)

+ CNS Depressants: Vistaril may potentiate the action of CNS depressants (e.g., barbiturates, opiates, etc.); conjunctive use with central nervous system depressants requires that their dosage be reduced. (1, 2) See Vistaril + Narcotics.
+ Demerol: see Vistaril + Narcotics
+ MAO Inhibitors: see Vistaril + Psychotropic Agents
+ Narcotics: Vistaril reduces the requirements by as much as 50% for narcotics such as meperidine, so that their use in preanesthetic adjunctive therapy should be modified on an individual basis. (1)
+ Opiates: see Vistaril + CNS Depressants
+ Phenothiazines: see Vistaril + Psychotropic Agents
+ Psychotropic Agents: Other psychotropic agents, particularly phenothiazines or monoamine oxidase inhibitors, that are known to potentiate the action of other drugs should not be given with Vistaril. (2)

VITAMIN B_{12} + Vitamin C: In solution, the Vitamin B_{12} stability will be destroyed unless solution is used at once. (1)

VITAMIN C: see listed under the following agents:

Anticoagulants
Coumadin
Feosol Elixir
Hedulin
Sintrom
Tromexan
Vitamin B_{12}

VITAMIN D + Rencal: see Rencal + Vitamin D

VITAMIN K: see listed under the following agents:

Anticoagulants
Coumadin
Dicumarol
Sintrom
Tromexan

VITAMINS, FAT SOLUBLE + Cuemid: see Cuemid + Vitamins, Fat Soluble

VIVACTIL: see Antidepressants, Tricyclic

VIVACTIL: see listed under the following agents:

Ismelin
Marplan
Parnate
Trisoralen

VIVACTIL

+ Aldomet: see Vivactil + Ismelin
+ Anticholinergic Agents: Anticholinergic, sympathomimetic compounds, and thyroid preparations may potentiate the effects of Vivactil. Vivactil has anticholinergic properties. (1, 97)

VIVACTIL (continued)

+ Antihypertensive Agents: Special care should be taken when Vivactil is used with other agents that lower blood pressure (e.g., phenothiazine compounds, thiazide diuretics, vasodilators). (97) See also Vivactil + Ismelin.
+ Diuretics, Thiazide: see Vivactil + Antihypertensive Agents
+ Drugs: When patients are receiving Vivactil in combination with other drugs, monitoring of liver function and blood cell count is recommended. (1)
+ Epilepsy: Vivactil may lower the seizure threshold and should be used with caution, if at all, in epilepsy. (97)
+ Ismelin: Tricyclic antidepressants, such as Vivactil, should not be given with Aldomet or Ismelin, since tricyclic antidepressants block the action of these antihypertensive agents. (1, 97) Mechanism: The tricyclic antidepressants will prevent the uptake of Ismelin and related drugs at the adrenergic nerve endings.
+ MAO Inhibitors: Monoamine oxidase inhibitor drugs may potentiate other drug effects, and such potentiation may even cause death. For this reason, when patients receiving an MAO inhibitor are to be treated with Vivactil, an interval should elapse between use of the two drugs to permit dissipation of the effects of the MAO inhibitor. This interval may be from several days to several weeks, depending on the MAO inhibitor involved, its dosage, and the duration of treatment. After the necessary time has elapsed, start therapy with Vivactil cautiously, increasing dosage gradually to the effective level. (1, 97) Mechanism: see MAO Inhibitor + Antidepressants, Tricyclic.
+ Norepinephrine: see Vivactil + Sympathomimetics
+ Phenothiazines: see Vivactil + Antihypertensive Agents
+ Surgery: Vivactil should be discontinued as soon as possible prior to elective surgery. (97)
+ Sympathomimetics: see Vivactil + Anticholinergic Agents. Caution is required if Vivactil is used in conjunction with sympathomimetics because of the possible development of unusual stimulatory side effects. (In common with other antidepressantsof the same category, Vivactil has autonomic effects, including ability to potentiate and/or prolong the various responses elicited by norepinephrine and by stimulation of sympathetic nerves.) (1)
+ Thyroid Preparations: see Vivactil + Anticholinergic Agents
+ Vasodilators: see Vivactil + Antihypertensive Agents

VLEM-DOME LIQUID CONCENTRATE

+ Iodides: see Vlem-Dome Liquid Concentrate + Mercurials
+ Mercurials: Do not use simultaneously with mercurials or iodides. (1)

VONTROL

+ Anticholinergics: Vontrol should be used cautiously in patients who are taking other centrally-acting anticholinergics concomitantly, since potentiation of the anticholinergic effects may occur. (1, 98)
+ Digitalis: see Vontrol + Drugs
+ Drugs: The antiemetic action of Vontrol may mask signs of overdosage of drugs (e.g., digitalis). (1)
+ Hypotensive Agents: Vontrol should be used cautiously in those patients receiving hypotensive agents. (98)

W

WARFARIN: see Anticoagulants, Coumadin, and Panwarfin

WARFARIN: see listed under the following agents:

- Alcohol
- Choloxin
- Euthroid
- Phenobarbital
- Questran
- Thyrolar

WATER: see listed under the following agents:

- Floropryl
- Surgicel

WATER for INJECTION: see listed under the following agents:

- Cardio-Green
- Ilotycin Gluceptate I.V.

WYAMINE: see listed under the following agents:

- Aramine
- Eutonyl
- Ismelin
- MAO Inhibitor
- Reserpine

WYAMINE + Thorazine: Wyamine, like epinephrine, is contraindicated in the treatment of hypotension induced by Thorazine, since the latter may reverse the usual pressor response and cause further lowering of blood pressure. (1)

X

XANTHINE DERIVATIVES: see listed under the following agents:

Anticoagulants
Asbron
Quibron
Theo-Organidin

XANTHINE PREPARATIONS: see listed under the following agents:

Brondecon
Choledyl
Lixaminol
Lixaminol AT
Lufylline
Theokon Elixir

X-PREP LIQUID + Diabetes: In diabetic patients, the physician must be be aware of the sugar content of X-Prep Liquid (60 gm per 2-1/2 fluid oz. dose). (1)

X-RAY THERAPY: see also Irradiation and Radiation Therapy

X-RAY THERAPY: see listed under the following agents:

Cosmegen
Cytoxan
Dicumarol
Mustargen
Panwarfin
Smallpox Vaccine Dried
Triethylene Melamine (TEM)
Uracil Mustard
Vercyte

XYLOCAINE

+ CNS Depressants: The effects of central nervous system depressants and cardiovascular depressing drugs may be enhanced when used simultaneously with local anesthetics. Accordingly, appropriate precautions should be observed even when using all of these drugs in recommended dosages. (1)
+ Drugs: Studies have revealed that many drugs currently employed alone or as combinations for preanesthetic medication exert a potentiating effect with local anesthetic agents. (1)
+ Drugs, Cardiovascular-Depressing: see Xylocaine + CNS Depressants
+ Eutonyl: see Eutonyl + Xylocaine
+ Metals: Local anesthetics react with certain metals and cause the release of their respective ions which, if injected, may cause severe local irritation. Adequate precautions should be taken to avoid this type of interaction. (1)

ZACTRIN COMPOUND + Anticoagulants: Salicylates (as part of formula) can add to the hypoprothrombinemic effect of anticoagulants. (1)

ZEPHIRAN

+ Soap: Zephiran Chloride is a cationic detergent. Soap is an anionic detergent; therefore soap should be rinsed carefully from tissue before Zephiran Chloride is applied. (1)

+ Tergemist: see Tergemist + Zephiran

ZYLOPRIM

+ Fluids: A fluid intake sufficient to yield a daily urinary output of at least two liters and the maintenance of a neutral or, preferably, slightly alkaline urine are desirable to (1) avoid the theoretic possibility of formation of xanthine calculi under the influence of Zyloprim therapy and (2) to help prevent renal precipitation of urates in patients receiving concomitant uricosuric agents. (1)

+ Imuran: In patients receiving Imuran the concomitant administration of 300 to 600 mg of Zyloprim per day will require a reduction in dose to approximately one-quarter to one-third of the usual dose of Imuran. Subsequent adjustment of doses of Imuran should be made on the basis of therapeutic response and any toxic effect. (1, 24)

+ Iron: Iron salts should not be given simultaneously with Zyloprim because animal studies suggested an increase in hepatic iron concentration. (1) Mechanism: Zyloprim is a potent inhibitor of xanthine oxidase. This enzyme is present in the liver and intestinal mucosa. It is involved in the mobilization of iron from liver stores and possibly in the regulation of iron absorption. Reduction of activity of this enzyme could effect iron absorption. (28)

+ Purinethol: Zyloprim may be utilized to inhibit the oxidation of Purinethol thus permitting use of smaller doses of Purinethol. The dose of the latter should be reduced to one-quarter to one-third of the therapeutic requirement when used alone, and then adjusted according to the observed effects. (1, 28, 55, 147, 148) Mechanism: Purinethol is biotransformed by the enzyme xanthine oxidase to the inactive 6-thiouric acid. Zyloprim inhibits xanthine oxidase thereby also inhibits the inactivation of Purinethol. (24, 69)

+ Uricosuric Agents: The concomitant administration of a uricosuric agent with Zyloprim may result in a decrease in urinary excretion

ZYLOPRIM (continued)

+ Uricosuric Agents (continued): of oxypurines as compared to their excretion with Zyloprim alone. This may possibly be due to an increased excretion of oxypurinol and a lowering of the degree of inhibition of xanthine oxidase. However, such combined therapy is not contraindicated and, for many patients, may provide optimum control. A report by Goldfinger et al. (149) on a patient treated with Anturane and salicylates in addition to Zyloprim did, however, show a marked decrease in the excretion of oxypurines, suggesting interference with their clearance at the renal tubular level. Although clinical evidence to date has not demonstrated renal precipitation of oxypurines in patients either on Zyloprim alone or in combination with uricosuric agents, the possibility should be kept in mind. (1)

BIBLIOGRAPHY

1. Pharmaceutical companies' package inserts, literature, etc.
2. New Drugs, 1966 ed., American Medical Association, Chicago.
3. S. A. Carter: Potentiation of the Effects of Orally Administered Anticoagulants by Phenyramidol Hydrochloride, New Engl. J. Med., 273:423 (Aug. 19, 1965).
4. J. E. Eckenhoff and R. K. Richards: Pharmacologic Limitations of Analeptic Therapy, Physiol. Pharmacol. Physic., (April 1966).
5. A. C. DeGraff: Guanethidine and Local Anesthetics, Am. Family Physic., 103 (Aug. 1965).
6. L. C. Jenkins and H. B. Graves: Potential Hazards of Psycho-Active Drugs in Association with Anesthesia, Canad. Anaesth. Soc. J., 12:121-128 (March 1965).
7. Analgesics and Monoamine-Oxidase Inhibitors, Brit. Med. J., 4: 284 (Nov. 4, 1967).
8. E. Twrdy, W. Weissel, and E. Zimmerman: Interactions of Coumarins and Phenobarbital, Munchen. Med. Wschr., 109:1272-1275 (June 9, 1967).
9. Current Guidelines to Anticoagulant Therapy, J. Am. Med. Assoc., 201:877-878 (Sept. 11, 1967).
10. H. M. Solomon: Pitfalls of Drug Interference with Coumarin Anticoagulants, Hosp. Practice, 51-55 (July 1968).
11. G. E. McLaughlin, D. J. McCarty, Jr., and B. L. Segal: Hemarthrosis Complicating Anticoagulant Therapy, J. Am. Med. Assoc., 196:202-203 (June 13, 1966).
12. T. B. Van Itallie: Treatment of Familial Hypercholesteremia, J. Am. Med. Assoc., 202:172 (Dec. 4, 1967).
13. E. S. Orgain, M. D. Bogdonoff, and C. Cain: Effect of Clofibrate (Atromid) with Androsterone upon Serum Lipids, Arch. Internal Med., 119:80-85 (Jan. 1967).
14. R. L. Katz: Clinical Experience with Neurogenic Cardiac Arrhythmias, Bull. N. Y. Acad. Med., 43:1106-1118 (Dec. 1967)

15. O. M. Spurny, J. W. Wolf, and G. S. Devins: Protracted Tolbutamide Induced Hypoglycemia, Arch. Internal Med., 115:53-56 (1965).

16. S. A. Cucinell, L. Odesky, M. Weiss, and P. G. Dayton: The Effect of Chloral Hydrate on Bishydroxycoumarin, J. Am. Med. Assoc., 197:144-146 (Aug. 1, 1966)

17. J. M. Weller and P. E. Borondy: Effects of Benzothiadiazine Drugs on Carbohydrate Metabolism, Metabolism, 14, 708 (June 1965).

18. L. K. Christensen, J. M. Hansen, and M. Kristensen: Sulfaphenazone-Induced Hypoglycemic Attacks in Tolbutamide-Treated Diabetes, Lancet, 2:1298-1301 (1963).

19. J. J. Schrogie, H. M. Solomon, and P. D. Zieve: Effect of Oral Contraceptives on Vitamin-K Dependent Clotting Activity, Clin. Pharmacol. Therap., 8:670-675 (Sept.-Oct. 1967).

20. Pickled Herring and Tranylcypromine Reaction, J. Am. Med. Assoc., 192:726-727 (May 24, 1965).

21. R. J. Cavallaro, L. W. Krumperman, and F. Kugler: Effect of Echothiophate Therapy on Metabolism of Succinylcholine in Man, Anesth. Analg., 47:570-574 (Sept.-Oct. 1968).

22. J. A. Winer and S. Bahn: Loss of Teeth with Antidepressant Drug Therapy, Arch. Gen. Psychiat., 16:238-240 (Feb. 1967).

23. J. Koch-Weser: Quinidine-Induced Hypoprothrombinemic Hemorrhage in Patients on Chronic Warfarin Therapy, Ann. Internal Med., 68:511-517 (March 1968).

24. G. W. Santos: The Pharmacology of Immunosuppressive Drugs, Pharmacol. Physic., (Aug. 1968).

25. A. Soffer: Digitalis Intoxication, Reserpine, and Double Tachycardia, J. Am. Med. Assoc., 191:777 (March 1, 1965).

26. J. C. Krantz, Jr.: The Problem of Modern Drug Incompatibilities, Curr. Med. Digest, 1951-1956 (Dec. 1966).

27. J. Menzel and F. Dreyfuss: Effect of Prednisone on Coagulation Time, J. Lab. Clin. Med., 56:14-20 (1960)

28. Allopurinol May Effect Iron Metabolism, J. Am. Med. Assoc., 200:39 (May 15, 1967).

29. H. M. Solomon and J. J. Schrogie: The Effect of Phenyramidol on the Metabolism of Diphenylhydantoin, Clin. Pharmacol. Therap., 8:554-556 (July-Aug. 1967).

30. H. Kutt and F. McDowell: Management of Epilepsy with Diphenylhydantoin Sodium, J. Am. Med. Assoc., 203:167-170 (March 11, 1968).

31. When Drugs Interact, Hosp. Practice, 72-77 (Oct. 1966).

32. J. B. Field et al.: Potentiation of Acetohexamide Hypoglycemia by Phenylbutazone, New Engl. J. Med., 277:889-893 (Oct. 26, 1967).

33. H. Beckman: Dilemmas in Drug Therapy, W. G. Saunders Co., Philadelphia, 1967

34. J. A. Oates: Antihypertensive Drugs that Impair Adrenergic Neuron Function, Pharmacol. Physic. (June 1967).

35. R. M. H. Kater, F. Tobon, and F. L. Iber: Increased Rate of Tolbutamide Metabolism in Alcoholic Patients, J. Am. Med. Assoc., 207:363-365 (June 13, 1969).

36. R. L. Katz and A. J. Gissen: Neuromuscular and Electromyographic Effects of Halothane and its Interaction with d-Tubocurarine in Man, Anesthesiology, 28:564-567 (May-June 1967).

37. R. H. Kessler: The Use of Furosemide and Ethacrynic Acid in the Treatment of Edema, Pharmacol. Physic. (Sept. 1967).

38. F. Robert Fakety, Jr.: Clinical Pharmacology of the New Penicillins and Cephalosporin, Pharmacol. Physic. (Oct. 1967).

39. J. Stuart Soeldner and J. Steinke: Hypoglycemia in Tolbutamide-Treated Diabetes, J. Am. Med. Assoc., 193:398-399 (Aug. 2, 1965).

40. J. E. Doherty and M. L. Murphy: Recognition and Management of the Intermediate Coronary Syndrome, Med. Times, 95:391-401 (April 1967).

41. H. Podgainy and R. Bressler: Biochemical Basis of the Sulfonylurea-Induced Disulfiram Syndrome, Diabetes, 17:679-683 (Nov. 1968).

42. E. A. Abramson and R. A. Arky: Role of Beta-Adrenergic Receptors in Counter-Regulation to Insulin-Induced Hypoglycemia, Diabetes, 17:141-146 (March 1968).

43. A. M. Anlitz, M. Tolentino, and M. F. Kosai: Effect of Butabarbital on Orally Administered Anticoagulants, Curr. Therap. Res., 10:70-73 (Feb. 1968).

44. Drug Interactions, Med. Sci., 27-28 (May 1967).

45. Heavy Drinking Accelerates Drugs' Breakdown in Liver, J. Am. Med. Assoc., 206: 1709 (Nov. 18, 1968).

46. J. R. Mitchell, L. Arias, and J. A. Oates: Antagonism of the Antihypertensive Action of Guanethidine Sulfate by Desipramine Hydrochloride, J. Am. Med. Assoc., 202:973-976 (Dec. 4, 1967).

47. E. Rubin and C. S. Lieber: Hepatic Microsomal Enzymes in Man and Rats: Induction and Inhibition by Ethanol, Science, 162:690-691 (Nov. 8, 1968).

48. M. G. MacDonald, D. S. Robinson, D. Sylwester, and J. J. Jaffe: The Effects of Phenobarbital, Chloral Betaine, and Glutethimide Administration on Warfarin Plasma Levels on Hypoprothrombinemic Response in Man, Clin. Pharmacol. Therap., 10:80-84 (Jan.-Feb. 1969).

49. R. A. Arky, E. Veverbrants, and E. A. Abramson: Irreversible Hypoglycemia, J. Am. Med. Assoc., 206:575-578 (Oct. 14, 1968).

50. H. F. Morelli and K. L. Melmon: The Clinician's Approach to Drug Interactions, Calif. Med., 109:380-389 (Nov. 1968).

51. S. Gitelson: Methaqualone-Meprobamate Poisoning, J. Am. Med. Assoc., 201:977-979 (Sept. 18, 1967).

52. H. A. Perkins: Concomitant Intravenous Fluids and Blood, J. Am. Med. Assoc., 206:2122 (Nov. 25, 1968).

53. J. C. deVilliers: Intracranial Hemorrhage in Patients with Monoamine Oxidase Inhibitors, Brit. J. Psychiat., 112:109-118 (Feb. 1966).

54. P. G. Dayton, Y. Tarcan, and M. Weiner: Influence of Barbiturates on Coumarin Plasma Levels and Prothrombin Response, J. Clin. Invest., 40:1797 (Oct. 1961).

55. Drug Interactions That Can Affect Your Patients, Patient Care, 1:33-71 (Nov. 1967).

56. M. Kristensen and J. M. Hansen: Potentiation of Tolbutamide Effect of Bishydroxycoumarin, Diabetes, 16:211-214 (April 1967).

57. H. M. Solomon and J. J. Schrogie: Effects of Phenyramidol and Bishydroxycoumarin in the Metabolism of Tolbutamide in Human Subjects, Metabolism, 16:1029-1033 (1967).

58. V. C. Dubach: Influence of Sulfonamides on the Blood-Glucose Decreasing Effect of Oral Antidiabetics, Scjweiz. Med. Wschr., 96:1483-1486 (Nov. 5, 1966).

59. M. E. Jarvik: "Drugs Used in the Treatment of Psychiatric Disorders" in L. S. Goodman and A. Gilman The Pharmacologic Basis of Therapeutics, 3rd ed., The Macmillan Co., New York, 1965.

60. Alcohol, General Anesthetics Influence Level of Anticoagulant Dosage, J. Am. Med. Assoc.,187:34-35 (March 7, 1964).

61. J. B. Field, M. Ohta, C. Boyle, and A. Remer: A Potentiation of Acetohexamide Hypoglycemia by Phenylbutazone, New Engl. J. Med., 277:889-894 (Oct. 26, 1967).

62. R. E. Tranquada: Diuretic for Patients Taking an Oral Hypoglycemic Agent, J. Am. Med. Assoc., 206:1580-1581 (Nov. 11, 1968).

63. W. A. Pettinger, F. G. Soyangco, and J. A. Oates: Inhibition of Monoamine Oxidase in Man by Furazolidine, Clin. Pharmacol. Therap., 9:442-447 (July-Aug. 1968).

64. J. E. Goss and D. W. Dickhaus: Increased Bishydroxycoumarin in Patients Receiving Phenobarbital, New Engl. J. Med., 273:1094 (Nov. 11, 1965).

65. J. Hellemans: Factors Influencing the Action of Coumarin, Belg. T. Geneesk, 18:361 (April 15, 1962).

66. T. E. Eiderton, O. Farmati, and E. K. Zsigmond: Reduction in Plasma Cholinesterase Level after Prolonged Administration of Echothiophate Iodide Eyedrops, Canad. Anaesth. Soc. J., 15:291-296 (May 1968).

67. G. Goldstein: Gamma-Globulin and Active Immunization, J. Am. Med. Assoc., 193:254 (July 19, 1965).

68. P. M. Aggeler and R. A. O'Reilly: The Pharmacological Basis of Oral Anticoagulant Therapy, Thromb. Diath. Haemorrh., 21:227-256 (1966).

69. E. Frei and T. L. Loo: Pharmacologic Basis for the Chemotherapy of Leukemia, Pharmacol. Physic. (May 1967).

70. S. Wessler and L. V. Avioli: Propranolol Therapy in Patients with Cardiac Disease, J. Am. Med. Assoc., 206:357–361 (Oct. 7, 1968).

71. Asthma Medication "Often Misused," J. Am. Med. Assoc., 206: 2639 (Dec. 16, 1968).

72. O. D. Gulati, B. T. Dave, S. D. Gokhale, and K. M. Shah: Antagonism of Adrenergic Neuron Blockade in Hypertensive Subjects, Clin. Pharmacol. Therap., 7:510–514 (July–Aug. 1966).

73. J. M. Hansen, M. Kristensen, L. Skovsted, and L. K. Christensen: Dicumarol Induced Diphenyhydantoin Intoxication, Lancet, 2:265–266 (July 30, 1966).

74. J. Roberts, R. Ito, J. Reilly, and V. J. Caivoli: Influence of Reserpine and beta TM 10 on Digitalis Induced Ventricular Arrhythmias, Circ. Res., 13:149–158 (Aug. 1963).

75. E. A. Abramson, R. A. Arky, and K. A. Woeber: Effects of Propranolol on the Hormonal and Metabolic Responses to Insulin-Induced Hypoglycemia, Lancet, 2:1386–1388 (Dec. 24, 1968).

76. K. L. Pines: The Pharmacologic Basis for the Use of Oral Hypoglycemic Agents in Diabetes, Physiol. Pharmacol. Physic. (Feb. 1966).

77. J. Marion Bryant: Monoamine Oxidase (MAO) Inhibition - A Therapeutic Adjunct, Med. Times, 95:420–434 (April 1967).

78. J. Barsa and J. C. Sanders: A Comparative Study of Tranylcypromine and Pargyline, Psychopharmacol., 6:295–298 (Oct. 14, 1964).

79. R. L. Dixon, E. S. Henderson, and D. P. Rall: Plasma Protein Binding of Methotrexate and its Displacement by Various Drugs, Fed. Proc., 24:454 (March–April 1965).

80. C. T. Dollery: Physiological and Pharmacological Interactions of Antihypertensive Drugs, Proc. Roy. Soc. Med., 58:983–987 (Nov. 1965).

81. E. H. Dearborn, J. T. Litchfield, Jr., H. J. Eisner, J. J. Corbett, and C. W. Dunner: The Effect of Various Substances on the Absorption of Tetracycline in Rats, Antibiot. Med. Clin. Therap., 4:627 (Oct. 1957).

82. J. C. Starke: Photoallergy to Sandalwood (Sandela) Oil, Arch. Derm., 96:62–63 (July 1967).

83. D. S. Robinson and M. G. MacDonald: Effect of Phenobarbital Administration on the Control of Coagulation Achieved during Warfarin Therapy in Man, J. Pharmacol. Exp. Therap., 153:250-254 (Aug. 1966).

84. Warfarin Plus Griseofulvin May Lower Prothrombin Time, J. Am. Med. Assoc., 197:37 (Aug. 1, 1966).

85. D. B. Hunninghake and D. L. Azarnoff: Drug Interactions with Warfarin, Arch. Internal Med., 121:356-360 (April 1968).

86. Evaluation of a New Antipsychotic Agent, Haloperidol (Haldol), Council on Drugs, J. Am. Med. Assoc., 205:577-578 (Aug. 19, 1968).

87. A Convulsant Agent for Psychiatric Use, Flurothyl (Indoklon), Council on Drugs, J. Am. Med. Assoc., 196:149 (April 4, 1966).

88. Evaluation of a New Antibiotic, Sodium Cephalothin (Keflin), Council on Drugs, J. Am. Med. Assoc., 194:182-183 (Oct. 11, 1965).

89. Evaluation of a New Oral Diuretic Agent, Furosemide (Lasix), Council on Drugs, J. Am. Med. Assoc., 200:979-980 (June 12, 1967).

90. A Non-narcotic Analgesic Agent, Methotrimeprazine (Levoprome), Council on Drugs, J. Am. Med. Assoc., 204:161-162 (April 8, 1968).

91. Evaluation of a New Antibacterial Agent, Cephaloridine (Loridine), Council on Drugs, J. Am. Med. Assoc., 206:1289-1290 (Nov. 4, 1968).

92. Current Status of Measles Immunization, Council on Drugs, J. Am. Med. Assoc., 194:1237-1238 (Dec. 13, 1965).

93. Evaluation of a New Antipsychotic Agent, Thiothixene (Navane), Council on Drugs, J. Am. Med. Assoc., 205:924-925 (Sept. 23, 1968).

94. Evaluation of a New Antipsychotic Agent, Butaperazine Maleate (Repoise Maleate), Council on Drugs, J. Am. Med. Assoc., 206:2307-2308 (Dec. 2, 1968).

95. An Agent for the Amelioration of Vertigo in Meniere's Syndrome, Betahistine Hydrochloride (Serc), Council on Drugs, J. Am. Med. Assoc., 203:1122 (March 25, 1968).

96. Evaluation of Two Antineoplastic Agents, Pipobroman (Vercyte) and Thioguanine, Council on Drugs, J. Am. Med. Assoc., 200:619-620 (May 15, 1967).

97. Evaluation of a New Antidepressant, Protriptyline Hydrochloride (Vivactil), Council on Drugs, J. Am. Med. Assoc., 206:364-365 (Oct. 7, 1968).

98. Evaluation of a New Antiemetic Agent, Diphenidol (Vontrol), Council on Drugs, J. Am. Med. Assoc., 204:253-254 (April 15, 1968).

99. J. Fulton and I. Willis: Photoallergy to 8-Methoxsalen, Arch. Derm., 98:445-450 (Nov. 1968).

100. R. L. Baer and H. Harris: Types of Cutaneous Reactions to Drugs, J. Am. Med. Assoc., 202:710-713 (Nov. 20, 1967).

101. R. L. Baer and L. C. Harber: Photosensitivity Induced by Drugs, J. Am. Med. Assoc., 192:989-990 (June 14, 1965).

102. L. R. Erickson and E. S. Peterka: Sunlight Sensitivity from Oral Contraceptives, J. Am. Med. Assoc., 203:980-981 (March 11, 1968).

103. M. G. MacDonald and D. S. Robinson: Clinical Observations of Possible Barbiturate Interference with Anticoagulation, J. Am. Med. Assoc., 204:97-100 (April 8, 1968).

104. A. S. Nies and K. L. Melmon: Recent Concepts in the Clinical Pharmacology of Antihypertensive Drugs, Calif. Med., 106:388-399 (May 1967)

105. L. I. Goldberg: Monoamine Oxidase Inhibitors, J. Am. Med. Assoc., 190:456-462 (Nov. 2, 1964).

106. J. H. Jaffe: "Narcotic Analgesics" in L. S. Goodman and A. Gilman: The Pharmacologic Basis of Therapeutics, 3rd ed., The Macmillan Co., New York, 1965.

107. P. I. Adnitt: Hypoglycemic Actions of Monoamine Oxidase Inhibitors, Diabetes, 17:628-633 (Oct. 1968).

108. G. H. Mulheims, R. W. Entrup, D. Paiewonsky, and D. S. Mierzwiak: Increased Sensitivity of the Heart to Catecholamine: Induced Arrhythmias Following Guanethidine, Clin. Pharmacol. Therap., 6:757-761 (Nov.-Dec. 1965).

109. D. P. Oakley and H. Lautch: Haloperidol and Anticoagulant Treatment, Lancet, 2:1231 (Dec. 7, 1963).

110. J. W. Dundee: Clinical Pharmacology of General Anesthetics, Clin. Pharmacol. Therap., 8:91-123 (Jan.-Feb. 1967).

111. J. Murdock Ritchie: "The Aliphatic Alcohols" in L. S. Goodman and A. Gilman: The Pharmacologic Basis of Therapeutics, 3rd ed., The Macmillan Co., New York, 1965.

112. Hospital Formulary: American Hospital Formulary Service, American Society of Hospital Pharmacists, Washington, D. C. (Monograph on Atarax).

113. Hospital Formulary: American Hospital Formulary Service, American Society of Hospital Pharmacists, Washington, D. C. (Monograph on Ponstel).

114. S. Consolo: Desipramine and Amphetamine Metabolism, J. Pharmacol., 19:253-256 (April 1967).

115. F. Sjöqvist: Psychotropic Drugs (2): Interactions Between Monoamine Oxidase (MAO) Inhibitors and Other Substances, Proc. Roy. Soc. Med., 58:967-978 (Nov. 1965).

116. P. Brazeau: "Inhibitors of Tubular Transport of Organic Compounds" in L. S. Goodman and A. Gilman: The Pharmacologic Basis of Therapeutics, 3rd ed., The Macmillan Co., New York, 1965.

117. M. F. Cuthbert, M. P. Greenberg, and S. W. Morley: Cough and Cold Remedies: Potential Danger to Patients on Monoamine Oxidase Inhibitors, Brit. Med. J., 1:404-406 (Feb. 15, 1969).

118. B. B. Brodie: Displacement of One Drug by Another from Carrier or Receptor Sites, Proc. Roy. Soc. Med., 58:946-955 (Nov. 1965).

119. C. G. Zubrod, T. J. Kennedy, and J. A. Shannon: Studies on the Chemotherapy of Human Malaria VIII: The Physiological Disposition of Pamaquine, J. Clin. Invest., 27:114-120 (May 1948).

120. J. J. Burns and A. H. Conney: Enzyme Stimulation and Inhibition in the Metabolism of Drugs, Proc. Roy. Soc. Med., 58:955-960 (Nov. 1965).

121. M. Weiner et al.: Effects of Steroids on Disposition of Oxyphenbutazone in Man, Soc. Exper. Biol. Med., 124:1170-1173 (April 1967).

122. M. Barber: Drug Combinations in Antibacterial Chemotherapy, Proc. Roy. Soc. Med., 58:990–995 (Nov. 1965).

123. H. L. Price and R. D. Dripps: "General Anesthetics I. Gas Anesthetics: Nitrous Oxide, Ethylene, Cyclopropane, and other Gaseous Anesthetics" in L. S. Goodman and A. Gilman: The Pharmacologic Basis of Therapeutics, 3rd ed., The Macmillan Co., New York, 1965.

124. M. D. Milne: Influence of Acid-Base Balance on Efficacy and Toxicity of Drugs, Proc. Roy. Soc. Med., 58:961–963 (Nov. 1965).

125. L. K. Garrettson, J. M. Perel, and P. G. Dayton: Methylphenidate Interaction with both Anticonvulsants and Ethyl Biscoumacetate, J. Am. Med. Assoc., 207:2053–2056 (March 17, 1969).

126. H. L. Price and R. D. Dripps: "General Anesthetics II. Volatile Anesthetics: Diethyl Ether, Divinyl Ether, Chloroform, Halothane, Methoxyflurane, and other Halogenated Volatile Anesthetics" in L. S. Goodman and A. Gilman: The Pharmacologic Basis of Therapeutics, 3rd ed., The Macmillan Co., New York, 1965.

127. W. C. Cutting: Handbook of Pharmacology, The Action and Uses of Drugs, 2nd ed., Appleton-Century-Crofts, New York, 1964.

128. L. F. Walts and W. McFarland: Effect of Vagolytic Agents on Ventricular Rhythm during Cyclopropane Anesthesia, Anesth. Analg., 44:429 (July-Aug. 1965).

129. S. A. Cucinell, A. H. Conney, M. Sansur, and J. J. Burns: Drug Interactions in Man. 1. Lowering Effect of Phenobarbital on Plasma Levels of Bishdroxycoumarin (Dicumarol) and Diphenylhydantoin (Dilantin), Clin. Pharmacol. Therap., 6:420–429 (July-Aug. 1965).

130. A. W. D. Leishman, H. L. Mathews, and A. J. Smith: Antagonism of Guanethidine by Imipramine, Lancet, 1:112 (Jan. 12, 1963).

131. G. B. Koelle: "Neuromuscular Blocking Agents" in L. S. Goodman and A. Gilman: The Pharmacologic Basis of Therapeutics, 3rd ed., The Macmillan Co., New York, 1965.

132. Pesticide Poisoning May Appear Anywhere, Calif. Med., III:68–69 (July 1969)

133. M. Shepherd: Psychotropic Drugs (1) Interaction Between Centrally Acting Drugs in Man: Some General Considerations, Proc. Roy. Soc. Med., 58:964–967 (Nov. 1965).

134. S. Gershon, H. Neubauer, and D. M. Sundland: Interaction Between Some Anticholinergic Agents and Phenothiazines. Potentiation of Phenothiazine Sedation and its Antagonism, Clin. Pharmacol. Therap., 6:749-756 (Nov.-Dec. 1965).

135. J. Murdoch Ritchie, P. J. Cohen, and R. D. Dripps: "Local Anesthetics: Cocaine, Procaine, and Other Synthetic Local Anesthetics" in L. S. Goodman and A. Gilman: The Pharmacologic Basis of Therapeutics, 3rd ed., The Macmillan Co., New York, 1965.

136. R. C. Northcutt, J. N. Stiel, J. W. Hollifield, and E. G. Stant: The Influence of Cholestyramine on Thyroxine Absorption, J. Am. Med. Assoc., 208:1857-1861 (June 9, 1969).

137. L. Weinstein: "Drugs Used in the Chemotherapy of Leprosy and Tuberculosis" in L. S. Goodman and A. Gilman: The Pharmacologic Basis of Therapeutics, 3rd ed., The Macmillan Co., New York, 1965.

138. D. A. Hussar: Therapeutic Incompatibilities: Drug Interactions, Hosp. Pharm., 3:14-24 (Aug. 1968).

139. E. W. Davidson, J. H. Modell, F. Moya, and O. Farmati: Respiratory Depression Following Use of Antibiotics in Pleural and Pseudocyst Cavities, J. Am. Med. Assoc., 196:170-171 (May 2, 1966).

140. A. Radler, P. Schmidt, and J. Metzl: Muscle Paralysis Induced by Intraperitoneal Neomycin, Orv. Hetil., 107:2384-2386 (Dec. 11, 1966). FDA Abstr., 20:347 (Oct. 24, 1967).

141. G. E. Kinyon: Anticholinesterase Eye Drops - Need for Caution, New Engl. J. Med., 280:53 (Jan. 2, 1969).

142. Adverse Effects of Topical Antiglaucoma Drugs, Med. Lett. Drug Therap., 9:92-93 (Nov. 17, 1967).

143. T. A. G. Tarda et al.: Interaction of Neuromuscular Blocking Agents in Man; Role of Hexafluorenium, Anesthesiology, 128:1010-1019 (Nov.-Dec. 1967).

144. R. L. Katz: Neuromuscular Effects of Diethyl Ether and its Interaction with Succinylcholine and d-Tubocurarine, Anesthesiology, 27:52-63 (Jan.-Feb. 1966).

145. P. Calabresi and A. D. Welch: "Cytotoxic Drugs, Hormones, and Radioactive Isotopes" in L. S. Goodman and A. Gilman: The Pharmacologic Basis of Therapeutics, 3rd ed., The Macmillan Co., New York, 1965.

146. W. L. Way, B. G. Katzung, and C. P. Larson, Jr.: Recurarization with Quinidine, J. Am. Med. Assoc., 200:153–154 (April 10, 1967).

147. G. B. Elion, S. Callahan, H. Nathan, S. Bieber, R. W. Rundles, and G. H. Hitchings: Potentiation by Inhibition of Drug Degradation: 6-Substituted Purines and Xanthine Oxidase, Biochem. Pharmacol., 12:85–93 (Jan. 1963).

148. R. C. DeConti and P. Calabrisi: Use of Allopurinol for Prevention and Control of Hyperuricemia in Patients with Neoplastic Diseases, New Engl. J. Med., 274:481–486 (March 3, 1966).

149. S. E. Goldfinger, J. R. Klinenberg, and J. E. Seegmiller: The Renal Excretion of Oxypurines, J. Clin. Invest., 44:623 (1965).

150. J. Adriani: Anesthesia Problems in Small Hospitals, Postgrad. Med., 45:116–123 (Feb. 1969).

151. M. M. Ghoneim and J. P. Long: The Interaction Between Magnesium and Other Neuromuscular Blocking Agents, Anesthesiology, 32:23–27 (Jan. 1970).

152. FDA Current Drug Information, Bull. on L-Dopa (June 1, 1970).

153. FDA Current Drug Information, Bull. on Lithium Carbonate (April 1970).

154. D. L. Azarnoff and A. Hurwitz: Drug Interactions, Pharmacol. Physic., 4 (Feb. 1970).

155. E. M. Sellers and J. Koch-Weser: Potentiation of Warfarin-Induced Hypoprothrombinemia by Chloral Hydrate, New Engl. J. Med., 283: 827–831 (Oct. 15, 1970).

156. M. Weiner: Drug Interactions, New Engl. J. Med., 283:871–872 (Oct. 15, 1970).

GENERAL INDEX

GENERIC NAME INDEX